Environmental Hazards - Are You Exposed?

Finding Hazards Where You Live

By Fred Siegel

with contributions by Anthony Quaglietta

To request permissions, contact the author at fred@toxictom.com

Paperback: 978-0-9816165-9-9
Ebook: 978-0-9816165-8-2

First paperback edition October 2020
First Ebook edition October 2020

Edited by Giffy Giffoniello
Cover art by Nabin Karna
Cover Photo by Famartin

Group 27
650 E. Palisade Ave 2-270
Englewood Cliffs, N.J. 07632

www.toxictom.com

This book is dedicated to:

To my patient and loving wife Vicki who tolerated listening to me all of the years talking about one day finishing this book.

To my two children Pam and Zach who had to listen to their father pointing out all the landfills, toxic waste sites and other hazards whenever we took family rides and trips.

To my wonderful granddaughter I call Miss Heidi who at the age of ten is already doing what she can to save the environment.

Heidi's favorite quote from Dr. Seuss's book "The Lorax"

"Unless someone like you cares a whole awful lot,
nothing is going to get better. It's not."

Table of Contents

Introduction

There is a 90% chance you are living near at least one environmental hazard and have no idea. Tony and I have been researching pollution and environmental hazards for thirty years. Based on our experience, we know environmental hazards exist in many places most people would never think to look.

For many years before I became involved in site assessments, there were things right in front of me I never saw. It wasn't that I couldn't see them. Of course, I could have. What I was unable to do was see behind them. Or underneath in my case. Because of my inability to see that way, for years I lived on top of an old landfill filled with potentially hazardous wastes and toxic chemicals. Many people living in the area had miscarriages and got cancer later in life. The information that the dump existed was there for me to see if I had known where to look or even knew I needed to review our property for potential hazards. Once I learned that many things in my everyday life were not as they appeared, I developed a new way of seeing.

Environmental hazards usually do not have giant flashing neon signs screaming, "Toxic waste here! Toxic waste here!" They rarely have chain-link fences topped with barbed wire and signs saying "Keep out--Danger." Most of them are simply normal looking factories and businesses. They are regular looking buildings, vacant parcels of land, and rural property that most people drive by, never noticing. It could even be a place people visit regularly, like a gas station or a dry cleaners. They look like many other everyday places as people pass. Most of these potential hazards do not look like they are a threat to your health. That is one of the biggest dangers of these sites.

I personally began checking for environmental hazards back in the early 1980s. That was right after I discovered my housing development had been built on top of an old landfill where all types of waste were disposed of, including hazardous industrial waste. Nobody informed us what was underneath our homes: not the developer, not the real estate people, and not the town. Wishing to never repeat that experience, my friend, Tony Quaglietta and I started to research potential environmental hazards nearby that could affect our families' health. Quite a few of my friends also expressed interest in knowing what was near them as well. Our new company, Site Searchers, was born.

Back then there was no internet, and all of our searches were done by driving around areas and spending many hours in town offices and state and federal agencies reviewing files and records. Over the years that my partner and I worked together, we have done many searches and added many new hazards that we check for to our list. While both Tony and I have been involved in other projects, we have continued over the years to do hundreds of searches for people, friends, and family members who asked us to help out.

Since learning about all the hazards back then, I have seen things differently than most people. When I'm driving down the highway and people say, "Oh, look at that pretty green hill over there," I see a closed landfill that was left there and covered up with grass. When I'm driving outside of a city and see the skyline, I also see the brownish haze caused by hundreds of thousands of cars on the highway pouring hazardous particulate matter into the air. Don't get me wrong. I can appreciate a grassy hill or the cityscape, but I also see what else is there.

The hidden dangers of many of these hazards is why Tony and I wrote this book. The government is not going to tell you about all of these hazards. Real estate developers are certainly not going to either. The chemical industry and all the other manufacturing sectors pouring millions of tons of chemicals into our air, water, and soil certainly will not be telling us anything either. So Tony and I are. We want to share what we have learned about many places that present a danger to you and your family and many you would usually never suspect. This book will first take you through each of the hazards we check for when doing a site assessment. We will start with the most obvious ones and work our way into many others that usually aren't looked at but can present threats to you and your health.

As you read along, you can do your own check of environmental hazards where you live. With the internet as a tool, it is so much easier. This book will point you in the right direction and show you where and how to look for the threats that might be lurking in your own backyard. There are many more hazards you should be aware of than you would think of.

Each of our chapters is laid out fairly much the same. We explain some basics about the topic and then lay out some of the specific risks each hazard has for you and your family in addition to the overall environmental impact. We share stories of the people and places impacted by these issues. All of the people and places are real. These stories have all been covered by reputable news sources, but these stories don't get much national attention. Unfortunately, there are thousands more just like them we see as well.

None of these chapters covers entirely all of the facts and the deeper story behind the hazard we are talking about. That is impossible as many of these subjects could be an entire book on their own. Some even have many books written about them. The main purpose of our short chapters is to plant seeds in your mind about seeing that topic from a different angle. What something looks like may be entirely different below the surface once you know more.

As you look through the chapters in the book at the individual hazards we talk about, hopefully you see many that have absolutely nothing to do with you. However, in the thirty plus years Tony and I have been reviewing sites, we have never found any place where there was not at least one potential area of risk. Some of the potential hazards we found were not a risk after we checked them out, but there are almost no places anywhere completely devoid of something needing to be checked out closer. This book is a tool to help you determine what is and what is not a risk to you and your family. If you are already renting or bought your house, this book can tell you where to look to see if these hazards exist and you are in danger of any of them. If you are about to buy a new home or move to a new place, this book will help you scout out your potential home.

The book doesn't stop at identifying dangers, though. Most of us do not have the ability to pick up and move, and this book explores ways to minimize the negative impact and exposure to some of these hazards.

You may be thinking you may not want to know if there is something nearby that can harm you and your family. You may be tempted to close this book and not read it. Unfortunately, what you don't know can absolutely hurt you. In many cases, it can hurt you badly. The consequences could be devastating: illness, potential breathing issues, and increased risk of cancer.

Tony and I have seen with our own eyes housing developments being built dangerously close to active and closed landfills. We met with and spoke to many people who were poisoned from many of these hazards in order to hear what their personal stories were. We see news stories every day about people and towns across the country whose wells are being contaminated from toxic waste sites and leaking gasoline tanks. Smokestacks belch out millions of tons of hazardous material. There are hundreds of huge factories literally pouring millions of gallons of toxic wastes into our waterways, most with the permission of the government.

There is one important fact Tony and I have learned that is indisputable in nature. There is no such place as "away." When the garbage trucks pick up your trash, it does not really go away. When you flush your toilet, it does not really go away. When trash incinerators burn garbage or factories pour toxic smoke into the sky and water, it does not go away. It all goes to someplace: other people's air, other people's water, and other places in the environment. It will end up somewhere, and it will release its toxic dangers.

SECTION ONE

Every one of the following hazards in this section is dangerous. There are imminent risks associated with living or working within the exposure zone for each of these. We have no doubt that you will be placing yourself at risk in some way.

- Superfund sites
- Trash incinerators
- Hazardous waste sites
- Power plants
- Coal ash dumps
- Hazardous waste incinerators
- Cement plants
- Military bases and government installations
- Medical waste incinerators
- Hazardous waste disposal and transport
- Factory farms
- Nuclear reactors
- Radioactive waste
- Landfill
- Airports

Chapter 1. Superfund Sites

Tony and I have been dealing with places of environmental risk and destruction over the past 30 years. The places we see the most long-term devastation and harm to humans and the environment are Superfund sites. These toxic waste sites are the absolute worst of the worst in the entire country. They have been hand-picked and selected for inclusion in this toxic fraternity because of how dangerous they are. Many times the pollution at the site is not under control and there is imminent danger to people and the environment in the area.

We have seen stories and met people who were totally unaware they were living near a Superfund site. We aren't entirely surprised by this. It is in a town's best interest not to advertise a Superfund site because communities with Superfund sites usually get a reputation as bad places to live, so they try to avoid negative press and publicity. Some states have disclosure laws regarding telling prospective buyers about certain hazards in the area, but many places do not require it. Despite this, residents living in a town with a Superfund site usually have some idea about its existence. It may be from a newspaper article of hearing talk about it from neighbors.

Even though they are immediate threats to the area, Superfund sites sit for a long time with nothing at all happening to clean them up. The government prefers a program of containment much more than actual remediation. Obviously, it is much less expensive and also allows the EPA to reclassify it. Once a Superfund site no longer poses an imminent threat, the EPA can technically move it off of the Superfund list and rename it a regular hazardous waste site. The name change doesn't make things any safer for the people, but the EPA can say it has made progress.

Dealing with a Superfund site is not like dealing with other environmental hazards. A Superfund site is already known to be of extreme danger for any human or animal that would come in contact with it. A Superfund site is a location that has been determined to contain chemicals and/or materials that are classified as hazardous waste, a material that is known to have a highly detrimental effect on the human body and the environment. The effects of these materials range widely, depending on the toxicity of the waste compound. Almost all the waste contained in a Superfund site has the potential to cause damage. Nerve and liver damage, various types of cancers, abnormally high rates of miscarriage, intellectual disabilities, gene and chromosome mutations--all are possible. Cancers in communities nearby to Superfund sites are usually seen at much higher than normal rates. They also tend to have more of the rare types of cancers as well. These sites contain a potpourri of the most deadly compounds known to humanity.

Many of the places that became Superfund sites got their starts between the 1950s and 1970s. After the end of World War II and the rapid expansion of the petroleum, pharmaceutical, and chemical industries, the volume of toxic waste skyrocketed. In that era, at the end of the manufacturing process, whatever waste and useless by-products remained were usually just disposed of by the least expensive methods available. Disposal practices varied. Sometimes, thousands of drums containing gallons of waste were dumped into unlined landfills or even simply disposed of in a ditch or an empty, abandoned canal, such as occurred in Love Canal, New York. Sometimes, lagoons were created that were filled in with a witch's brew of bubbling, oozing, toxic substances, such as the Stringfellow acid pits in California. Old mines were a favorite place to pull up a tanker truck, put the hose into the hole, and empty the truck out, just like they did in Ringwood, New Jersey.

At the time, there were almost no regulations as to what anyone was supposed to do with toxic waste. Most of the time, this toxic and hazardous material ended up in the town landfill along with the regular solid waste. People didn't care because it was away from them. Many Superfund sites are old landfills used by towns to accept industrial waste in addition to residential garbage. We now know, too late, that there is no such place as "away."

Disposal of this waste wasn't even regulated by the government until the year 1976. The public outcry was getting too loud to ignore any longer, and Congress had to act. Congress was finally pushed and prodded into passing our first real laws controlling disposal of municipal, industrial, and hazardous waste. Love Canal was the first toxic waste site to receive Superfund status. The government likes to claim that they were responsible for enacting the Superfund legislation in 1980. Maybe they enacted it, but as Tony and I proudly state, it was driven by the environmental activism of two of our heroes: Lois Gibbs from Love Canal and Rachel Carson author of Silent Spring. The credit, according to us, goes to them because Lois and Rachel, started the modern environmental movement.

While this legislation addressed some of the problems moving forward, many communities had already sustained damage from hazardous waste at a monumental scale. Before the passing of the law, millions of gallons of waste were regularly disposed of simply by dumping it into rivers, sewers, landfills, and any other place it was economically feasible. Sadly, that practice is still happening in many places, often illegally. Sometimes it is being done in the middle of the night, other times it is being done openly with some participants looking the other way, such as at the Oregon landfill and radioactive waste disposal.

These continued, illegal practices are generating new Superfund sites. For instance, in Chicago, a 67 acre site reached Superfund status when large piles of slag and waste were illegally dumped over years. Federal and state officials have found the site is contaminated with cyanide, mercury, manganese, lead, and other contaminants. The site was added to the Superfund list in 2019, and although the toxic chemicals threaten the health of the 10,000 people living near it, the site remains uncovered and protected.

It's not just new Superfund sites that are a threat, though. Our landfills, deep wells, lagoons, and pits have been coming back to haunt us for decades now. In Plumstead Township, New Jersey, there are several Superfund sites in the middle of this predominantly farming area. Years ago, unethical waste disposers paid farmers to allow them to dump their industrial waste on the farmers' property. With the rapid expansion of the suburbs, we are starting to realize the great extent to which farmland dumping has occurred over the years.

At the time we wrote this chapter, there were 1,344 sites listed on the Superfund list. The sites that get placed here are considered as the National Priorities List. Superfund sites can affect those living nearby through all three primary exposure routes.

1. Air Contamination

Most of these sites sit open and exposed to the rain and the winds. Particles from the piles of hazardous materials blow off the waste site and are carried into the nearby areas. Neighborhoods are exposed to these toxic chemicals, and the residents breathe them. Over time, the continued small exposures begin to take a toll on the health of people living nearby, especially children and women who are pregnant.

Way back in 1983, the EPA listed 21 square miles in the Silver Valley in Northern Idaho as a Superfund site. The area was used for mining by the Bunker Hill Mining Company. The area was contaminated by significant amounts of mine waste from lead, silver, and zinc mining. Tailings, or mine waste, were simply dumped into rivers and streams for disposal, or stacked in enormous piles for the wind and weather to blow and wash away. Lead levels were so high that the residents in the area said they could taste the metal in their mouth. Hundreds of residential yards needed decontamination, as well as roads and other surfaces. There was a coating of lead everywhere.

In 2002 the EPA expanded the size of the Superfund site from 21 square miles to 1,500. It was one of the largest, most polluted places in the country. Over 35 years since it was first listed as a Superfund site, clean up is still ongoing, and the long-term effects of the contamination have been handed down for several generations. At one time a quarter of all the children had dangerously high lead levels in their blood. Even now, the dust still blows lead all around the communities in the region, contaminating the air they breathe and the soil the children play on and gardeners grow food in. The EPA estimates a minimum of 30 years just to partially clean up a portion of the Superfund site. The entirety may not be cleaned up for deaconess more, leaving the entire area a continued risk to all.

2. Groundwater Contamination

Water is a resource that everyone depends on. Whether public water systems or private wells, a safe, dependable water supply is needed for everyday existence. There are quite a few communities where the groundwater has been contaminated beyond the point of remediation, at least for the present time. Truckloads of water are being brought in to fill large tanks to supply the homes with safe water. In many of these places, this has been life for years. Other sites are using water that the city or government tells the people is safe now until up pops another toxic chemical to add to the poison list.

In Tucson, Arizona, the Tucson International Airport had a 10 square mile area declared a Superfund site back in 1983. The contamination came from electronics manufacturing, toxic fire suppression foam used at many airports and military bases, and unlined landfills where hazardous materials were disposed of. The groundwater in the area was contaminated with toxic chemicals. In 1994, 11 years after being listed as a Superfund, a water treatment plant was finally installed to handle some of the contamination.

In 2019, new toxic chemicals, PFOA and PFOSs, were discovered in the supposedly safe water. While the water treatment plant is adding equipment to deal with the newly found toxin, the presence of a previously undetected chemical adds to the concerns Tony and I have about what else has been in the water that went untested or may yet be waiting for discovery. 1983 to 2019 is 36 years, and counting. The groundwater remains an issue.

In Santa Clara County, California, semiconductor chip manufacturing has been a major industry since the 1980s. This area, better known to us as Silicon Valley, was at the heart of the U.S. computer industry's explosive growth. While the business of most Silicon Valley companies may have shifted away from manufacturing semiconductors and microprocessors to developing software, the legacy of its manufacturing days lives on. Santa Clara County boasts 23 Superfund sites within its boundaries. That puts them at number one in the county.

The groundwater beneath the area is contaminated with volatile organic compounds and thousands of gallons of trichloroethylene, toxic chemicals used in the semiconductor manufacturing process among others. Trichloroethylene levels were found as high as 130,000 parts per billion (PPB), and the EPA safe level is only 5 PPB. For the past 30 years, efforts have been made just to get the spread of the contamination in the groundwater under control. These efforts don't appear to be successful as, at last count, there are 518 individual plumes of contaminated water spreading through the groundwater like little fingers. Efforts to clean up these waste sites may never be completed. Homes in the area have vapors intruding upward from the soil, bringing high levels of toxic gases into their homes.

In Ringwood, New Jersey, a site known as the Ringwood Mines/Landfill Site was listed for Superfund status and cleanup in 1983. The property was used as a dumping ground for paint sludge from a Ford plant operating nearby in Mahwah, New Jersey. The sludge and other materials were dumped in the forests and old mine shafts from closed mines in the area.

Many homes in the area were on private wells, and in the 1980s were exposed to the chemicals by using on-site spring water for household use. According to an EPA report, public water was not available to all residents until the early '80s. Locals used to collect water from streams and brooks for drinking and cooking. The EPA estimated 20 years of exposure to these chemicals was from water they used. Children rode bicycles through the woods and in an area nicknamed Sludge Hill. Streams and ponds in the area also register toxic chemicals in the waters. Lead, beryllium, and chloroethane are other toxic chemicals found with levels potentially able to cause serious health effects in people such as kidney and liver damage and nervous system disorders.

The cleanup is still ongoing after 50 years with more areas of additional toxic chemicals such as 1,4 Dioxane, benzene, and arsenic first being discovered in 2016. Exposure to the chemical 1,4 dioxane is likely to cause cancer, the EPA said, adding to the health risks already present from prior contamination.

Much of the toxic material still left to clean up will simply be covered with asphalt and left. Unfortunately, the site will continue to contaminate the groundwater much of the region depends on. Ford Motor Company is saving millions of dollars by covering up the waste instead of removing and disposing of it. There will be 166,000 tons of this hazardous waste left onsite forever.

The government of the Town of Ringwood isn't blameless in this situation. They allowed the dumping to occur, giving Ford's contractors permission to use the land and mines for disposal. They are also on the hook for liability and the arrangement to cover the remaining waste saves the town millions of dollars as well despite the persistent hazards to the locals.

In Phoenix, Arizona, sits an underground plume of contamination that has 120,000 houses on top of it. Known as the Motorola 52nd Street Superfund Site, it stretches for eight miles from the factory Motorola used for manufacturing for years. The groundwater is contaminated with the chemical TCE, which is known to cause cancer and raises the risks in babies having heart defects. The worst part about this contamination is that TCE is known to change into a vapor and rises from the groundwater through the soil and into the buildings on top. Vapor intrusion as it is known is threatening the health of many people in the 120,000 homes affected. The EPA has known about the problem for years, but many of the people say they are dragging their feet when it comes to taking any sort of action. For now, as in many other contaminated places, the people wait and hope they are not getting sick by the toxic contamination the government knows is there.

3. Soil Contamination

Many times contamination is carried away from the polluted area by various methods including physically moving contaminants to other areas, winds blowing old tailings and other materials away from one site and settling in another area. The contamination is deposited into nearby neighborhoods. The soil in these areas many times will have hazardous levels of materials that could affect the health of the people and animals living there. In the years previous to a site being declared a Superfund, toxic materials have been accumulating in the soil. Discovery of the problems comes well after the damage may have been done.

In Chattanooga, Tennessee, a group of neighborhoods sits within what is known as the Southside Lead Superfund Site. This is an area which is plagued with lead poisoning from industrial activities. The EPA is slowly testing people's yards in the area, and out of the first 1,100 yards tested for lead contamination, 450 are needing to have the top two feet of soil removed due to high lead levels. Lead is toxic, especially to children, causing brain and nervous system damage in addition to causing learning disabilities. In adults lead can cause kidney damage and cancer. The EPA is especially concerned about the hundreds of children living in the area who are exposed to lead when they play outside. Parents have been urged to test their children for lead as it is going to take years to test and remediate all the properties that need it. In the meantime, hundreds of people live in conditions that expose them to toxic levels of lead.

In Dewey Humboldt, Arizona, the Iron King Mine/Humboldt Smelter (IKHS) Superfund site sits nearby. Designated as a Superfund site in 2008, the area was home to mining operations for gold, silver, lead, and zinc. Mine tailings totaled four million cubic yards of waste piles. For years, tailings with high levels of arsenic and lead have been blown about the area.

For over 40 years, until 1937, there was also a smelting operation located there that crushed and melted copper and lead ore in furnaces. The smelter spewed out huge amounts of lead and other metals from its smokestacks that were carried away by the winds. The lead and other contaminants settled all over the town. Lead does not go away. It stays in the soil, and over the years continuously exposes people to it.

Finally, the EPA decided to test some properties for lead. The area does have naturally occurring levels of lead and arsenic in the soils, but after all these years since the smelter and the mines closed, dozens of yards showed high levels of lead. They lived like this for generations. The children played in the yards and the dirt. People grew vegetables in gardens using lead-contaminated soil and ate them.

The EPA is going to dig up a foot of contaminated soil and replace it with clean fill. As for the other properties that did not test high enough for remediation, the EPA will do nothing. The lead is still there, but activity on the properties over the years simply mixes the lead into a greater amount of soil. The additional amounts from decades of industrial pollution are just being washed under the soil.

The lead remains. The gigantic piles of lead-contaminated mine tailings still remain to blow about in the wind and wash away during storms. The problem has only had a band aid solution for now.

In Charlevoix, Michigan, residents there learned the hard way that just because the government and EPA says the site is cleaned up does not necessarily mean it is. In downtown Charlevoix, there was a Superfund site back in the 1990s where dry cleaning operations and a couple of industries had contaminated the groundwater of a substantial portion of downtown. The city moved the wells, which were the main source for public drinking water, and banned any wells from being drilled in the area. The EPA thought the contamination in the water would clean itself after a period of time. Residents thought the Superfund had been cleaned up and went about their lives. Unfortunately, the EPA was wrong, and toxic fumes from the saturated soils underneath their building had penetrated them and vapors of toxic chemicals were seeping in and mixing with their air. Meanwhile, the health effects from the chemical contamination in the homes is being downplayed. The EPA is installing venting systems in the affected homes to clean the air. There are also plans to demolish several buildings and remove the highly contaminated soil beneath. How much exposure to the chemicals were the residents of these homes subjected to because the EPA closed the clean up way too early? Nobody knows yet at this time, but, as expected, the residents are frightened.

It is fairly easy to learn if you are currently near a Superfund site or there may be one in the area you are considering moving to. The advantage to being on the Superfund list is it makes it easy to see where these waste sites are. It has been estimated, according to EPA figures, as of 2016, approximately 15 million people live within one mile of Superfund sites (roughly 5% of the U.S. population). Many of these sites the EPA lists as "Not Under Control." This means that human exposure, contamination of the groundwater, or both are still a sizable risk.

There are quite a few websites, including the EPA's itself, with a comprehensive list of current Superfund sites. In addition to telling you the location, there are also profile information about the known and suspected wastes at the site, descriptions of clean-up measures to date, and planned future ones. Using this information and the internet, you can easily find a large number of news stories and people connected with these sites. In the past, Tony and I have spoken to dozens of people to help us in gathering information on how the cleanup was progressing and discussed health issues the Superfund site caused to people.

The category of Superfund sites demands using extreme caution and alertness if you live near one or are considering a move close to one. The potential risks from exposure to the chemicals and hazardous substances in these sites can't be stated too strongly. Do not in any way think living near a Superfund site is safe now or will be in the future. Nobody, not even scientists or government experts, can ever guarantee the area is totally clean and safe.

There truly is no one safe distance to be from a Superfund site. There always is the prospect of chemicals that were never discovered to still be present and working their way out through the air or infiltrating the water underneath. Fumes still may permeate buildings and homes in the area for many years to come. We suggest a minimum of two miles away from the Superfund site as a safer place to be.

Chapter 2. Trash Incinerators

Trash incinerators have been marketed to the public as the environmentally safe way to dispose of our trash instead of sending it to those horrible landfills. In some places the industry has done such a good job at convincing politicians of this, several have even included these incinerators creating electricity from garbage as part of their push to meet clean energy goals. The industry likes to call them "resource recovery units," but they have other names, like "trash incinerator," "garbage incinerator," or "solid waste incinerator." Regardless of what they're called, these incinerators are all the same thing. There is nothing resource-recovering about these giant trash-burning monstrosities.

The trash incinerator industry likes to refer to themselves as "waste management and treatment companies." They like the public to believe they are doing a great service for the environment and mankind. All trash incinerator companies take the waste produced by industry, commercial, and residential customers and burn it. That is the simplest way to describe what they all do. What else they all have in common is they emit toxic compounds like mercury, lead, and cadmium, to list just a few.

I admit that I am not a fan of the incinerator industry and most likely will never become one. Since the mid-1980s, I have been involved in one way or another with the incinerator issue, always on the side that incineration is not an answer. It just shifts the problem from one place to another, mostly into the air.

The belief or idea that anything is able to be incinerated and disappear makes absolutely no sense if looked at from a scientific view. Basic science teaches us that nothing is ever destroyed completely, only changed to another form. The definition of "incineration" from an environmental viewpoint is "to burn something and create different products more dangerous than the one originally put in." The burning of trash in an incinerator simply changes the structure of much of the materials into other ones.

Trash incinerators have been around since the late 1800s. They started having widespread usage in the 1920s. It seemed to be a simple solution to get rid of trash. Many people had backyard incinerators, which were nothing more than drums with holes punched in them. I admit that I remember going to my grandfather's house and sitting in the backyard with him while he threw everything from the household trash bins into his incinerator and burned it. My job was to clean out the ashes at the bottom after the fire was out and put them into the garden as fertilizer. If I only knew back then what we were doing.

The giant, specialized, modern-day trash incinerators still have some things in common with the backyard burn barrels. They may have fancy bells and whistles, but they still continuously emit toxic chemicals. The problem now is, due to their size, the amount of toxins and pollution they dump on the surrounding community is magnified thousands of times. Incinerators reduce the volume of what they burn, but at the same time, they multiply the toxicity by concentrating the materials that can't be destroyed through incineration. The incinerators end up with large amounts of hazardous, highly toxic ash that needs to be specially handled and disposed of. Some of it is more dangerous than what was incinerated in the first place due to chemical changes during the incineration process.

The term "municipal solid waste" is the one frequently used for the trash or garbage that the trucks are picking up in town. Incinerators use natural gas or other fuel to burn the paper, food waste, plastics, metal, and whatever is in the trash stream. The EPA estimated plastic made up 16.4% of the materials incinerated in 2017. The largest problem with burning plastics is that, when it is burned, it creates new toxic compounds much deadlier than the waste itself.

When the trash is placed into the incinerator kiln and burned, the original volume of trash is reduced by about 90%. The other 10% is made up of heavy ash residue and incompletely burned items, as well as fly ash, which is the lighter ash that contains heavy metals and toxic chemicals, such as dioxins, furans, and PCBs. The fly ash also contains "products of incomplete combustion," which are new chemicals formed by unburnt ones recombining in new, more dangerous combinations. Many of those go out the smokestack and into the air people breathe.

No matter how they're marketed, incinerators are not green renewable energy. If the goal is to reduce waste, these materials, which otherwise would be recycled into new usable items using less energy than creating them again, are instead turned into waste and pollution. The companies that run incinerators also discourage recycling by the simple fact that the materials most easily recycled are their best fuel.

The appeal these incinerators make to the communities is typically to promise to share the wealth. The company speaks about all of the money that can be made in fees for the trash taken into the incinerator. They also will point at how much less will have to be landfilled and mention how all the energy from the burning of the trash is clean, renewable energy. What usually is failed to be discussed are more than a few critical factors.

One often overlooked fact is that there needs to be a minimum amount of trash for the incinerator to burn just to cover operating expenses. In the case that a city or county owns the incinerator, there is the additional expense of the bond and interest payments for the incinerator itself. So, if a certain amount of trash is needed just to break even and the more trash it burns the more revenue, trash reduction and recycling programs usually take a hit. After all, more trash is more money, so there's a financial incentive not to reduce waste. In many cases incinerator operators still did not have enough trash to burn to pay their bills, so outside trash from other counties and states is being shipped in.

All trash is not equal either. Most incinerators use natural gas as the primary fuel for burning trash. The higher the temperature, the better the burn, and the less volume left at the end to dispose of. Everything generates heat when burned. How much is generated is said to be its BTU value. Wet trash, mostly made up of food waste, does not have any real burnable fuel in it, or, as experts would say, it has a low amount of BTU value.It needs additional BTU's in the form of extra fuel to generate enough heat to burn.

The best trash is the stuff with high BTU value. Incinerator companies prefer trash that has a high percentage of plastics and paper, the very items that are the easiest to remove for recycling. Those materials allow the incinerator operators to use less natural gas and can burn more trash--less expenditure, more income. So, there's also a financial incentive for them not to recycle anything.

There is a big problem with burning the items with the most BTU value for the operator: it creates the most harmful pollution for the air and people living nearby. One material with a high BTU value is some metals that when subjected to the high temperatures of the incinerator will melt and sometimes burn. Metals create a lot of heat when burned, as does PVC. However, metal is not destroyed by incineration; it goes up in smoke and is mostly released out the smokestack or remains in the incinerator ash. PVC, when burned, creates dioxins and furans. If chlorine is in a product that is incinerated, the chemical reaction also will create dioxin. Chlorine is used in manufacturing a lot of items, so it is fairly common to have this issue as well.

The biggest problem with trash incinerators is not what goes into the facility even though hazardous materials that should not be burned there are frequently found. The biggest issue is what comes out of the incinerator in the smokestack emissions. I can't emphasize enough that the danger to the community from these emissions is so dangerous both from specific toxic chemicals in the air to tons of tiny particles too small for human lungs to screen out. They get into the bloodstream and attack the body from the inside. Emissions from incinerators typically contain heavy metals; lead and mercury are just two of the metals found in smokestack emissions. They also are two of the most harmful metals to be exposed to continuously.

The emissions also contain a plethora of other potentially harmful chemicals: sulfur dioxide sulfur, nitrogen oxides, dioxins, furans, particulate matter, carbon monoxide, and more.

Some of the nastiest chemicals are formed when the incinerator does not burn material completely. These partially combusted materials recombine with others to form new, more dangerous compounds (like the previously mentioned dioxins), many of which scientists are unable to fully determine based on the lack of extensive monitoring and wide air sampling .

All incinerators pollute and release dangerous toxic compounds into the air, even new incinerators featuring the latest technology in pollution control devices. Regardless of how advanced pollution controls are, they fall short. Modern incinerators may be able to trap some of the pollutants, but dioxins, furans, and other hazardous emissions of small to ultra-fine size are not, allowing them to escape. The EPA neither regulates nor monitors particles of this size even though they are the most dangerous to the human body.

Some of the issues with trash incinerators need a closer examination.

1. Dioxins and Furans

Incinerators naturally create emissions as materials inside burn. Some of the most hazardous materials known to humanity, dioxins and furans, are created by the burning process itself when chlorinated wastes such as wood, plastics, and various types of paper are incinerated. When plastic waste is burned, especially PVC, dioxins are only some of the toxic chemicals created in the incinerator. New toxic compounds are created at all the stages of the incineration process. Toxic compounds are created in the initial combustion of the trash. Toxic compounds are created when chemicals not destroyed in the incinerator combine together in the smokestack.

Toxic compounds are created at every step in between. Smokestack filters only reduce and remove a certain amount of pollutants, and some are not filtered out at all. Dioxins and furans are almost always present and found in the materials emitted from the incinerator smokestack.

The particle size of much of the dioxins and furans measured in smokestack emissions is the smallest category, the ultra-fine size. Not only are many particles of that size not captured by the filters and pollution devices in the smokestack, they are also the most dangerous to breathe. Those are the particles the lungs are unable to filter out as well, and they go deep in the lung's air sacs and into the bloodstream. A study published in the *Journal of Air and Waste Management* that analyzed measurements of dioxins in various size emission particles has shown that when the particle size decreases, the quantity of dioxins found increases. The National Institute of Health also showed the same results when testing samples of the dust in the air.

According to the National Institute of Health, much of the formation of dioxins is thought to take place in the exhaust gases as the emissions begin to cool. This is one of the primary reasons for the toxic chemical being released along with other emissions from the incinerator. While some are produced prior to that point in the process and partially filtered or deposited in the fly ash, a significant quantity makes it out of the smokestack and into the atmosphere.

2. Truck Traffic

The incinerators are huge and stay hungry for as much trash as they can be fed. Sometimes an incinerator may be closely located to a major highway, but usually the trash trucks need to travel through local streets and neighborhoods to reach the incinerator. Then, they dump their load and travel right back out again through the local streets and neighborhoods. Besides the noise they make, they also stir up dust and dirt and burn diesel fuel in the truck engines, adding more fine particulate matter for the residents nearby to breathe and the diesel fumes contain chemicals considered known carcinogens.

3. Particulate Matter

The particles in incinerator emissions contribute to serious health issues. When these are inhaled, the particles and other pollutants are often too small to be filtered by the lungs and little cleansing hairs in the upper airways. They make their way deep into the lungs and eventually the bloodstream. The amount of particles emitted from one of these trash incinerators is measured in tons per year.

The incinerator operators brag about how effective their pollution control devices are and how they eliminate much of the pollution. Regardless of how well they work, there are still tons of fine and ultra-fine particulate material too small to be completely filtered out of the exhaust gases. Many of these escape and wind up covering the local areas, waiting to be breathed in and trapped in the lungs where they can do significant damage.

4. Toxic Air Releases

State of the art filters and scrubbers do decrease the amount of pollutants released from these incinerators, in some cases tremendously from 20 years ago. However, for many of the pollutants and toxins currently emitted, the amounts are still far above harmless levels. In fact, according to the EPA, exposure to even very small amounts of toxins like mercury, lead, and dioxin cause irreparable damage, especially in children.

Furthermore, the information the incineration industry touts almost entirely relies on testing done under 100% optimal and perfect conditions. They do not address the startup phase in which incomplete burning is common, nor do they address upsets, which happen frequently in the combustion chamber, releasing pollution in far greater quantities. Occurring quite frequently, upset conditions are when the incinerator is not running properly, a state that can be caused by several reasons. There could be an imbalance in the various types of materials being burned. There could also be an issue with the temperature. If the incinerator is too hot, the emissions controls need to be bypassed for safety reasons to avoid an explosion or other issue. If the incinerator is not hot enough, it cannot completely burn what is in the incinerator and more waste is vented partially burned into the air.

In Baltimore, a trash incinerator emitted 300 pounds of lead into the surrounding neighborhoods and communities. Lead is a dangerous heavy metal known to harm children's developing brains and pregnant mothers' unborn babies. The city is trying to get the incinerator to upgrade its pollution control devices and enacted more stringent air pollution regulations in an attempt to force the operators to comply or close. There is a court battle over whether the new rules are legal. The incinerator was responsible for emitting over 1,100 tons of nitrogen oxides in a single year.

In Montgomery County, Maryland, the trash incinerator is one of the major polluters in the county. The facility located in Dickerson emitted over 390,000 tons of pollutants according to the group Chesapeake Climate Action Fund.

In a heavily industrialized area in the city of Baltimore, according to figures from the EPA, in 2014 one incinerator was responsible for 64% of nitrogen oxides and 82% of sulfur dioxide inside Baltimore city limits.

In Saugus, Massachusetts, the trash incinerator owned by Wheelabrator is an older incinerator, and the cost to upgrade the pollution control devices to meet new Massachusetts standards for nitrogen oxide is very costly. The operator is requesting an exemption from having to meet the new standards and be allowed to continue polluting at the more dangerous levels it had been. Health effects from high levels of nitrogen oxides include irritation of the respiratory system, nausea, coughing, difficulty breathing, skin and eye irritation, increased symptoms of asthma, and respiratory infections. There are 150,000 residents living in the area the incinerator emissions impact.

If the incinerator is not allowed to continue releasing the current amounts, they are asking the state to allow them to purchase emission credits to offset the extra pollution. Emission credits are created by companies who actually reduce their emissions and get credits for the accomplishment. The rules of the game, however, allow those companies to sell the credits to bad polluters like Wheelabrator who uses them to continue emitting non-compliant levels of pollution. Residents are asking the state to deny the request and force the company to comply or close.

5. Uncontrolled Releases

Regularly occurring situations in the incinerator cause large amounts of hazardous chemicals to vent to the atmosphere. This venting of hazardous chemicals is much more common than the industry likes to admit. Particularly when the incinerator is starting up or shutting down, materials placed in the incinerator may not get an optimum operating condition to completely destroy them. Releases of heavy metals, dioxins, and mercury in excess of what should be emitted are quite possible. There are other reasons this could occur, such as a malfunction in the equipment, the operators not conforming with propers standards, and even unpredictable coincidences.

In Newark, New Jersey, the trash incinerator sent out pink smoke clouds. Local television and social media showed pictures of the smoke billowing out. The incinerator spokesman said there was nothing to see; the unusual color must have come from a load that had too much iodine in it. They then blamed it on the customers. In Lancaster County, Pennsylvania, the trash incinerator sent out billows of pink smoke too. The spokesperson repeated the same story as Newark. The EPA lists iodine as an inhalant with a serious risk to health. Iodine can irritate the lungs, and in high exposure, could cause a build-up of fluids in the lungs. In the meantime, it can cause headaches, nausea, vomiting, and diarrhea.

6. Accidents

The trash incinerator industry has a track record of having accidents. Considering they literally set things on fire in their incinerators, accidental fires are far from a rare occurrence. When fires do happen, all of the emissions from whatever is on fire goes directly into the air. Any heavy metals, dioxins, and fine particulate matter blanket nearby areas, and residents are subjected to breathing the polluted air.

In Fairfax County, Virginia, an incinerator that burns one million tons of trash per year caught on fire. The smoke billowing out contained unknown chemicals and substances. The firemen were unable to control the fire for a while because of limited access. The surrounding community is being subjected to whatever is in the smoke and ashes raining down.

Six firefighters needed medical attention from issues related to fighting the blaze. While the county claims there should not be any long-term effects to the residents, the inhalation of the smoke and fine particles led to irritation of the eyes and throat, difficulty breathing, and chest pain in some residents. The county said that because the fire lasted a relatively short time, there should not be long term health issues but obviously if there are complications, people should continue to see their doctor.

There have been three previous fires at this facility.

In Dickerson, Maryland, another of Covanto's trash incinerators caught fire. When fire personnel arrived, there was an 85 foot high pile of trash engulfed in flames. This fire burned for 11 days before it was able to be extinguished. This facility also had two other fires earlier that year. The company says everything is under control and they are in compliance with regulations. During the fire, Montgomery County issued a stay in place health advisory for people who have respiratory issues due to high amounts of particulate matter in the air. The emissions from the open trash fire released emissions that may have contained dioxins, lead, and mercury. Despite the plant owners telling the county and public everything was fine, during 2016, the plant was shut down for 105 days for what the operators described as unscheduled outages in a seven month period.

7. Non-Enforcement of Regulations

Violations of environmental laws often go unenforced, and so the incinerator operators routinely exceed permitted emissions--a significant problem. Sometimes, holes are written into the laws to allow for these emissions.

In the city of Detroit, an environmental group watches the incinerator run by Detroit Renewable Energy. They reviewed emission violations between 2013 and 2017 and found that the incinerator had exceeded allowable limits almost 800 times. The total number of times the state Department of Environmental Quality issued fines was eight. The regulations were written so the incinerator can exceed pollution limits anytime the equipment is starting up or shutting down.

The state enforcement also used their own arbitrary guidelines for other emissions as well. The rule says the facility can't exceed carbon monoxide emission for over 60 minutes. However, the state only enforces that regulation when they exceed emissions for over two hours, or double the exposure time. The spokesperson for Detroit Renewable Energy said "We are trying to be a better neighbor."

8. Illegal Dumping and Discharges

It costs money to dispose of the leftover ash and unburned materials from the incinerator. Proper management of the toxic ash is a problem for incinerators. The level of toxicity often requires special disposal methods to be used. The costs of disposal may skyrocket. To avoid this, it is a common practice to blend highly toxic ash with less toxic material to reduce the level of toxicity, making it easier to dispose of in less expensive landfills as solid waste.

Ash management is a problem for incinerators, and serious cases of mismanagement have occurred in recent years. In Massachusetts, an incinerator owned by Wheelabrator was caught not properly treating and disposing of the toxic ash leftover after burning trash. They also were caught releasing 8,000 gallons of ash sludge and contaminated water into the streams, rivers, and wetlands surrounding three of their facilities. The incinerator operator was fined $7.5 million dollars. Depending on how many other times they were never caught doing the same thing, the amount of the fine is likely small compared to the amount of money they earn.

Living near a trash incinerator subjects people in the area to constant exposure to hazardous emissions and large amounts of particulate matter. Trash incinerators are certainly not safe to have in a residential community. The risks both to human health and the community are many.

In spite of this, in two dozen states the lobbyists for the waste industry have managed to get approval for tax credits for trash incineration companies. Covanta's incinerator in Oregon is trying to get it approved there now. The company donated $35,000 to political campaigns in 2018 as the decision was nearing.

In 2019 in the state of Maryland, lawmakers passed a bill that mandated a large increase in the percentage of energy which must be from green and renewable sources. Through the use of the powerful waste industry lobbying groups, the law also now considers trash incineration renewable energy. Burning of large amounts of plastic and other recyclable materials does not meet any reasonable definition of renewable energy.

There are no upsides to trash incinerators. From the day one opens its doors for the first truckload of trash until the day it closes, it will be one of the largest polluters and sources of hazardous emissions in town. Stay as far away from them as you can manage. Leave at least two miles in between you and the incinerator. More if possible.

Chapter 3. Hazardous Waste Sites

One of the worst ways to be exposed to hazardous chemicals is to spend years living on or near a hazardous waste site without knowing it's there. But it can't stay hidden forever. Eventually, someone in the family gets sick or the neighborhood starts having three, four, or more cases of a rare cancer on the same block. Exposure to a hazardous waste site is one of the many frequently preventable hazards that Tony and I see.

In all our research, Tony and I quickly learned one important truth: there is no such thing as a typical waste site.

It's hard to say anything collectively about their sizes because they come in all different sizes. It may be as small as an old gas station with a few abandoned drums of waste, or it could cover hundreds of acres like some polluted industrial parks where textile mills and chemical companies operated for years. These sites could be located in highly populated areas or could be in a very rural area such as a group of three farms in central New Jersey. All located in Plumstead Township, they were used for the disposal of hazardous waste and polluted a large area of the township. Known as Goose Farm, Prijak Farm and Spence Farm, all were listed as Superfund sites.

It's also hard to say anything collectively about their contents because these sites could contain a single type of hazardous waste or many different types all dumped together, creating a toxic ooze.

It's even hard to say how many hazardous waste sites there are. In an attempt to find the number of hazardous sites in the United States, Tony and I contacted quite a few sources. We received numbers from the federal EPA, individual states, environmental groups, and even Alexa. Counts varied widely, as did opinions of what types of waste sites should be included in the numbers each counted.

One group, Infrastructure Report Card, an organization that does annual assessments of different critical aspects of American infrastructure, counted 18,000 sites. The EPA talks about 450,000 sites in the brownfields category alone with half of those coming from leaking underground storage tanks, most from gas stations. The National Research Council did a count some years back looking at sites with groundwater pollution from industrial operations. Their number was more than 126,000. New Jersey alone counts almost 14,000 hazardous waste sites of varying levels of danger in their small state.

There is really only one thing though all hazardous waste sites do have in common: they all contain hazardous waste. That is the only commonality that the hundreds of thousands of sites all have.

According to the EPA, "Hazardous waste is waste that is dangerous or potentially harmful to our health or the environment. Hazardous wastes can be liquids, solids, gases, or sludges. They can be discarded commercial products, like cleaning fluids or pesticides, or the by-products of manufacturing processes."

That definition covers a lot of ground. Hazardous waste can be almost anything if it has the potential to be harmful to human health and the environment. This is disturbing because there are an estimated 22 million acres of land considered to be part of the primary hazardous waste management program administered by the EPA and other responsible agencies. Enormous parcels of land have been laid to waste by mining operations stretching out over thousands of acres.

Many of the sites became contaminated by people and companies who improperly handled toxic waste or disposed of hazardous waste carelessly in a manner that was going to harm other people and the environment. The majority of our toxic waste sites are the result of such carelessness. The industry for years discarded their toxic waste, treating this poison as insignificant, throwing it onto the ground or into a ditch. Millions of tons of that hazardous poison were simply discarded.

There were many shady companies who specialized in helping industries dispose of their waste. These companies sometimes bought land in rural areas and turned it into lagoons filled with the most toxic poisons on Earth. These companies found farmers who allowed trucks to bring barrels of these toxic chemicals and pile them somewhere on their properties. Tanker trucks would pull down dirt roads and unroll their hoses to empty their liquid into a ditch or creek. Town landfills routinely allowed hazardous waste to be dumped alongside household trash and then covered over.

Accidents really do happen. Although many accidents could have been prevented with better maintenance or attentiveness, there are times when an accidental release of toxic substances happens. There are also leaks that may remain unknown for prolonged periods of time. A perfect example of preventable accidents are many of the toxic waste sites that are former gas stations. Many of these sites slowly leaked gasoline out of rusty old underground storage tanks. Had someone paid closer attention to the numbers pumped out with the numbers put in, they may have noticed a consistent difference and caught the leak much earlier.

When the location of a hazardous waste site is obvious, or at least suspicious and likely, we can avoid the area. The larger problem is that sometimes the chemicals from a hazardous waste site migrate off the contaminated area and into a nearby residential area. Many of the risks from living near hazardous waste sites are in the contaminants that travel off the site and away.

Some of the same groups who estimate the number of hazardous waste sites also estimate the number of people who live near one. Looking specifically at Superfund sites, the EPA estimated 16% of the population of our country is within three miles of one of these. In a one mile radius, 15 million people, or 5% of the U.S., is living that close to a Superfund site. These numbers certainly are not uplifting to see. By some definitions and estimates, over 75% of the population is within five miles of a hazardous waste site.

At many hazardous waste sites, the poisons sit, uncovered and exposed. Some have sat for decades since their discovery, leaking their toxic waste into the ground and soil and blowing away in the wind.

The three main ways people living near hazardous waste sites can be exposed to pollution are, as always, water contamination, soil contamination, and air contamination.

1. Water Contamination

Almost every household is hooked up to some type of a water supply. Public water supplies about 87% of the population with the other 13% of the country using private wells. Both sources are vulnerable to pollution from hazardous waste sites. Public or private, hazardous waste is an equal opportunity polluter. Toxic chemicals can infiltrate and contaminate either water supply. The only difference is that public water supplies are supposed to be regularly tested for toxins before they deliver water to their customers.

Underground plumes of chemicals from hazardous waste sites contaminate water, which we then drink, not knowing it is full of toxic chemicals. Many times the pollution may be there for years before it is even tested for and detected.

On Long Island, New York, the town of Bethpage is finally learning all the facts about the chemical leak Grumman Aerospace Company has known about since the 1970s, possibly even earlier. Decades of drinking contaminated water has many residents wondering if the cancers and illnesses they have had over the years are from the poison that Grumman knew was in the water but warned nobody. Investigative reporters uncovered information the state actually knew about the health risks once the toxins were found in the first well back in the 1970s but allowed people to continue to use the water as the poison spread. The state government kept private the health risks associated with drinking water even though officials from the federal Environmental Protection Agency told state officials that the water was unsafe to drink, but the state did not publicly admit until 2019 that it may have posed a health risk. The plant had been responsible for leaking and dumping dangerous chemicals such as trichloroethylene and volatile organic compounds and PCE, 1-4 Dioxane, and radioactive isotopes into the soil, all of which has extensively contaminated the groundwater. Eleven municipal wells supplying water to residents needed to be shut down, and about 17 more are currently threatened. The plume of contamination stretches 4.3 miles long 2.1 miles wide and is 900 feet deep. The EPA knows it will continue to expand.

In Paden City, West Virginia, residents are finding out their water from the city water department is contaminated with the toxic chemical Perc. An industrial solvent used for years in dry cleaning was found in the aquifer that supplies public water to the town's four wells. The EPA considers Perc a likely carcinogen, and studies link the chemical to increased risk of cancer, neurological issues, and reproductive and developmental problems. The city water has three times the federal limit considered safe. The town is telling residents that it's okay to drink the water for now, but doing so could lead to long-term health problems. The EPA took samples from a test well close to the dry cleaner that was suspected of leaking the Perc for over 20 years; the well had levels of Perc eight times the federal limit. Nobody yet can say with certainty how long the residents have been drinking the contaminated water.

In Prairieville, Louisiana, a hazardous waste site thought to have been cleaned up thirty years ago turned up more surprises. Very high levels of additional chemicals such as vinyl chloride; 1,1,2-trichloroethane; and other chlorinated solvents and volatile organic compounds have turned up, indicating there is much more toxic waste still on the site. Testing of the groundwater found 13 toxic chemicals exceeding exposure standards. One of them, 1,1,2-trichloroethane, was measured in the groundwater at 44 parts per million, nearly 1,050 times greater than the DEQ's level of safety. The EPA said the chemical only causes cancer in laboratory mice, not humans. The CDC however calls this chemical a possible carcinogen also affecting the liver and nervous system. The state claims even though the drinking water aquifer under the site is contaminated, none of the water is currently being used for drinking water. There is a plume of chemicals stretching under the area to the end of the neighborhood. Shockingly, the state environmental agency has no plans to fully clean up the area. They only want to remove enough of the toxic chemicals to bring the contamination down to governmentally acceptable levels even though these levels are far from acceptable to the people living there.

In Phoenix, Arizona, a plume of contaminated water in the aquifer below the city stretches out, connecting to other plumes of water all contaminated with industrial chemicals and waste. For decades, many factories in Phoenix's industrial areas dumped toxic chemicals on the grounds around their plants. The interconnected plumes cover 15 miles beneath the city streets and neighborhoods.

Parts of the plume have been known since the late 1980s. The chemicals in the water cause cancer, lead to birth defects, and affect neurological systems. The water is not being used for drinking, which is good, but vapors from the large concentration of chemicals are suspected of traveling through the soil into the neighborhoods above, continuously for a long time. The city also installed air strippers to help clean up some of the water pollution, but they are vented directly out on the streets in the neighborhoods too. Maybe someday people will know what they are being exposed to, and it may eventually be cleaned up. That day most likely will not come until the fight over who owns the water gets resolved. The bigger issue for now is who is going to be able to have the rights to sell the water once it is finally clean.

In Belmont, Michigan, PFAS were found in neighborhood wells near a waste dump used by the shoe company famous for Hush Puppies. State health officials found levels of PFAS almost

90,000 parts per trillion when the EPA advisory limit is 70 parts. The people there had been drinking dangerous levels of PFAS for years. One woman's husband died from liver cancer, and her blood contained PFAS levels 750 times the national average. She now has thyroid problems and wonders what is next. The EPA says liver and kidney problems may develop from exposure as well as thyroid issues.

In 2017, more industrial waste was discovered in wooded areas hidden in the overgrowth where the neighborhood children play and hang out, exposing themselves to the chemicals. The landfill the company used was not the only place these chemicals were disposed of as they are being found in additional areas around the old site and in the neighborhood.

In addition to the water issues and waste outside the complex, excavators working inside the site have uncovered massive amounts of soil so contaminated with different types of chromium some in such high levels it appeared like bright blue soil at first. Multiple types of chromium waste in addition to PFAS and lead are being dug up at levels over ten feet deep.

2. Soil Contamination

Soil around and underneath many manufacturing sites contain levels of hazardous chemicals. Many hazardous waste sites consist of liquids and solids that have been poured onto the ground or burned on the ground.

In Honolulu, Hawaii, the discovery of lead in soil that is believed to have come from a long-gone fishing weight company is due to be removed. The hazardous waste site is only about 6,400 square feet, but almost 2,000 people live less than a tenth of a mile away. The level of lead was over 100 times what the regular level should be. It was so bad that the EPA said people could potentially have been exposed to the lead simply by walking near the waste site.

In the Overtown neighborhood in Miami, Florida, a crew doing some work found ten rusty barrels along the rear of an apartment building, buried under only two feet of dirt. The barrels were removed, and testing of the soil around the barrels showed a mix of toxic chemicals including barium, arsenic, Dieldrin (a now-banned, highly toxic pesticide), and other chemicals known to cause cancer.

Six months after the barrels were found, the area was fenced off using a cheap plastic fence and had no signs warning about the danger. None of the residents were told about the toxic chemicals in the soil because, as the county employee said, "There was no requirement to do so." One observer said the fence had been trampled down, so it is hard to say with certainty how many people walked through the area.

Some of the illegal disposal is being done by several of the large corporations such as AT&T, Walmart, and Crystal Geyser Water Company. All have been fined millions of dollars after being caught illegally disposing hazardous wastes at regular landfills.

Other instances include one in Savannah, Georgia, when a waste company transported hazardous waste and, instead of taking it to the disposal facility, dumped it on the ground in a nearby neighborhood. Luckily, local hazmat teams were able to respond quickly and clean it up.

In a section of the Coconino National Forest outside Flagstaff, Arizona, hikers stumbled onto an area where midnight dumpers disposed of drums of highly toxic cleaning solvents. Specialized hazmat teams needed to remove the contaminated soil to prevent the chemicals from reaching a nearby river where they were dumped.

The Forest Service says this is not an isolated instance but one which is becoming all too frequent. South of Seattle, Washington, in the Mount Baker-Snoqualmie National Forest, the hazmat coordinator said dumping of hazardous materials has increased with 20-30 toxic waste sites discovered annually.

3. Air Contaminants

All types of materials are found in toxic waste sites. Depending on whether they are liquids or solids, wastes may travel off the site through the soil, water or air. Airborne particles blown by the winds contaminate nearby areas. Some materials found at hazardous waste sites are perfect for being blown away with the breeze. Large piles of mine tailings containing lead and arsenic catch in the wind and float away, toxic ash from giant coal ash dumps comes down like gray snowflakes coating cars and shrubs, and dried particles of various toxic waste--many small enough to be trapped in the lungs--get breathed by people living near the open waste site.

In Chicago, Illinois, a toxic waste dump sits next to crowded residential neighborhoods. About 10,000 people live close enough to be at risk from the waste sitting open and unprotected. Testing has shown lead, mercury, chromium, and cyanide to name only a few. The dump was used by a steel company to dispose of its waste for years.

One material especially stands out as a danger to the nearby people. Dust from the special furnaces at the steel mill was dumped on the site. This dust is collected from filters on the furnace to prevent it from being emitted from the smokestack. It contains many toxic metals in particle sizes so tiny they are able to penetrate people's lungs and get into the air sacs and bloodstream. No covers or any protection prevents the dust from blowing into the neighborhoods nearby.

To call our country a toxic wasteland is far from the truth, but with millions of people living near hazardous waste sites, drinking water containing chemicals leached from these waste sites, and the diminishing possibilities or prospects for getting these cleaned up any time soon, many people live in their own toxic wasteland. Love Canal, Times Beach, and thousands more places exist. Despite environmental rules, despite the government spin, hazardous waste sites are one of the biggest threats to the health of the country's population.

Use any of the databases available to check if there is a hazardous waste site near you. The state environmental office almost always has the best and most current listings. Check with your county environmental office as well. Sometimes the town hall where you live is a good resource, but Tony and I have been more disappointed than pleased at their knowledge of what is in town and the risk it presents.

Living in the area of a known hazardous waste site presents multiple risks of exposures to toxic substances. Many sites are still considered uncontained or uncontrolled, words the government uses to mean the people and the environment are at high risk in the area. Because hazardous wastes travel in multiple ways such as through the air, soil, and water, determining a specific safe distance from any of these sites is can be difficult and varies depending on individual sites. Factors depend on your answers to various questions. Are you using a private well or public water? Is the site open and the wind blowing the poisons into the surrounding areas such as the thousands of hazardous waste sites containing toxic mine tailings? Are there any bodies of water such as creeks or ponds near the site? Is the site currently in any stage of being cleaned up and what is the plan? Is there even a clean up plan in place at all? Your answers to these questions affect how far from these hazardous waste sites you should live. Impacted areas from many of these sites extend for one to two miles in many cases. The sheer number of these sites in some locations makes this even more difficult for you to keep distance between yourself and the sites.

Living near a hazardous waste site is taking a chance we do not suggest. We have never found any of these sites not to present a risk at some level that did not concern us. As a rule of thumb, pathways of exposure would suggest keeping at least two miles away. If you depend on well water, regularly have the water tested at least annually and immediately if you suspect any strange tastes or smells.

Chapter 4. Power Plants

The United States is undergoing a change in the power generation industry. While clean energy resources such as wind, hydroelectric, and solar remain only a small part of the current power supply, they will have an increasing share in the coming decades. The big changes at this time are coming in the number of coal-fired plants closing and the number of new gas-fired power plants opening.

For years, coal-fired power plants have generated the majority of the electricity this country needed. The cost of coal as a fuel has always been quite low, and the availability of the supply is abundant. Coal-fired power plants are called the dirtiest way to make electricity because of the large amounts of chemical emissions and particulate matter the plants put out through the smokestacks. The communities that the power plants were located in usually had very little say in pollution control. Many plants were located in poorer communities that were happy to have the jobs these plants brought to town. Accepting the fact that coal-fired power plants continuously blanketed their towns and neighborhoods in dust and soot was a way of life for many people. These dirty burning coal plants are closing, not because they have become more expensive to run. Rather, gas-powered ones have become that much cheaper due to the fracking boom and the immediate supply of natural gas.

The era of the coal-fired power plant is slowly winding down. The closing of these plants can't come soon enough for many people. Fortunately for thousands of people, almost half the existing plants may be closed and phased out over the next 10-20 years, possibly even sooner, depending on how the industry handles the transition. At the time this was written in May 2020, oil and gas prices are at the lowest levels they have been in decades. Perhaps that may hasten the closings of these pants even faster.

In 2018, there were 336 coal-fired power plants still operating. That is significantly down from the 598 that were running back in 2008. Still, the amount of pollution they produce is significant. According to the federal government statistics for all types of power plants, the amount of pollution generated from coal-fired power plants is measured in the millions of tons for some of the chemicals emitted.

According to the U.S. Energy Information Administration (EIA. gov), carbon dioxide emissions alone from only coal-fired power plants was 1.15 million metric tons in 2018. That accounts for 65% of all the emissions from power plants. Considering the fact coal-fired power plants only accounted for 23.5% of all the power generated, that is an enormous amount of pollution compared to the small percentage of the power it created.

There are quite a few hazards people living near these power plants are subjected to everyday. Over time, the continued exposure to high levels of air pollution that they breathe, including chemicals and toxic metals from the exhaust, takes a toll on the body and health. It is impossible not to be affected by the assault of these toxic emissions just wearing down the immune system and respiratory system day after day.

Tons of chemicals are emitted every single day into the air near the plant. In addition, poisons from the irresponsible disposal of the waste products left after the coal was burned contaminates and poisons many communities' water supplies. A closer look at the risks from these monstrosities will show areas people living near still operating, or even closed, coal-fired power plants should be aware of.

1. Air Pollution

The amount of pollution and the specific chemicals and metals that coal-fired power plants emit make living near these plants a hazard. There is no dispute over that fact. The emissions cause a very wide range of potential negative health effects. Out of 187 air pollutants hazardous to health, coal-fired power plants are known to emit 84. According to the National Toxicology Register, many of these may cause cancer.

The hazardous air pollution emitted by coal-fired power plants is responsible for a wide range of heart and lung diseases. Other hazardous pollutants attack the nervous and respiratory systems. Kidney issues are also associated with these emissions as are some memory, learning, and behavioral issues in children.

While it would be tiresome to go over all 84 of the pollutants released into the air by coal-fired power plants, it's important to know some of the most prominent ones and their potential hazards.

A. Nitrogen Oxide

Nitrogen oxide visibly hangs like a fog in the air around communities near power plants. When inhaled, the particle irritates the linings of the lungs and inflames them. It causes the worsening of asthma symptoms in people already having a difficult time. Chronic respiratory diseases such as pneumonia and influenza are more likely. Power plants as a whole emitted 1.02 million tons of nitrogen oxide in 2018.

B. Sulfur Dioxide

This gas is created when sulfur reacts with oxygen in the combustion chamber. It then goes on to combine with other chemical elements, creating tiny and highly acidic particles able to get inside our lungs. Bronchitis and asthma are just a couple of the diseases caused by breathing in this chemical. Over 1.26 million tons were emitted by all power plants in 2018.

C. Carbon Monoxide

High concentrations of carbon monoxide such as those typically found near coal power plants are known to cause headaches in many people. It also places a lot of stress on the heart, particularly with people who already have heart issues. Over 1.15 million tons of carbon monoxide were released by these plants in 2018.

D. Mercury

Coal-fired power plants are the biggest source of mercury pollution in the country. For the year 2015, the EPA said coal-fired power plants were responsible for half of all the mercury emissions in the country. Mercury is a heavy metal that is very toxic. It does not get destroyed when burned in a power plant but does get emitted into the exhaust emissions or into the toxic coal ash. While better pollution control devices have been installed at many coal-fired power plants, the quantity of mercury still being emitted continues to put people nearby at risk.

Exposure to mercury is linked to issues with the neurological system and developmental problems. Simply put, exposure to even tiny amounts of mercury can lead to brain damage and birth defects in unborn children. According to the EPA, just 1/70th of a single teaspoon of mercury would contaminate a 25 acre lake so badly the fish all would contain too much mercury to eat. In the year 2014, EPA records show that coal-fired power plants emitted 45,676 pounds of mercury into the air.

E. Arsenic

This chemical is also frequently found in power plant emissions in high enough quantities to be a concern to surrounding areas. It is extremely toxic, and as few as 50 parts per billion is enough to cause cancer. Power plants pumped 77,108 pounds into the air in 2014.

F. Lead

One of the most highly toxic heavy metals found in significant amounts in plant emissions is lead. There were more than 41 tons of lead emitted in 2014 alone. Like mercury, it only takes a tiny amount to cause harm to people's brains, especially those of young children, which then affects their development. Lead affects the nervous system, decreases kidney function, hampers the immune system, and can lead to cardiovascular issues in adults.

G. Other Toxic Heavy Metals

According to the EPA, coal-fired power plants emit between 20-60% of all the toxic metals emitted by industry in the U.S. Even small amounts of some metals such as cadmium have large effects on the body and organs. Continuous, low-level exposure can lead to

cardiovascular disease and the early onset of dementia. Almost 10,000 pounds of cadmium was emitted in a single year by these plants. Thallium is another heavy metal frequently found in coal-fired power plant emissions, affecting the nervous system, heart, liver and lungs. While there are technologies available to further reduce some of the emissions, the industry has been slow to add them.

H. Radioactive Contaminants

Much of the coal burned contains some amount of radioactive isotopes. This is not a huge issue, but certainly an area of concern. When burned in the power plant, emissions contain uranium, thorium, and other radioactive elements that are then released into the atmosphere in concentrated forms. The amount is not negligible; over time and continued exposure, there is enough to accumulate in the bodies of people living nearby. The radioactive elements stay in the body and cause damage to the cells they attack, gradually causing the cells to mutate.

I. Particulate Matter

This is one of the largest single causes of health issues connected to coal-fired power plants. Particulate matter is simply small particles that contain a variety of substances, sometimes also called soot. This is the dirty coating that accumulates on cars and other surface areas in coal plant communities. The particles vary in sizes, and the smaller the particle, the more dangerous it is. When inhaled, the smallest ones get trapped in the lungs and are unable to be exhaled.

They attach to the tiny air sacs, and some end up traveling into the body through the bloodstream and affect the organs. Some of the known effects of continued exposure are severe asthma, chronic bronchitis, heart attacks, and other cardiovascular issues. Particulate matter in the lungs makes the body work much harder to compensate for the reduced air flow, putting extra strain on several critical body systems. Tons of this particulate matter are emitted by coal-fired plants every day.

2. Coal Ash - Water Pollution

Power plants generate a toxic product so hazardous it is considered one of the greatest threats to people and the environment. This threat is doubly hazardous due to the quantities of coal ash being generated, which is measured in the millions of tons annually. The toxicity of the ash is almost like the contents of a hazardous waste dump itself. This pollutant is so toxic that it gets its own chapter later in this section, but we included coal ash in this chapter to ensure our readers understand the threat it presents. Coal ash has been disposed of in thousands of unlined dumps over decades and threatens many communities across the country everyday.

Even though a coal plant closes and is no longer actively releasing hazardous materials into the air, the threats posed by coal-fired plants don't end there. There are the hazards that remain.

Even after these coal plants are closed decades from now, there are still other legacies of their days that will live on possibly for hundreds of years.

The single biggest issue threatening the quality of water in the country today is coal ash disposal.

Coal ash is the hazardous waste that remains after the coal has been burned. It is the ash from the bottom of the plant's furnaces as well as the ash caught in the pollution control devices if the plant has any. The amount produced annually is well over

100 million tons, and the amount stored in various places over the years is in the billions of tons.

A report compiled by the Environmental Integrity Project using data submitted by the power plants themselves detailed the extent to which toxic coal ash has contaminated the groundwater near these plants. It shows over nine in ten already have polluted the groundwater with unsafe levels of pollutants from a wide range of toxic and hazardous chemicals leached from the fly ash disposal.

The data showed the groundwater of 242 power plants out of the 265 that sent in reports for review had at least one pollutant found in coal ash in unsafe levels. More than half of the 242 had unsafe levels of arsenic, which is linked to cancer and is very toxic in even small amounts. Lithium, a metal usually associated with batteries, was found in high levels at 60% of the plants' groundwater. Neurological damage is associated with lithium exposure.

Many communities are already reaping the ill effects of coal ash contamination in their water.

In Memphis, Tennessee, a dump used for coal ash disposal for years is leaking toxic chemicals into the groundwater. Arsenic, among other pollutants, has been found in tests of the aquifer underneath the plant, an aquifer that is used as a drinking water supply for over 650,000 people in the community. There is no known way to prevent the current contamination from moving into the water needed for drinking. The town plans to clean up the site, but it may be too late since the leaking has been going on for decades.

In Belmont, North Carolina, letters were sent to residents advising them the wells they were using were contaminated. Tests showed high levels of several pollutants, including chromium-6, exposure to which can cause kidney and liver damage, pulmonary issues and cancer. The water was unsafe to drink or use for cooking, and over 900 homes affected had to rely on using bottled water.

Every power plant in the state of Texas has contaminated the groundwater according to documents the companies submitted to the government that were reviewed by the Environmental Integrity Project. Toxic chemicals have contaminated the groundwater near the plants with lithium, cobalt, arsenic, chromium, boron, manganese and beryllium. One family ranch near a power plant has water contaminated with 12 different chemicals, some at levels 100 times the safe level set by the EPA. People across the state are exposed to health effects such as kidney and bone damage, as well as a variety of cancers.

The most important way to combat potential water contamination is to routinely test your water for any pollutants (If you are using well water, an annual test should be done no matter where you live). If unsafe levels of pollutants are found, you should call the local or state health department. They can offer advice on how to reduce the levels using a filter to limit any exposure. The last alternative would be to use bottled water instead for drinking and cooking. Bathing may release chemicals into the air, which you would then breath, so this presents another problem.

Regardless where you get your water from, don't just rely on tests performed by local authorities. Public water systems and private wells are increasingly being contaminated from nearby coal ash disposal sites. There have been many cases where the public water was supplied to customers for years before tests revealed it had been polluted the whole time. Doing a test of your water on your own may determine there is an existing problem needing action on a higher level.

There are thousands of people living near power plants that have been polluting the local community's water supply for many years. The extent of the widespread problem continues to grow like the plume of poisons spreading from the coal ash leaking into the water below. It is only going to get worse. More communities are finding their water supplies polluted. Too many communities are facing situations without many solutions to solve this problem.

Sadly, it falls on the individual to keep their own water safe. The government is doing way too little way too late. They knew this was a major threat to the water supply but allowed the sloppy disposal of coal ash to continue.

As a whole, no matter where, you should avoid being in an area within four miles from one of these power plants. The huge amount of pollution from the smokestacks travels for miles. Even the ones that have already been closed down or are scheduled to be closed down have coal ash disposal areas nearby. The question is not if they will leak, but when. Based on the lack of any real structural requirements for these disposal areas, they most likely already are.

Chapter 5. Coal Ash Dumps

When I first heard about coal ash, it was because I read about an accident that occurred when a coal ash dump burst through its dam and emptied its contents into a nearby river. It contaminated nearby waterways and affected an entire community's water supply. At that time I had no idea what a coal ash dump was or, for that matter, what coal ash was anyway. Looking into this, I found out that the power plant industry creates millions of tons of coal ash every year and has to dump this hazardous waste product someplace. As a result, they need coal ash dumps, giant impoundments holding millions of tons of material that anywhere else in the world would be called hazardous waste, but aren't classified as such here in the U.S.

Coal ash and its disposal is not a well-known topic to most people. However, people who live near one of the known 735 giant slurry ponds or over 310 ash dump landfills, likely are already aware of the hazards associated with it. And, unfortunately, they may have been one of the thousands affected by accidents and exposure to the hazardous substances it contains.

Coal ash is the residue leftover from coal-fired power plants. According to the EPA, it is the second largest waste product in this country with over 130 million tons produced every year. Despite coal ash containing high levels of metals and toxic chemicals, the EPA after years of review decided not to regulate it as a hazardous waste. The labeling of the waste as hazardous would have made disposal costs for coal ash much higher. The decision is unsurprising as Andrew Wheeler, the current head of the EPA, is a former coal lobbyist.

Coal ash is not simply harmless leftovers from burning coal. Coal ash contains a toxic assortment of metals and chemicals known to cause a variety of health problems in people who are exposed over a period of time. The list of the toxic ingredients in coal ash include lead, mercury, and arsenic. Also found are aluminum, boron, selenium, chlorine, and barium. Known health effects from exposure to these include birth defects, cancer, heart damage, lung disease, reproductive issues, kidney disease, and others. Long-term exposure builds up the amount of these toxins in the body, slowly eating it away and causing damage. This toxic brew affects practically every part of the human body. As always, children and pregnant women are especially affected by these toxins.

Over the years, the industry has had loose regulations when any at all were enacted. Further complicating the matter, the states have much of the responsibility to handle regulation at their level. Coal states especially benefited from this arrangement. The power plant industry saved many millions of dollars not having to meet strict requirements that other industries had to when they disposed of hazardous wastes. The power plant industry was special, and politicians labeled their coal ash as "special waste." Even though it was toxic and hazardous, the special classification gave them the freedom to dispose of millions of tons in ways they never could have if they had not been exempted from hazardous waste regulations.

With anything that is produced in such massive amounts, dealing with its disposal is an equally massive undertaking. For years, power plants simply dumped millions of pounds of coal ash slurry in unlined storage ponds or in mountainous piles. The slurry ponds ranged in sizes from a few acres to lake-sized ones held in place by simple dams.

The industry also tried to find other uses for the ash besides waste disposal. They used some very creative ways to make this toxic material appealing to other industries and have them want some too. After all, it was a "special waste." Currently, almost half of the volume of coal ash is, as the industry puts it, "recycled." In spite of the toxic chemicals coal ash contains, millions of pounds are used every year in making cement, gypsum board, and other building materials. It is used as a filler and binder. While some cheer the effort to use coal ash in a beneficial way, others can't believe regulations allow the ash to be disposed of in this way. In essence, this so-called recycling is putting toxic coal ash in the walls and foundations of people's houses.

When I took a trip to the Navajo Nation in Arizona, I could not help myself asking what they do with all the coal ash from their huge power plant. The lady we were with on our tour proudly said they used it to build their houses. I explained to our driver and the others in the vehicle with us my view of that. Our driver looked at me and said she was certain if they were using it in their houses, then it must be safe. After she made that statement, I just closed my mouth for the rest of the ride.

There are several hazards coal ash creates. The situations where thousands of people live their lives affected by it every day are almost everywhere these dumps exist.

1. Accidents and Dam Breaks

The favored method for coal plants is to find somebody else to take their coal ash waste and not have to deal with it. When they are able to get someone else to take it off their property, they will. When they can't get rid of it, though, it needs to be dealt with by the plant. Considering the power industry produces an average of 130 million tons of this hazardous substance every year, this presents a few logistical problems. Instead of creating a system that handles the waste in an environmentally safe way that puts the safety of people first, the government allows the industry to dispose of it how they see best.

The coal industry long ago decided that the solution was to take the ash and create giant areas resembling huge lakes or reservoirs and just dump the waste there. After a while, in order to accomplish this, they needed a way to hold the waste in place, so they built dams. Many of the dams they built to hold back millions of gallons of toxic coal ash were made out of clay or were simple earthen barriers.

These dikes and dams were doomed to fail. No long-term thought was given to the possibility of failure. It was all about cutting costs to maximize profit. The government allowed this to continue for many years by letting the industry basically self-regulate. All of the years the industry managed to escape from public scrutiny and government oversight changed on December 22, 2008.

It was on that day the poorly constructed dam at one of the largest containment ponds in the country finally had enough and gave way. Spilling through the breach came over one billion gallons of toxic coal ash. It flooded many of the residential areas near the plant. The toxic waste eventually flowed into the nearby river, contaminating that as well. The cleanup was a massive undertaking. The cost was placed at $1.1 billion dollars, and it took over ten years to complete.

Up until this enormous disaster, the industry managed to slip through the loopholes of regulations for disposing all this toxic ash and monitoring all of the lagoons, pits, and the dams expected to contain them.

In reality, not much has changed. The industry talks about all the new ways they are protecting the environment and the groundwater such as creating more “beneficial uses” of recycling coal ash, which the industry sees as the answer for disposal. However, these improvements register more as talking points from the industry rather than real improvements with the goal of safety first. Dumping tons of toxic coal ash down mine shafts and calling it “beneficial fill” is not a solution, nor is placing it in the wall boards and concrete we use to build our home with.

2. Contaminated Groundwater

The contamination of groundwater is one of the most serious problems facing our country. Once a source of water becomes unsafe for use as drinking water, it takes years to restore it, if it can be at all.Today, the majority of power plants have already polluted the groundwater near them. Leaking storage ponds are the most widespread problem currently. In 2018 a study was conducted by the Environmental Integrity Project with assistance from Earthjustice. They used industry data that became available to the public for the first time in 2018 because of requirements in federal coal ash regulations issued in 2015. The report found that the groundwater near 242 of the 265 power plants with monitoring data contained unsafe levels of one or more of the pollutants in coal ash, including arsenic, a known carcinogen, and lithium, which is associated with neurological damage, among other pollutants. There was very little dispute over the findings as the chemicals found in the tests were the same as found in coal ash. There was no chance of a coincidence here at all.

For decades, the coal ash was disposed of in unlined pits and impoundments. This was exactly how toxic waste and landfills were also operated for years: dump it on the ground or find a hole to put it in.

All unlined coal ash dumps will eventually leak. This may seem like a strong claim, but the EPA stated this years ago and stands by the conclusion that all landfills eventually will leak into the environment. The U.S. Geologic Survey site corroborates this claim as well.

With nothing at all between the hazardous ash and the ground, it is only a question of how long it will be until the toxic brew reaches the groundwater underneath. Many of these ponds also leak from the side as well, running their contents into nearby streams and waterways.

In some of the worst examples of complete disregard for what the consequences could be for the water supply, some plants built their storage ponds and dumps below the water table. They actually dug down deeper than the level of the groundwater and then started throwing in their hazardous, toxic waste product to sit and mix like a soup with the groundwater sources until it all became unusable and, in many cases, untreatable.

A report issued by the EPA in 2007 detailed coal ash leaks that had contaminated groundwater. Almost 70 towns in 26 states had heavy metals in groundwater at unsafe amounts.

At a power plant near Pittsburgh in Lawrence County, Pennsylvania, groundwater under the ash site had arsenic levels of

3,720 parts per million. The EPA standard is a maximum 10 parts per million. The site also showed the groundwater had lithium in it at 54 times the federal guidelines.

In Yorktown, Virginia, gravel pits were used to dump coal ash from the power plant. Over 50 homes had their water supply contaminated with arsenic, chromium, manganese, and other hazardous pollutants. The underground aquifer these homes depended on for safe drinking water is now polluted.

In Gambrills, Maryland, Constellation Energy used an old sand and gravel pit to dispose of coal ash. The unlined pit leaked toxins into the groundwater, including lithium and arsenic, among others. Multiple residential wells were impacted, and residents' water was contaminated. Constellation Energy paid the residents $53 million in a settlement, suggesting there were some very serious health consequences the pollution caused.

In Colstrip, Montana, unsafe levels of pollutants from Colstrip Power polluted the water supply in homes nearby. The plant paid $25 million in damages and was required to provide clean water to five residents whose safe water supply was destroyed.

3. Coal Ash as Fill Dirt

Although the power plant industry has tried to recycle coal ash as a useful filler product, projects that actually used coal ash show just how toxic that use can be. The contaminants in the coal ash are not removed or neutralized but get transferred into whatever the new so-called beneficial use is. Whether used in creating a golf course or, as material in making gypsum board for building, the ash contains its toxic components.

In an effort to dispose of coal ash and try to claim it was used for a beneficial purpose, Duke Energy in Virginia dumped a million and a half tons of coal ash to be used as fill for a golf course project in Chesapeake, Virginia. Over two dozen drinking water wells were contaminated with high levels of lead and arsenic.

In Pines, Indiana, coal ash was landfilled from a nearby power plant in Michigan City. The DPW used some of the ash for projects, and the coal ash was made available to residents to take and use around their homes for landscaping or other projects. Now the EPA has over a dozen homes that have tested positive for contamination, and the area where the landfill was is a Superfund site. Fill was also used in the town park, which also became a Superfund site.

The story today is no better than it was in the past. In some respects, it has become worse as the EPA is currently headed by a former coal lobbyist. In fact, in 2018 the EPA proposed many changes in the rules regulating "Coal Combustion Residuals." Those who were informed enough to notice the vocabulary change saw this as an attempt by the industry to gut the already lax regulations. The EPA wants the states to have the power over the industry in their own state. This could result in states deciding there doesn't even need to be any monitoring at all. At risk of editorializing, this sounds like allowing the fox to watch the chicken house.

One thing was made clear in EPA statements, though. These new changes would save the industry tens of millions of dollars in annual compliance costs. Nothing at all was said about the negative impact this will directly have on the environment and the health of people already affected by contamination and exposure to this hazardous waste product.

The problems associated with coal ash disposal and storage threaten many people's safe drinking water supply. Collapse of dams and dikes threatens communities, streams, and rivers. Particulate matter containing high levels of toxic chemicals blows off store piles of ash and into the surrounding community to be breathed in by its residents.

If you live near one of these places coal ash is disposed of, a high level of caution is needed, especially with regards to your water supply. The government or water company may not be legally required to test for all the contaminants that you may be consuming--the laws vary from state to state. Those with private wells need even more caution about potential contamination to their water. Private wells need to be tested annually for the list of chemicals most likely found in coal ash. If you are able to place at least two miles between you and any known coal ash dumps, we highly suggest it.

Chapter 6. Hazardous Waste Incinerators

In theory, incineration of hazardous waste sounds good. Take a large volume of toxic waste, and burn it up in an incinerator at a very high temperature until it's gone. However, nothing can ever be destroyed. So, what exactly does "gone" mean in this context?

The United States found out far too late that these monstrous hazardous waste incinerators, which some people thought were a cure-all for our toxic waste disposal problems, have been creating bigger problems the whole time. We now know what they really do to the material that they are fed to destroy. Even as the U.S. moved ahead with these incinerators, there was a large number of people who said all along the problem really had not been solved and the toxic wastes were not really going away. Time has proven them right. Unfortunately, the people who can actually do something about this will not listen and act.

Since the early 1980s, across the country, the people being poisoned by these incinerators continue pushing and showing evidence these incinerators simply do not work. But, the industry built around these giant toxic waste incinerators is making too much money to give this up easily. When the overwhelming body of evidence is presented to the EPA and other environmental officials showing why these incinerators should be shut down, no action is ever taken, possibly because, in spite of the failures of the method currently in use, they have no alternative.

Hazardous waste incinerators are supposed to destroy 99.99% of the chemicals put in them. That level increases to 99.9999% for the most hazardous toxic chemicals that are put into them. That is the claim: all of them, not just the select few the operators monitor emissions for. The biggest danger of hazardous waste incinerators is that they really do not destroy the toxic materials to the efficiency level that on paper they are supposed to. That leaves a whole lot of those deadly chemicals pouring out of the smokestack into the nearby neighborhoods.

Adding to this problem, there are documented cases of incinerator operators turning off or disconnecting monitoring devices while huge amounts of waste are put into the kiln for destruction. Almost daily, large amounts of deadly chemicals are released from these facilities, and the general public are continually exposed to them.

Besides the release of all the toxic compounds that are not fully destroyed, there is another huge risk: products of incomplete combustion (PICs). When the incompletely destroyed toxic materials leave the combustion chamber of the incinerator and join together with other toxic chemicals that also were not completely destroyed, PICs form new, unknown, and possibly deadlier chemical compounds than those that were originally put in for destruction.

This consequence of hazardous waste incinerators has been known for decades. However, the emissions of PICs happen all the time, and there is nothing at all being done to stop them.

For almost 40 years now, people outside the industry have been calling for monitoring of the full spectrum of waste that is allowed to be burned. It makes sense that efficiency level monitoring takes place for each of the individual chemicals that is put into the facility for destruction, but these facilities are only required to monitor emissions for only a few dozen different chemicals instead of monitoring for each individual chemical placed in the incinerator. This is a pitiful attempt at a serious monitoring program because thousands of different types of chemicals go into these incinerators. If monitoring was expanded to cover all chemicals, incinerators would likely fail miserably and would cease to be an easy way to dump hazardous wastes. It is because of this lack of sufficient oversight, hazardous waste incinerators continue to be as dangerous today as they were 40 years ago.

The permit system as a whole needs to be overhauled. For instance, the permit application for Heritage Thermal in East Liverpool, Ohio, lists only a couple dozen toxic chemicals in the list of emissions it is going to release annually. This short list alone is frightening. The company will be able to emit 335 pounds of mercury, 4,170 pounds of nickel, 954 pounds of silver, 102 pounds of selenium, and 254 pounds of lead into the town of East Liverpool and the surrounding areas. Any one of these chemicals on its own is so toxic a small amount can cause tremendous damage to the human body. The potential effects of all these combined on the people nearby is unimaginable.

That is just the tip of the iceberg about what will be raining down on the city. The permit allows Heritage to store up to 855,475 pounds of toxic deadly hazardous wastes. In case of accidents, the permit uses terms like "clean up must be at the earliest practical time" or "in as timely a manner as possible." This, unfortunately, is the standard way these incinerators are allowed to operate by the state environmental agencies and the EPA.

Outside Arkadelphia, Arkansas, there is a huge hazardous waste facility where Alcoa Aluminum disposed of large volumes of toxic waste. Alcoa sold the 1,300 acre facility to the hazardous waste incinerator company Veolia who operates other incinerators in the U.S. and abroad. Veolia has a contract with the Department of Defense to incinerate the toxic chemicals PFAS and PFOA, which are found in water supplies from industrial operations and military bases everywhere. The government has no idea what toxic chemicals are being released from the incineration of PFAS compounds, and the public is trying to halt the incineration of them until further study has been done. As of 2020, the Veolia facility has already received over 121,000 gallons of PFOA and started to incinerate them. The people in Arkadelphia are basically the human guinea pigs subjected to what toxins are being poured out of the smokestacks and into the air they are breathing.

At Veolia's Sauget, Illinois, hazardous waste incinerator, community members are outraged that the EPA is loosening the permit requirements in order to allow toxic waste to be incinerated without monitoring for many of the hazardous metals that are commonly emitted. The equipment previously used for monitoring those emissions is being eliminated by the company so no information will be available on those emissions. The EPA points at the new permit in which pollution control devices and monitoring for mercury is now required. The community responded that mercury had been emitted into their neighborhoods for years poisoning them while the EPA did nothing all of that time.

Environmental groups and residents want to have the emissions of toxic metals monitored and have some standard set for the emissions the incinerator must comply with. The spokesperson for Veolia claims the technology is still unproven; the metals can't be successfully monitored for, so in the meantime they refuse to install the equipment. This response is another example of how dangerous these facilities are. There is technology available, but because the industry convinces the EPA it is not yet proven, the incinerator gets a pass.

Most of the commercial hazardous waste incinerators currently operating are used by all types of industrial waste producers in many different states. There are no restrictions on who can use the incinerators. Therefore, the incinerators take in wastes from any customer who needs a disposal company. However, there are some large producers of hazardous wastes big enough to operate their own incinerator on the property.

Much remains unknown about the practices of these private incinerators. Some of the time, people living in the area have no idea these incinerators are even there and even less of an idea what they are burning. There are instances of these private incinerators being manipulated to bypass emission controls and log books with falsified records to mimic compliance.

In Waterford, New York, a hazardous waste incinerator operated by General Electric was cited for air pollution exceeding their permits. The excess releases at the plant potentially included cancer-causing chemicals such as dioxins and furans.

The federal complaint stated General Electric manipulated the controls of the incinerator to allow hazardous wastes to continue burning even when the incinerator had exceeded automatic emissions cutoff levels. In a six-month period, employees overrode the cutoff system at least 1,859 times. During those times, excessive emissions were continually released from the plant in clear violation of permits.

In addition, records were falsified for pollution control entries almost 1,000 times during that period as well. There was some evidence of the possibility the incinerator's automatic shut-off system may have been regularly disabled thousands of times over a 15 year period leading up the federal complaint. The New York State Department of Environmental Conservation stated they did not have health studies on the residents affected as the incidents happened long before knowledge of the incidents was known. The DEC did state that there were cancer-causing chemicals released; however, conclusions with any present day issues could not be definitively made.

At the DuPont Experimental Station a few miles outside of Wilmington, Delaware, the company operated its own hazardous waste incinerator. Many other household names such as Kodak, Dow Chemical, 3M Corp, and Monsanto all operated private hazardous waste incinerators at facilities. According to the EPA records, every one of these facilities had at least one serious accident in addition to regular daily operations that had all the same issues as every other hazardous waste incinerator, private or not.

The business of hazardous waste incineration is worth millions every year in disposal fees. One company taking advantage of the potential for profit in the hazardous waste disposal industry is Clean Harbors. They claim to be a complete waste disposal solution company, doing everything from transporting the waste to disposing of it. Clean Harbors operates facilities across the country. They recently built a third incinerator at their El Dorado, Arkansas, site alongside the two older units. Environmentalists had hoped the two older incinerators would close, but Clean Harbors would have been foolish to do that. Because of EPA loopholes, those two incinerators can continue to emit hazardous chemicals at much higher levels than the new one. The savings from not having to improve emissions on the older units will help offset the costs of the new incinerator.

In addition to the hazards of emissions themselves, there is the danger of possible explosions of stored materials awaiting destruction. In the past, there have even been major explosions of hundreds of 55-gallon drums of toxic waste and ensuing deadly fires forcing evacuations of thousands of people surrounding these facilities.

In Elizabeth, New Jersey, a hazardous waste disposal company called Chemical Control had one of the most deadly toxic chemical accidents in history. A fire fueled by an estimated 50,000-60,000 drums of toxic waste exploded into flames. Millions of gallons of water used to fight the fire were contaminated with toxic waste and washed through local neighborhoods into the river. This was over 35 years ago, and the site still remains on the Superfund list.

More recently, in El Dorado, Arkansas, employees at the hazardous waste facility operated by Clean Harbors were transferring waste from a larger drum into a smaller drum. According to a spokesman for the company, they were following standard procedures when, suddenly, the waste spontaneously burst into flames. The fire created by the hazardous waste was unable to be controlled by company staff, and the area was evacuated. A nearby college was evacuated, and residents were told to stay where they were while toxic smoke drifted into some of the surrounding neighborhoods. The company placed the blame for the fire on a customer they said did not accurately disclose the hazardous waste they sent for disposal. Regardless of who was to blame, this incident shows a gaping hole in the procedures the company uses to ensure the safety of the operation and the health of the people in the community.

In upstate New York, the city of Cohoes is home to a hazardous waste incinerator run by Norlite. The incinerator facility is located in a densely populated area and is right next to a large public housing complex. A recent 2020 investigation by Bennington College in Vermont discovered the facility was emitting PFAS chemicals it was supposedly destroying in the incinerator. A spokesperson from the college said soil and water testing nearby the plant indicated it was not effective at destroying the compounds but was emitting them into the surrounding communities. A former EPA regional administrator said it was very troubling since the plant has been allowed to bun these compounds since 2018. The New York State Department of Environmental Conservation allowed the burning to start without stack testing. He went on to say the residents of the surrounding affected communities should not be exposed to unknown hazards while the incinerator company rakes in money.

In East Liverpool, Ohio, a hazardous waste incinerator operated by Heritage Thermal Services exploded in 2013. The explosion sent gas and toxic ash into the surrounding neighborhoods. Several fires also started as a result of the explosion. Shockingly enough, an elementary school was less than 1,200 feet away from the incinerator. There was an emergency plan the teachers were supposed to follow in the event of a situation at the nearby incinerator. They were told to close the windows and doors to the school and throw wet towels and blankets over the windows. That was supposed to be adequate in the event the toxic waste incinerator blew up. Unfortunately, the incinerator did just that. No report was available on the effects experienced by the school.

The incinerator, which had been operating for over 30 years, was known for numerous emissions of toxic pollutants. According to a report issued by the EPA, the incinerator in East Liverpool continuously emitted toxins to contaminate the air in the neighborhood over a four year period. The emissions exposed residents to chemicals linked to cancer, spontaneous abortions, miscarriages, and early death. In 2015 the EPA sent correspondence to the company outlining separate violations it had committed. The incinerator was charged with emitting high levels of toxic chemicals 195 times in a 175 day period between November 2010 and December 2014. In spite of the toxic waste incinerator being classified as a high priority violator, very little happened in the way of enforcement. The people living and going to school near the incinerator see the lack of action as the EPA condoning the violations.

In 2017 a study was conducted on children in the East Liverpool area. The results showed the presence of high levels of manganese. The manganese was also present in the area around the hazardous waste incinerator. The study results also showed a correlation between the high levels of the neurotoxin and lower IQ scores on tests. The area children also had a higher proportion of needing special education schooling.

In West Carrollton, Ohio, an explosion and subsequent fire at the Veolia ES Technical Solutions facility, seriously injured two workers and damaged 20 nearby houses and five businesses. The federal government's Chemical Safety Board conducted an investigation and concluded the accident never should have happened. OSHA cited the company for improper safety management practices but no other consequences followed.

Hazardous waste incineration is one area Tony and I have spent quite a bit of time on over the years. Actually, almost 30 years has passed since we first were drawn to this topic. If we believed this method actually was a safe working solution for disposing of hazardous wastes, we would be two of its biggest advocates. Instead, the actual results of incinerating these wastes only concentrates and creates new and more horrendous toxins. The industry, however, points at the reduction in volume and calls that a successful treatment method. The communities where these incinerators have set up shop have become, in a sense, a laboratory experiment, and the people living there the rats. The communities certainly feel that way.

These incinerators still operate in the face of scientific evidence showing their failures. Ultimately, these incinerators continue to operate because of the money involved in both the hazardous waste industry and especially the industries that create all of the waste that needs to be disposed of in the first place. There remain many unknowns about what exactly happens in the combustion chamber and in the smokestack, but so much of what is emitted can neither be measured or monitored. No sane person can look at the evidence compiled over the decades and say hazardous waste incinerators are remotely safe or destroy the wastes even close to how they initially were supposed to.

These incinerators are among the most dangerous hazards anywhere. It continues to amaze me that the agencies set up to protect the health of the people and of the environment only require minimal monitoring of such a limited number of toxic chemicals and let so many operating malfunctions to continue. There are only a handful of industrial operations Tony and I consider to be unsafe to be operated at all and should have been closed long ago. These hazardous waste incinerators are one of those few. Studies from every operating facility in the United States across the board have come to the same conclusion.

Choosing a safe distance to be from one of these facilities is difficult as the constant toxic emissions travel at minimum several miles from the smokestack. They travel even further when they are in greater quantities and the winds are blowing. If it is within your power to do so, at an absolute minimum, keep four miles away from any hazardous waste incinerator regardless of any information saying closer is safe.

Chapter 7. Cement Plants

The smoke plume rises in the air far ahead up the highway. You keep driving for miles, and the plume gets closer. Finally, you see the top of the giant smokestack the plume is billowing out of. As you start to pass the plant to the left of the highway, you notice a long, thin, covered wood structure that runs about a half mile from the other side of the highway and up to the building the smokestack is on. Welcome to the Capital District. Albany, New York, and home of the LaFarge Cement Plant in Selkirk.

For years, I used the plume of smoke as the sign I was almost back home. The long building transported coal to fuel the giant furnaces inside the cement plant. I would hold my breath as long as I could while I sped up to pass the plant as quickly as possible. That turned out to be a smart idea.

Cement plants in the U.S. churn out millions of pounds of cement every year. What else they churn out in the process is a whole other story. Historically, the cement industry is less regulated than other industries in the U.S. Perhaps some of the reason may be the support the industry gives to politicians who work in their favor. According to documents obtained from the Federal Election Commission, a single political action committee run by CalPortland Cement, had over $105,000 cash on hand in December, 2018 to use.

Cement plants are part of what supplies our country with needed materials for building and maintaining infrastructure while creating jobs for locals in those communities. Unfortunately, a cement plant makes for a horrible neighbor. Whether residents are employed by the plant or not, they and all their family members have to deal with the environmental risks and negative health effects the plants typically cause. Communities hosting cement plants get far more than their share of pollution.

1. Mercury Emissions

Many cement plants still use coal to fire their furnaces in the manufacturing process. Burning coal to power huge furnaces is a major source of mercury. Coal contains mercury and other hazardous air pollutants that are emitted when the coal is burned in cement plants. Mercury is one of the more toxic substances emitted by these plants. Mercury is toxic to the central nervous system. Inhaling mercury also harms the immune and digestive system as well as the lungs and kidneys. Pregnant women and children are especially at risk.

It was estimated by the EPA that 23,000 pounds of mercury was emitted by the industry in 2006. In Ravena, New York, the Lafarge Cement Plant was the second largest polluter of mercury in the state, emitting 400 pounds. In the nearby community, the people who were tested for mercury had levels in their bodies well above the national average. In fact, a study done by the Harvard University School of Public Health showed almost one person in every ten had elevated levels of mercury high enough to warrant a visit to their doctor

2. Cement Kiln Dust

The process of making cement creates an enormous amount of residue known as cement kiln dust (CKD). The main ingredients to make cement are calcium, silicon, aluminum, and iron in addition to other trace ingredients. For instance, the manufacturing of the most popular type of cement, Portland cement, produces CKD that may also contain small amounts of calcium oxide, magnesium oxide, potassium sulfate, and sodium sulfate, which are considered hazardous (and the case of crystalline silica, carcinogenic) as well as amounts below 0.1% of chromium salts or compounds (including hexavalent chromium, which is also considered carcinogenic) or other metals (including nickel compounds).

The industry group says, "Cement is so fine that 1 pound of cement contains 150 billion grains." All around the cement plant, CKD is in the air and covers everything in a fine coating. A person could drive their newly washed car in to work in the morning, and by the afternoon, it could need another wash.

CKD has certain components that make this dust especially harmful when inhaled, the most dangerous of which is crystalline silica, a known carcinogen. The workers at the plant are constantly covered in the dust, but neighborhoods nearby are subjected to much of the dust as well.

3. Alternative Fuel Burning

Many of the cement plants across the country are turning to other fuels to supplement coal. The entire reason is cost-driven. When the cost of alternate fuel is less than the coal, they will burn anything they can. There are very few types of waste the industry will turn away. Some cost them very little money, and others they get paid money to take. Sometimes, it is not about how much energy the plant can get from the fuel but how much can they get paid to take it from someone needing to get rid of it.

The industry likes to keep quiet about what it is burning at their plants because these items can be trash or even toxic. If people realized these materials were being burned, they would likely object. As long as the plant operator has secured the needed permits to burn what they want to, they do not need to inform the town what they are burning, and there is very little the community can do to stop them. What the community doesn't know in this case may very well be hurting them. There are several favorite alternative fuels the cement industry likes.

Tires are a popular alternative as they may even generate income from their disposal. Also, because of the high level of petroleum in tires, they make wonderful fuel. The fumes, however, are quite toxic.

Pollution equipment such as air scrubbers and filters do reduce the amount of pollutants released into the air. However, considering how much toxic emissions are produced, even after scrubbing, there are still alarming quantities emitted. For some chemicals such as cadmium and aluminum, the amount released is substantially above safe levels. Particulate matter, especially the smaller 2.5 micron size, the most harmful, is measured in tons. Dioxins are not entirely removed and sent into the air as emissions despite the fact that, according to the EPA, even the smallest exposure to dioxin is hazardous to humans.

Solid waste is another revenue stream cement plants use. With the difficulty of finding a site for new solid waste incinerators, cement plants have positioned themselves as an alternative disposal site. An increasing number have contracts with major cities and towns to burn their garbage. With landfill space at a premium and filling up, getting rid of trash to cement companies to burn as fuel is seen as a victory, except for the people living anywhere near the plant.

In Ravena, New York, LaFarge Industries, the same cement plant that spewed toxic chemicals including mercury over the community for years has quietly been talking with the city of Hartford, Connecticut, in an attempt to bring solid waste from Hartford to burn in the cement plant as fuel. LaFarge denied for months the idea was even on the table until a letter was released by the EPA showing LaFarge had been deceiving the public. The letter also showed LaFarge is seeking to burn old tires as well.

The same logic applies for sewage sludge. There are only a few options for sewage sludge disposal, all of them fairly costly. While the cheapest disposal method has always been land application, letting cement plants burn it as fuel is seen as a reasonable alternative.. The heavy metals contained and concentrated in the sludge increase the emissions for certain dangerous chemicals such as mercury and zinc. Sewage sludge may also have trace amounts of radiation from fracking waste now being disposed through the treatment plants. Cheap fuel, though, is the first consideration of the cement plant.

The most frightening alternative fuel is hazardous waste. The most toxic and potent mix of chemical soup is used as fuel for the cement plant. Incomplete combustion, inadequate air filters, and lack of emissions monitoring combine to create a dangerous situation. At the same time, this is one of the biggest money makers for the cement plant operator.

There are rules and regulations that control hazardous waste disposal. Somehow, cement plants were given a special classification that allows them to burn hazardous wastes and emit even more toxic chemicals and pollution into the air than an incinerator built specifically for burning hazardous waste is allowed to emit. As a result, cement plants are using industrial solvents, old chemicals, paint sludges, and pharmaceutical waste as fuel. Because there is a steady supply of over 20 million tons of hazardous wastes needing disposal each year, there is plenty more where that came from.

In the town of Chanute, Kansas, the residents live under a cloud of hazardous emissions pumped out by the cement plant. The plant is allowed to emit much more lead, cadmium, chlorine gas, and particulate matter than actual hazardous waste incinerators. They are not the only cement plant allowed to pollute like this either. Over a dozen others get a free pass from the EPA thanks to the loophole gifted to the cement industry.

For the most part, if something has BTU value like old oil, industrial solvents, or old millings, some cement plants will burn it.

In the town of Glens Falls, New York, the cement plant received their permits to start burning paper and plastic waste. Plastic creates dioxins when burned, and the emissions from the plant contain not only dioxin but additional toxic compounds as well. The state's Department of Environmental Conservation is allowing the cement plant to monitor and report their own emissions once every 30 months. The DEC actually stated their reason for allowing the additional pollution from the plant is the area is too clean and has room for more toxic air pollution before it reaches the point of an unhealthy level.

To reiterate, the state is allowing a company to pump more toxic pollution in the air because there is room for more. Unfortunately, this is not a typo.

4. Particulate Matter

In addition to CKD, which by itself presents inhalation hazards and respiratory issues, there are a number of other particles generated by the other materials the plant uses in its operation. For instance, toxic fly ash, dust kicked up from trucks and equipment moving around, is constantly present. Other materials added to the cement kiln such as lime, silica sand and fly ash, some of which are caustic and need to be handled carefully sit in piles around the yard and also create dust and particles as they too blow around.

Once the cement is finished, it needs to be bagged and packed. Clouds of dust rise from every corner and building of the site. Even after hours, if the plant closes overnight, there are still particles in the air. Studies done by Cook Children's Hospital have shown increases in asthma, especially in children who live in communities near cement plants. The EPA also says the emissions from these plants contribute to smog and can lead to childhood asthma. Exposure to particulate matter pollution and soot, is a well-known known cause for asthma. It can also make a child's asthma worse.

In Midlothian, Texas, a recent study analyzed data from a survey the Cook Children's Hospital previously conducted on regional childhood health issues. The results of the study confirmed the higher levels of childhood asthma were directly associated with the presence and concentrations of three Midlothian cements plants just over the county line. The people working on the study tracked the direction of the plume of pollution from the plants. The highest concentrations of the asthma cases recorded occurred in the neighborhoods and areas that were directly in the pathway. Those were where the highest levels of pollution and particulate matter from the cement plants passed over.

When a cement plant exploded in Boston, dust from the explosion covered a school bus yard next door. Thousands of pounds of slag dust sent 61 people to the hospital. Burning in the eyes and throat in addition to shortness of breath were reported by those admitted. The dust was also thought to contain beryllium, and those exposed may have been poisoned. Inhaling airborne beryllium may cause irreversible lung damage. Similar to silicosis, CBD scars the lungs, causing shortness of breath, chest pain, fatigue and other symptoms.

Heavy exposure to beryllium even for a short period of time can lead to acute beryllium disease, which causes symptoms similar to pneumonia or bronchitis

5. Toxic Gases and Metals

According to the EPA, the cement sector is the third largest industrial source of pollution, emitting more than 500,000 tons per year or 1 billion pounds of sulfur dioxide, nitrogen oxide, and carbon monoxide. As a major emitter of these three harmful chemicals the cement industry contributes substantially to the harm of people living near the facilities.

All three of these chemicals cause a variety of health problems. Nitrogen oxide can cause damage to lung tissue in people who work or play outside. Children especially are affected.

Sulfur dioxide when present in high concentrations can aggravate breathing issues in people with respiratory problems as well as cardiovascular problems. People with asthma, bronchitis and emphysema are especially at risk from increased attacks and symptoms.

Carbon monoxide reduces the amount of oxygen that gets to the body's organs and tissues and has adverse effects on the cardiovascular system and the body's central nervous system. Carbon monoxide especially increases ground level ozone, which affects breathing, especially in children and elderly people.

The EPA states emissions of metal compounds from Portland cement kilns include mercury, thallium, antimony, cadmium, lead, selenium, zinc, potassium, sodium, barium, chromium, arsenic, nickel, vanadium, manganese, copper, and silver. Add that to the already toxic emissions from normal kiln operations, the remnants of the alternative fuels, dioxins, and products of incomplete combustion, and the resulting emissions may be deadlier than the original chemicals burned in the first place. *Science Direct, referencing* a compilation of over 20 peer reviewed studies on cancer rates near cement plants, showed a higher incidence of respiratory system cancers were linked with proximity to cement plants.

What truly is scary is how the EPA sets emission levels. Many times, they use what the industry tells the government the current technology is capable of, not the threshold for toxicity for the surrounding community. Especially older cement kilns are allowed to emit much more based on their technology. The EPA does not require upgrades to equipment if it costs too much to the company. This policy needs as much upgrading as the cement plants.

When residents living close to the cement plant near Selkirk, New York, were asked about possible links between cancer and the plant, the answer was simple. The response was yes: there seemed to be many people with cancer. Ewing's Sarcoma is a rare bone cancer that affects usually one person out of 500,000. In Selkirk, four children are suffering from it in a county whose population is under 300,000.

The Justice Department's Environment and Natural Resources Division calls the cement industry a significant source of air pollution that poses real health threats to the communities they are in. They made this statement after settling with Cemex and five of their plants. Cemex agreed to spend $10 million to reduce pollution at those five plants and were fined an additional $1.67 million.

6. Power Plant Fly Ash

Cement plants use a toxic waste product from coal-fired power plants called fly ash made out of the residue left from unburned coal as an ingredient in cement. It contains heavy metals such as mercury and a host of toxins. Power plants often give it away because it costs money to dispose of legally. When fly ash is mixed with the cement and heated, it releases mercury into the air, adding to the mercury they're already emitting by burning coal. The use of fly ash in cement plants really amounts to taking mercury from one place and sending it to another to be emitted in the smokestack. The industry claims they emit much less than in prior years, but it still is in amounts in harmful levels of emissions.

Between 15-30% of the cement ingredients is toxic fly ash. The EPA considers this use of a toxic by-product an allowable use under the beneficial use clause as the cement plants are recycling an otherwise unusable material. Toxic fly ash is also allowed to be used in road construction, fill material for architectural projects, and basically any place they can dispose of it. Because the volume of the fly ash is tremendous, many cement plants create a storage area where tons of toxic fly ash sit until needed for the blending operation. The piles sit open to the winds blowing the particles all over.

In 2007, the EPA announced it was considering regulating coal ash as a hazardous waste. Since that time, a number of powerful industry lobbying organizations have opposed the label for fly ash and other forms of coal ash. After seven years and intense lobbying by the cement and construction industry along with their political friends, the EPA reclassified fly ash as non-hazardous.

7. Water Pollution and Emissions Violations

Many of the cement plants operating across the country have a track record of violating water pollution regulations. In some cases, highly alkaline water was discharged and caused damage to the ecosystem of the stream it was released into. Spills of acids and discharges of sewage have been documented. Sometimes the violations are simply described as "unidentified material discharged." What was actually released may never be known.

Chemicals in the form of emissions exceeding the amounts cement plants are allowed to release occur, and unfortunately, they occur too frequently. The release of dozens of toxic chemicals is already legally sent into the air from the smokestack, but thousands of pounds in additional pollutants are regularly sent out as well. Monitoring is very weak, and enforcement is usually much weaker.

On the occasions fines are issued, the amounts penalized are small in comparison with the revenues the companies earn, so the fines may not have the desired effect to force compliance with regulations. In fact, they might just be chalked up as the cost of doing business.

Cement plants make for the very worst type of neighbor. The pollution they emit can't be escaped. If the amount of dust and particulate matter was the only issue, maybe it would not be so bad. However, with the amount of highly toxic chemicals that are released, negative health effects from living near a cement plant may only be a matter of time. Children deserve to breathe air that is not filled with mercury and other deadly chemicals. Allowing cement plants to burn alternative fuels broadens the list of toxic emissions to include almost all the deadly chemicals of many other industries combined.

In no way would we suggest living in any community that has a cement plant operating or nearby. Even a small plant can emit very large amounts of hazardous pollution. It also too easily could expand into a much bigger operation if allowed. If it expands, it may wish to burn anything it can for fuel. Stay away and look for another place no matter what. All of the other communities that already host one are rife with problems that should be a signal to you.

A minimum of four miles from the cement plant needs to be given to distance yourself from the particulates and other materials emitted. Many cement plants have smokestacks much higher than other factories that allow for the dispersal of the toxic emissions over a much greater distance and area from the plant. If possible, even further away is preferred.

Chapter 8. Military Bases and Government Installations

Our military is the largest polluter in the world. The amount of hazardous waste the United States military produces overshadows the amount made by corporations like Dupont and Exxon. The military has produced incredible amounts of waste from manufacturing bombs, defoliants, chemical weapons, fuel, and so much more over the years, and it still continues to do so. Thousands of acres of land are contaminated with nuclear waste, and millions of gallons of underground water are contaminated with solvents, hazardous wastes, gasoline, and much more. All this contamination was caused by activities at these military installations and the contractors who work for the Department of Defense.

The health problems caused by all this toxic contamination is not confined to the bases themselves but go beyond the guarded entrances to the communities they are near. Women with miscarriages, children born with birth defects, different kinds of cancer, and blood diseases are all just some of the results of exposure to the air, water, and soil contaminated by the military.

From radioactive waste dumps to water wells contaminated with dry cleaning fluids, the military has a toxic legacy on these lands. A news analysis by *TruthOut* magazine reviewed public data on the military and toxic waste back in the 1980s. Based on the government's own data, it was estimated the government was creating a ton of toxic waste per minute. They were the largest polluter in the nation. This figure was not even including the Department of Defense nuclear energy plants or any of the Pentagon's civilian contractors, who have their own toxic legacies as well. Outside of Denver, Colorado, the Army Corp of Engineers christened the 100 acres at Rocky Mountain Arsenal, "the Earth's most toxic square mile." That area was used as a dump for storing waste from the production of nerve gases, pesticides, and a host of unknown chemicals from the 1940s through the '70s.

For the past 90 years, perhaps longer, the United States Defense Department has created thousands of toxic waste sites across the country. Many of these are located on military bases, research facilities, and current and former manufacturing locations. The waste ranges from high-level radioactive waste that will be active for centuries to come to old dumps used to dispose of unused mustard gas, toxic nerve agents, and other weapons of war.

Military bases are in every state in our country. They range in size from a few hundred acres up to many square miles in size. The one thing all these bases have in common is the public knows nothing about what is happening inside the fences. This lack of transparency has led to many of the bases having toxic waste sites where previous military or disposal activities have taken place.

The focus of this chapter is hazards for both people living on and near the base. The biggest obstacle in the discussion is the lack of accountability and transparency of the polluters. While much has already been exposed and made known about specific sites, there is plenty more that has not been found yet. This is further complicated by the wide range of toxins currently used and those used in the past. Tracking such a large amount of information, much of which is unknown, is exceedingly difficult. Thankfully, there are several outstanding databases and websites that list locations and suspected contaminants. By being aware of what already exists, we have an idea of what to look for. Our concern focuses both on avoiding exposure to the known hazards as much as possible and avoiding the ones that may turn up later.

According to recent figures, in 2017 there are 130 sites that are military bases and installations on the National Superfund List. The sites placed on this list are the worst of the worst. There are about 1,317 Superfund sites in total, so the military has created 10% of them. Again, that is only counting sites known to date. Many military installations and proving grounds are completely off limits.

The lack of transparency creates long-term situations that outside of a military setting may have been taken care of. A good example is Camp Lejeune in North Carolina. For decades the water at Camp Lejeune was contaminated by cancer-causing chemical solvents, including known carcinogens such as trichloroethylene, perchlorate, and benzene among others. Exposure to these chemicals are associated with miscarriages, birth defects and a range of cancers.

Who knew what and when they knew it is still in dispute, but, there were over 900,000 people both on and off the base between the 1950s and 1990s who drank, cooked with, and bathed in the water, never told it was contaminated with all these toxic chemicals. There are terrible stories of cancers, miscarriages, and children born with physical and mental disabilities. Moreover, children born there later suffered childhood diseases and cancer that could not otherwise be explained. Eventually, the base acknowledged the issue, but not until a large number of people were affected for life.

A website from the Veterans Administration lists diseases which may have been a result of living on the base for 30 days or longer during a period of years exposure was uncontrolled. They are leukemia, bladder and kidney cancer, liver cancer, multiple myeloma, non-Hodgkin's lymphoma and parkinson's disease. There are quite a few more as well.

Part of the problem is people in the service have to unquestioningly follow orders. The opportunity to avoid a situation is sometimes out of individuals' control regardless of what they know about the risks they are taking.

The biggest environmental threats from these bases comes from actions the bases fail to take. Once known about, the waste can usually be monitored to ensure the contamination is not spreading farther and contaminating water resources or additional lands. The unknown, hidden, or ignored damage is hardest to address. Many of these contaminated sites were created with full knowledge of the substances and what the results would be from handling and disposal of the wastes.

An old report from the Federal General Accounting Office to the Department of Defense stated that they had a problem and expected it to be taken care of. The report back then said, "Over 15 million acres in the United States are suspected of being or known to be contaminated with military munitions. These sites include ranges on closing military installations, closed ranges on active installations, and formerly used defense sites." Much of this has yet to be taken care of despite the Department of Defense's responsibility to do so.

There are many thousands of people affected by the tremendous extent of military base pollution. On the base or in the surrounding areas, very little is left untouched by the toxic legacy the military created. The main areas of danger from the known hazardous waste sites are similar to other toxic dumps. Because of what is not being told to the public and what is being hidden, the potential risks increase exponentially.

1. Contaminated Water Supplies

On base or nearby, water supplies for tens of thousands of people have been contaminated with everything from radioactive materials to dry cleaning fluids and PFOS. Other chemicals disposed of improperly include toxic wastes illegally disposed in unlined pits and old nerve and mustard gas burned out in the open. There seems to be no end to the ways the military has polluted and contaminated the drinking water supply, not only for their own bases, but for many towns and cities nearby. There also seems to be no end to new discoveries of places where toxic waste has turned up to add another town to the growing list.

The dangers of known and unknown toxic contaminants in water supplies make living on or near a military installation a questionable gamble. Many of these military bases and neighboring communities are at risk from widespread use of a substance group known as PFAS. This stands for "perfluorinated alkyl substances." Since this chemical substance has been invented, it has worked its way into almost every human body. This substance is one of a group of chemicals called "forever chemicals." They never break down and persist in the environment and human body. Originally, PFAS were proclaimed a miracle substance as one of its original widespread uses was non-stick cooking pans and pots.

Many other uses for these substances have been developed over the years. One of the other widespread uses was for fighting fires. PFAS was added to firefighting foam because it excelled at quickly smothering and extinguishing fires. Unfortunately, this chemical is now suspected to be cancer-causing, contaminating water supplies at many military bases and the communities they were in. In a report from the Pentagon's PFAS Task Force released Feb. 13, the Department of Defense had previously identified 651 active or former military bases where perfluorooctane sulfonate or perfluorooctanoic acid, also known as PFOS and PFOA, may have been used or released. In fact, at the former England Air Force Base near Alexandria, Louisiana, PFAS concentrations of 20.7 million parts per trillion were found. The EPA sees 70 parts per trillion as the threshold level.

The Pentagon further reported a minimum of 126 military sites had potentially dangerous levels of PFAS in the drinking water. The Department of Defense tested 2,668 wells that depend on local groundwater and found 61% of them exceeded the EPA's safe level for PFAS. The CDC issued a report that this substance can be hazardous at 1/10th the level the EPA's legal limit.

In Newburgh, New York, the toxic chemical PFOS was found in private wells and the public water supply, Lake Washington. Residents for years used the water for drinking, cooking, and bathing. The town says the source of the pollution is the Stewart Air National Guard Base nearby. The base is also the site of a Superfund.

In Fallon, Nevada, near the Naval Air Station there are high rates of certain rare diseases and cancers, including childhood leukemia. The population of Fallon is 8,000 total. In 2000, nine children were diagnosed with a rare leukemia. Sixteen more were diagnosed over the next six years, ultimately marking the town as having the most significant cluster of childhood leukemia cases in the history of the country. Possible links to a leaking pipeline carrying jet fuel (which contains PFAS) and groundwater possibly contaminated with radiation were investigated as they may have been a cause. Other experts pointed to high arsenic levels in the water and levels of tungsten. No conclusive cause has ever been found, only suspected possibilities, most traceable back to the military base.

On beautiful Cape Cod, Massachusetts, the Massachusetts Military Reservation is believed to be the source of perchlorate found in the water supply for the town of Bourne. The chemical is used in rocket artillery., and the contaminated well is located one mile from the base. Tests showed 70% of the town's water supply had been contaminated. Ingesting perchlorate over a period of time has an effect on the thyroid gland. The exposure may reduce thyroid activity leading to a serious condition known as hypothyroidism. This condition can have adverse effects on the nervous, pulmonary and the cardiovascular system. It also is known to affect the liver, kidneys, muscular and reproductive system.

In South Carolina, houses located near Fort Jackson had their water source test positive for RDX. The chemical RDX is an ingredient in making hand grenades. The affected wells have been closed, and the Army is supplying water to the houses while they figure out what to do. While the CDC does not know what effects drinking water contaminated with RDX has on humans, they have determined it is a possible human carcinogen. In addition, rats and mice who ingested RDX showed signs of liver and kidney damage.

Outside Spokane, Washington, people living around Fairchild Air Force Base in the Airway Heights area drank water contaminated with PFOA and PFOS. This chemical was used in firefighting foams, probably for decades, before the contamination was discovered or at least made public. The city is trying to find an alternative source of clean water for the residents. In the meantime, the residents are on their own to find alternatives to drinking more of the contaminated water. Many wells in the area exceed the EPA's safety standards dozens of times over.

Research into all the health effects are ongoing, but the chemicals have been linked to lower birth weights in infants and birth defects. Increased risk for certain cancers is also associated with high exposure to these compounds according to the Agency for Toxic Substances and Disease Registry.

At a Naval station in Virginia Beach, Virginia, there was a spill of 94,000 gallons of jet fuel. The fuel spilled into a nearby waterway less than one mile from the ocean. Two local creeks were contaminated by the fuel, and there were 200 homes in the neighborhoods impacted by the spill. The Navy paid to relocate 48 families to other housing while the fuel spill was being cleaned up. Over 1,700 tons of contaminated soil needed to be removed and disposed of. While the Navy said no people were directly affected by the spill, there was concern. Exposure to jet fuel may cause damage to the liver and impaired neurological functioning, and if the skin is exposed to the fuel, the chemicals will be absorbed through the skin. People exposed in other cases have shown chemicals in their blood, fat, brain, lungs, and liver.

2. Contaminated Soil

Buried underground, sometimes deep, sometimes just below the surface, is waste carelessly disposed of on and near military installations. The way this waste was disposed of shows severe disregard for the environment and the people who live in the areas they polluted.

Near Frederic, Maryland, next to Fort Detrick, a developer was hoping to build a planned community. The discovery of biological, radioactive, and chemical waste changed these plans. For years, the army routinely buried this waste on the base in unlined pits as many others did. Leaking radioactive and chemical waste contaminated the groundwater underneath and created a plume that spread off-site to the areas outside the base. One cancer-causing chemical, trichloroethylene, was detected at a level 42 times the maximum set by the EPA. The EPA says this solvent is associated with central nervous system disorders, immunological and developmental effects, and some cancers.

San Diego, California, is home to a massive Navy base. The Navy has the distinction of being named the largest polluter in the San Diego area. Over the past 80 years, 100 toxic waste sites have been credited to them. The naval base currently has eight active hazardous waste sites listed as undergoing some type of remediation. The chemicals include dioxins, arsenic, PCBs, radioactive elements, heavy metals and explosives. The contamination may extend beyond the boundaries of the base onto private and public lands in addition to contamination of the groundwater. According to *Propublica* in California overall, the military has 598 military locations with at least one hazardous waste site.

In Paducah, Kentucky, radioactive waste was found outside the fenced area where a Department of Energy facility produced enriched uranium to be used in nuclear weapons, commercial power plants, and the Navy's nuclear submarines. The latest find of more radioactive waste was located in a lot covered up by a layer of soil.

Recently, evidence of radioactive metals in the groundwater and radioactive ooze was discovered by workers near the plant who noticed the ground oozing a tar-like substance after they drove their truck through an area outside the plant boundary.

Thousands of tons of radioactive waste have routinely been buried in ditches and in piles inside the plant property. Workers at the facility alleged radioactive waste was also being dumped outside the plant in the woods and areas easily accessible to the public.

Workers from the plant also said radioactive waste was sent to a nearby landfill that was not authorized to accept any hazardous waste. The Department of Energy disputes that and claims no knowledge of any waste not disposed of properly. The department produced records from their private hauler that show only uncontaminated trash and garbage being taken to the landfill. In

2018 workers at the plant took multiple truckloads of contaminated soil to a nearby residents home who was looking for fill dirt.

The groundwater near the plant and the site itself has been contaminated with trichloroethylene, which was also found in some residents' wells along with traces of uranium. The groundwater also is contaminated with radioactive technetium-99 and PCBs. There are two plumes of contamination approximately 2-4 miles long traveling out from the base. Billions of gallons of TCE contaminated water has been pumped up and disposed into a nearby creek running past neighbors' homes and into the Ohio River. One of the creeks being used run through a 120 acre property belonging to a long time resident. Signs are posted all along the creek warning not to contact the water. The well which used to supply the property with water has been padlocked due to TCE contamination from the DOE facility. TCE causes damage to the central nervous system and causes kidney cancer according to the National Cancer Institute. It also could increase risks of liver cancer and non-Hodgkin lymphoma. The family living there has severe gastrointestinal issues they attribute to the water prior to finding out it was contaminated.

In San Francisco, California, radioactive items were found less than a foot from the surface of the ground at the upscale neighborhood known as the San Francisco Shipyard. The area features condos selling for close to $1.5 million. In this project, which some call the area's $8 billion neighborhood, residences were built over what used to be a top secret nuclear testing facility run by the Navy. One of the many functions they performed there was to examine ships and equipment that had been exposed to atomic bomb explosions. The radioactivity from the objects found would expose anyone in contact with it to radiation over six times the limit for nuclear power plant operators.

There is now concern the converted former shipyard may be holding more radioactive items beneath the surface. One past employee who worked at the shipyard during its Navy days claimed radioactive objects similar to the one discovered were routinely just thrown away. In the '90s, when people began to move into the housing, chemicals including DDT, lead, and dioxin were discovered under the area of the schoolyard, daycare center, and in some of the homes' yards. The residents were told by the Navy not to worry but to not garden or disturb the soil. A few years later, lead, dioxin, and benzene were also found under the schoolyard. One sample of lead registered 22 times the level deemed safe by the EPA. DDT levels were found at 31 times EPA levels considered safe. One count had a total of 1,289 radioactive items that have been removed.

Rashes, asthma, thyroid issues, and ovarian cysts have all been reported by people living in these properties. Records reviewed by news agencies confirm complaints were made to the state about asthma, lumps, children losing hair, and cancers. At one point the state of California advised residents not to allow pets or children to track any soil from the yards inside the house.

The issues are ongoing at this project, and there are more toxic and hazardous materials expected to be found. Nobody really has any type of solution or knows the additional health effects on residents from exposure to the materials already found and yet to be discovered.

3. Air Emissions

The spread of contamination from these installations is challenging to determine. With other operations, such as a trash incinerator, researchers are able to do independent testing and review paperwork and records (with varying levels of trustworthiness) to determine what is coming out of the smokestack. The military, on the other hand, disposes of and stores hazardous wastes and materials in a variety of different ways. This confusing array of practices creates additional exposure, but just what this exposure is and where it is located isn't readily apparent. It is hard to tell what to even look for.

Despite this level of confusion, many of the obvious risks posed by military bases are airborne pollutants. There are installations where open air burning of old bombs, ammunition, and hazardous materials send thick, billowing clouds of toxic materials over all the neighboring areas. There are open and uncovered waste dumps where leaking and half-full drums of chemical defoliants and other chemicals sit, allowing their toxic contents to evaporate into the air to float someplace where they are inhaled. There are fuel spills large enough to fill the tanks of dozens of vehicles, possibly occurring on an almost daily basis. Furthermore, the military has been known to utilize mobile incinerators to burn everything from hazardous waste to contaminated soil. Nobody ever really knows what is happening, and that it is very scary for those who live near one of these bases.

In West Valley City, Utah, at the ATK Bacchus Facility Naval Industrial Reserve Ordnance Plant, there is an area set aside for open burning of hazardous waste. In spite of there being safer alternatives than open air burning with no pollution control devices at all, the base continues the practice. No place in the entire United States is allowed to burn hazardous wastes open air like this except on military bases. Sixty one of them according to Propublica. Across the entire country there are bases just like this one burning hazardous waste in the open without regard for the environment and human health.

We honestly do not know for certain how much pollution is sent into the air and what it may contain. We do, however, know for certain this happens, and it happens in many more ways than the several examples we used. By her own admission, Maureen Sullivan, the person in charge of the Pentagon's environmental program, has to deal with 39,000 different contaminated sites under her jurisdiction.

Tony and I are continually stunned when we hear that yet another discovery of hazardous waste has been made. We really should not be, but we are. To us, it is incomprehensible how so much has been destroyed by our government. Much of the pollution is ongoing to this day. There are still about five dozen or so military sites where they burn hazardous waste in open pits. As documented by the EPA, there currently are sites operated by the Department of Defense or their contractors that annually burn and destroy millions of pounds of waste, including explosives, not in specially constructed furnaces or disposal units to prevent any pollution from escaping, but in the open. Everything that goes up in smoke also goes into the air and the environment.

It was announced in 2019 that the Radford Army Ammunition Plant in Radford, Virginia, will be constructing an incinerator to be able to burn waste onsite. The hope is much of the waste that is currently burned in open pits with no pollution controls at all will now be incinerated in equipment offering some type of reduction in toxic emissions. The Army is, however, still keeping the possibility of open burning, depending on future needs. It is also entirely possible that the base may start receiving hazardous waste from other military bases around the country, making this area the epicenter for hazardous waste disposal in the country.

In Arizona where I currently live, there is an Air Force base. It has a hazardous waste site on it that has contaminated some of the underground water supply. The distance between the base and me is about 10 miles. Because it is an Air Force base, there are many jets taking off and landing. There are many days when they deafeningly fly directly overhead. However, when we moved into the area, we were told about this, signed a knowledge of disclosure, and are simply fine with watching our military pilots in the air. The point is, we were told and knew.

If you have a choice in how close you locate near a government or military installation, we suggest placing as much distance between you and the base as feasible. Plumes of contaminated water have been known to travel distances of 4-5 miles (such as the Gruman plume in Bethpage, Long Island, New York). Giving yourself at least four miles is reasonable. Almost without a doubt, there are going to be more public discoveries of new hazardous waste situations at almost every one of the places.

Chapter 9. Medical Waste Incinerators

Incinerators specializing in burning medical waste are not as common as they once were. In the past, almost every hospital had an incinerator onsite and disposed of their own waste. Toxic emissions poured out of smokestacks into neighborhoods all the time. Since environmental rules and regulations became stricter, many of these facilities closed down and large companies constructed larger incinerators to handle the flow.

Typically, the medical waste disposed of at the incinerator include a range of bloody bandages, used needles, empty intravenous bags, narcotics, and other pharmaceuticals. Add to this list dead fetuses, removed body organs, biopsy discards, amputated limbs, and cancer treatment byproducts. Many similar items from veterinary offices are sometimes allowed to be accepted and sent for disposal as well.

According to a paper from Columbia University, there were over 2,400 medical waste incinerators operating around the country in 1996. The year 2000 saw new regulations required by the EPA to reduce some of the toxic emissions and odors the incinerators were pumping out of their smokestacks. The cost of the upgrades was more than some of the companies were willing to spend, so they operated up until either their permit ran out or the deadline for the installation of upgrades came. Either way, many of these facilities finally closed their doors.

The ones that stayed and upgraded saw a big increase in business for the most part. Small operations in towns became medium and large ones, filling the gap left by the closed incinerators. On the national level, the volume of waste was increasing. The companies would like to see their investment in new equipment pay dividends as soon as possible. In 2014, there was another change in the regulations that went into effect, again requiring better emission controls to meet the new Clean Air Act standards the government had set.

However, just because the government has standards and says they are stricter than previously required, that does not mean that poisons are still not being emitted into the air. Mercury and other heavy metals, dioxins, and furans are still emitted. "Clean" is a relative term.

For instance, in some cases, the regulations actually allow smaller, rural incinerators to emit up to 15 times the amount of toxic metals and chemicals the larger ones are. Tony and I are unsure what the logic is. All we have are questions. Does the EPA think people in rural areas are worth less and, therefore, have less protections? Does the EPA think smaller incinerators don't burn as much, so, therefore, they should be allowed to pollute more? Or, is it that the EPA doesn't want to make the smaller operators have to upgrade their pollution controls because the cost is too high for small operations? It could even be all three. These different levels of compliance based on the size and location of the operations do not seem to be based on scientific reasons but financial ones.

Apparently, because medical waste disposal is worth billions of dollars a year for those who can get a permit to dispose of it, many of the companies operating trash incinerators are moving into the medical waste incineration business too. They're fighting over the medical waste business because it's more than just additional revenue; it's triple the money of burning garbage. The incinerator companies say they only burn certain medical wastes, but the waste that they burn is almost always received in sealed boxes from the hospital and never opened before putting it into the incinerator. As they do not x-ray or examine these packages, it is absurd to believe that they actually know what they are receiving and burning.

Medical waste is a broad category that includes a range of materials: everything from relatively harmless product wrappers to extremely harmful radioactive or pathogenic waste. Disposal of the waste and the separation and handling prior to disposal is important to protect people from being exposed to these pathogens and toxic components. Proper handling of the waste at the source when done correctly separates the waste needing special handling to avoid risks until reaching the final disposal point.

Waste is usually separated into different categories and boxed for storage in the medical facility until picked up by the disposal company. Pathological or infectious waste contains blood, body fluids, and tissue. Many times there are quantities of body parts, such as organs or amputated limbs, as well as bloody bandages, gloves, towels, and other items soiled with blood and body fluids. These wastes are infectious and have the potential to spread diseases or contaminate the areas the waste comes in contact with. Extra special care in handling is needed to ensure the waste stays secure. The receiving facility is supposed to check the type of waste that comes in to ensure they are placed into the incinerator at the proposer temperature and with the correct amount of fuel. Unfortunately, many of the boxes are never checked to see what the contents are and if the labels are correct. The lack of checking may lead to waste that needs higher temperatures in the incinerator or additional levels of fuel based on the contents being wetter not being properly incinerated, leaving much more contaminants and unburnt materials to go out of the smokestack.

The operation of the incinerator poses dangers to the surrounding area in the community from a number of areas.

1. Heavy Metals

Many metals are basic elements, and one of the main issues of a medical waste incinerator is that much of these metals put in for destruction are not incinerated. Materials such as lead, mercury, and cadmium simply do not disappear or get destroyed. Quite simply, they take another form after incineration and either become part of the incinerator ash and dumped in a landfill or, in many cases, are emitted from the smokestack to the surrounding areas. This isn't even secret, hidden information; this information is on the permit application from the incinerator operator to the state environmental agency and the EPA.

Medical waste incinerator operators may make claims such as they reduce emissions of toxic heavy metals by 94% using the latest state of the art technology. We respond by saying that this is only under perfect, ideal operating conditions. On top of that, the 6% that will still be emitted under ideal conditions is still toxic and deadly. Usually, the amount of toxic chemicals being emitted on their permits amounts to quantities quite dangerous to be exposed to. Furthermore, the emissions are ongoing, and the exposure will be cumulative.

Since 1989 North Salt Lake City, Utah, has been plagued with a hospital waste incinerator first run by BFI and then sold to Stericycle in 1999. There was controversy starting when the incinerator first opened in 1989. The clean-running operation the town thought it was approving never materialized. Most people had the impression the medical waste incinerator would only burn wastes from Utah. Instead, it brought in materials from as many as eight other states. Residents in the area noted there were many times thick black smoke coming from the incinerator smokestack. At times, pungent odors drifted through the areas.

For whatever reason that no one to this day understood, the town allowed the rezoning of land up to several hundred yards away from the incinerator to allow building single-family homes. Many of the residents and first-time buyers had no idea they were moving next to a medical waste incinerator that actively burned hazardous materials.

Regulators suspected irregularities in late 2011. Throughout 2012, they conducted a series of three stack tests to determine the level and nature of pollutants released from the plant. These tests are supposed to be conducted at the maximum production or combustion rate and reflect normal, operational variances. The tests showed hazardous pollution beyond the allowed rate, violating their permit.

The company first attempted to blame a flawed laboratory analysis for the test results. After the division obtained additional information, a stack test conducted on December 27 and 28, 2011, found that the incinerator exceeded levels for hazardous pollutants as well as nitrogen oxides, or highly reactive gases. Regulators also said they believed the company's logs were manipulated to show compliance with operating conditions.

In 2014, the Utah Division of Air Quality fined the company $2.3 million, the largest fine in the history of the division, for misreporting emissions from its North Salt Lake incinerator and violating emissions limits set by the federal Clean Air Act. Finally after 25 years of the community trying to stop the incinerator from poisoning them, the company is closing down the incinerator facility in Utah and moving the disposal facility to North Las Vegas where they will open a larger facility, handling all the medical waste they process in the Western states. The company says they will use the latest technology at this facility. The people in North Las Vegas, however, did not want Stericycle to relocate there and stopped the company from building a new facility. After trying since 2016 Stericycle pulled their permit from consideration in 2019 after receiving opposition not only from the residents but from officials who had reservations about the safety of the incinerator and the pollution it would emit. Experts from North Salt Lake City offered information and stories about their first hand experience with Stericycle in their community and what the incinerator inflicted and subjected people in their community to for years. Stericycle has now announced they are going to push to site a new incinerator in the Reno, Nevada area. They are hoping to convince another small community of the benefits they will bring. If the community has honest and thorough people at the county level the project here will also eventually be not approved.

2. Upset Conditions

An upset condition is when the incinerator furnace is not running properly, so steps need to be taken to correct it. A common upset condition is when the furnace has become too hot and needs to be cooled quickly. The operator will bypass certain steps that control the pollutants and vent directly into the atmosphere until the proper temperature has been reached. Clouds of thick smoke, usually black (but other colors too depending on what is in the incinerator), come out of the smokestack. While the operator is required to log and report upset conditions, many times they go unreported due to penalties the operator could be subject to.

Upset conditions are far more frequent than any facility will ever admit. Many incinerators have a documented track record of being cited and fined for them. Based on the way these incinerators operate, it is impossible to eliminate these occurrences. Sometimes they happen because of a human error, such as using the incorrect temperature for what was placed in the incinerator. Sometimes an upset condition happens because the mix of materials was incorrect, so there was an unexpected reaction. Whatever the reason, they can and do occur. At the Stericycle incinerator in North Salt Lake City, neighbors recalled one incident when the power had gone out in their neighborhood and for 20 minutes, thick black smoke billowed out of the incinerator's smokestack as all the emissions bypassed the pollution control devices and vented into the neighborhood. The state of Utah kept a record of this incinerator: it had 10 instances of these bypass events in the previous 12 month period.

Accidents as well are frequent occurrences, and sometimes entire buildings are destroyed. In Petersburg, Virginia, an explosion destroyed most of the medical waste incinerator operated by Curtis Bay Medical Waste. While employees and the company said no contamination was released, outside emergency personnel noted there were medical waste and contaminated items such as hypodermic needles and blood-soaked bandages strewn around the field of debris. It was determined a boiler in the building had exploded, causing the damage. Environmental crews worked to secure the area by setting up netting and dikes to keep any hazardous substances and materials from being washed off by predicted upcoming rains.

3. Dioxins and Furans

Dioxins are one of the deadliest materials known to mankind besides radioactive isotopes. Dioxins are made from the process of combustion. Nature can also make dioxins when there is a forest fire. That is true; however, the amount of dioxin created from a medical waste incinerator is far greater. The EPA states medical waste incinerators are one of the fourth leading causes of dioxins. Other major industrial sources are cement plants, power generation stations, and metal smelters. Once dioxin enters a person's system, it never leaves. According to the World Health Organization, dioxin has a half life of 7-11 years, so even though some degrades, there will be dioxin still present. Continued exposure to the source maintains this cycle. This material accumulates in your body and damages cells and organs.

In Brooks, Oregon, the incinerator run by Covanta is importlng medical waste from outside the state to burn at its facility. The incinerator was actually an ordinary trash incinerator whose revenue had dropped, so they convinced the state of Oregon to allow them to burn medical waste at the facility. Up to 15,000 tons of the waste per year are incinerated there. Previously, the incinerator was bringing in medical waste from Canada until people found out the plant was burning human body parts and dead fetuses to generate electricity at the plant. The very idea of this infuriated locals from all across the political spectrum who banded together and stopped the importation of the wastes.

Plastic waste is another issue. Large amounts of plastics are used and disposed of as hospital waste. A trade organization, Plastics Today, estimated that, of the total amount of plastic disposed by hospitals, a third of it is PVC plastic. According to Practice Greenhealth and the Healthcare Plastics Recycling Council, plastics make up between 20 - 25% of medical waste sent out for incineration. Whichever of these two numbers is most accurate, a sizable chunk of medical waste is plastic, especially PVC plastic.

While PVC is great for the incinerator operators because it has a high BTU value and saves money in fuel cost, it is very toxic and releases highly toxic compounds. Cadmium and lead are common elements released by the combustion of PVC, as are dioxins. Rather ironically, the actual medical waste bags used to collect and transport the waste contribute even more to the toxicity and in some cases contain more toxic chemicals than the contents they hold. These bags are mostly made from polyethylene, and burning that compound created dioxins. Another item in particular adds to this problem: blue wrap. Blue wrap is used on surgical instruments and accounts for up to 19% of the plastic medical waste at one large California hospital. While recycling programs for blue wrap are available, many hospitals continue to dispose of the polyethylene plastic wrap into medical waste bins, and this too will emit dioxins when it is burned. The smokestack at medical waste incinerators is only capable of filtering some of these dioxins out. The rest get emitted, and more are formed after the smoke has passed the filters because toxic chemicals can bond and create them in the way out too.

4. Odors

Burning human flesh and body parts gives off an unforgettable smell. Along with all the plastics a hospital disposes of, this creates a painfully memorable smell. Of course, the incinerator has air scrubbers and other pollution control devices on the emissions stacks, but only so much odor can be suppressed or masked over.

In Mathews, North Carolina, a medical waste incinerator operated for years. During that time, local residents could tell when human waste was being burned from the smell that permeated their backyards as black smoke came out from the smokestack. Over the years, the residents fought to have the incinerator closed, citing health concerns from mercury and heavy metal emissions, also citing the incinerator's non-compliance with environmental regulations. Sometimes the area residents could not even hang their clothes out back as black ash and soot would collect on them from incinerator emissions. The town pressured the facility and eventually got the EPA to require the incinerator to add even more pollution control equipment and reduce emissions further. Under all these conditions, which were very costly to the operator, the incinerator finally closed down. The operator denied that was the reason behind the closing.

There are new medical waste incinerators being proposed throughout the country. A great example of one is in West

Warwick, Rhode Island, where people wanting to try the industry's latest experiment are looking to set up a new operation. As always, the industry's latest experiment has to be tested someplace. This mostly untested method cooks the medical waste and even turns it into electricity. The people looking to operate this waste-cooking business are calling the process "clean and green." The plans call for processing 70 tons of medical waste from all over New England. They want to operate seven days a week, 24 hours a day. None of the medical waste will be inspected or opened, but they will be checked for radiation. The lack of oversight is self-evidently suspicious and dangerous.

Staying away from a known medical waste incinerator is simply good sense. At the same time, be on the lookout for other places that currently or in the future could burn medical waste as well. Permits need to be issued for facilities to legally accept medical waste. Tracking applications to your state environmental agency gives you a heads up before it becomes too late. Be aware that trash incinerators may also be proposing to burn medical waste because it brings much more revenue to the operators. Places like the Covanta incinerator in Brooks, Oregon, is a perfect example of one which now burns both types of trash.

Tony and I suggest locating a distance of at least four miles or more away from the area. It should be beyond where many of the heavier particles will be falling. There will still be contaminants in your air; however, they should not be at a level so high as to be an immediate hazard or cause long term health issues.

Chapter 10. Hazardous Waste Disposal and Transport

There are three main ways hazardous waste is disposed of: incineration, injection wells, and landfilling. Each one of those methods has drawbacks and presents unreasonable risks and dangers. In addition to disposal companies, there are hundreds if not thousands of businesses involved in the transportation and storage of these hazardous waste.

1. Hazardous Waste Landfills

There are landfills that specialize in hazardous waste. Solids, liquids, and sludges all get dumped here. Some of these landfills were specially built to hold hazardous waste. Others were simply repurposed trash landfills meeting minimum standards of safety to get a permit.

Many hazardous waste landfills may not currently be operating, but the toxic chemicals are still present, working their way through into the surrounding soil and the groundwater beneath. For instance, in Andover, Minnesota, an old hazardous waste landfill sits waiting for cleanup. Long ago closed, its toxic contents ate away the liner underneath, and thousands of barrels of waste disposed there worked their way through the soil and contaminated an aquifer underneath the site. How large the plume of toxic waste in the aquifer is or how far it will spread is unknown.

In 2011, polychlorinated biphenyls (PCBs) were discovered leaking from the hazardous waste pit.

In Williamsburg, Ohio, sits another of those hazardous waste landfills closed long ago. There have been thousands of tons of hazardous waste disposed of there over the years. This landfill also accepted startling amounts of toxic waste from several Superfund sites that needed another place to dump their toxic chemicals. The hazardous waste landfill is located upstream from the drinking water source the people use. Workers at the landfill were found directing wastewater from the landfill into a nearby creek used for drinking water for the town. In addition, one of the walls in part of the landfill caved in because that part had been placed in an aquifer and was below the water table.

Hazardous waste landfills that are presently operating receive tens of thousands of tons of hazardous waste every year. While many do not allow radioactive waste to be disposed of at the facility, some may have a permit for that as well. However, research has shown that a lack of permits does not always mean these landfills do not accept something they should not.

In a strange twist of fate, a hazardous waste landfill that operated near the Niagara River in Model City, New York, just 12 miles from the infamous Love Canal waste site, which started much of the environmental movement, wants to restart its operation, which was closed in 2015. For 31 years the hazardous waste landfill accepted PCBs and toxic waste from 38 states at their facility. For years, they discharged into the Niagara River. They called the contaminated discharge "treated wastewater," but it had PCBs, dioxins, and other toxic materials. The landfill reached capacity in 2015, holding over 30 years of the most hazardous substances known to mankind. With the huge amount of money to be made in toxic waste disposal, the company is applying not just to reopen but to expand the landfill. The new capacity would be able to accept half the toxic waste made in the entire country. It may also give the landfill operators the right to discharge 20 million gallons of PCB-contaminated wastewater into the river as well.

In New Jersey, hazardous waste landfills were constructed by Dupont and Monsanto to be used for disposing of waste generated at their manufacturing facilities. In both cases it was under six months before leaks were detected in the landfills. Dupont argued the fluids detected were from rainwater, but the compounds in the fluids showed levels of dissolved solids, the presence of phenol compounds and colors indicating the liquid was not rainwater but contaminants leaking. other contaminants. The Monsanto landfill leak was investigated for four years by an outside consultant, and the source of the leak was unable to be located. In both cases the landfills were built using double liners and leak-detection equipment. In both cases there was a breach in the liner and leaks occurred.

At the U.S. Ecology hazardous waste landfill outside Detroit, Michigan, the company says everything is fine and that they operate a state of the art facility. However, records from the Michigan Department of Environmental Quality show otherwise. The liner under the landfill, according to the waste industry, will never leak. The primary barrier to prevent toxic and hazardous waste from escaping containment has been verified to be leaking. There is documentation that toxic leachates from the landfill have been spilled into surface water near the facility. Improper venting and monitoring of hazardous wastes stored underground have been found, which could lead to leaks and possibly explosions for interactions of non-compatible chemicals.

The company violated regulation by disposing of hazardous wastes improperly in a non-hazardous waste landfill, cutting costs by using this cheaper and illegal method, ultimately creating a dangerous situation. And, by failing to properly control reactions between chemical waste they were handling at the facility, they caused nine fires in a period of nine years. One of them was an explosion that burned for three days and forced 900 homes to be evacuated. Fires in the landfill have created thick black smoke containing unknown amounts of fumes from hazardous waste to drift over the surrounding areas. Recently, U.S. Ecology announced that it wants to start taking radioactive fracking wastes from out of state. And, the company had been seeking to be allowed to take in PCBs, one of the deadliest compounds known to humans. They are planning on doing all of this in a landfill known to be leaking.

2. Injection Wells

These wells are used to literally inject millions of gallons of toxic wastes into the ground at high pressure. Although not as popular a disposal method for hazardous waste as incineration, it is still used and has had serious problems in the past.

The EPA is responsible for regulating these Class I Hazardous Waste Disposal Wells. This class of injection is primarily used by the petroleum, pharmaceutical, and chemical industry to dispose of their toxic waste and other by products considered hazardous. The wells are up to two miles deep in some cases and need to be designed to ensure the waste does not escape from the injection zone. In fact, the EPA requires companies operating these wells to show evidence the hazardous waste will not move from the place it is injected into for a period of 10,000 years.

The industry prefers not talking about this method, so the public for the most part doesn't even realize what it is the industry is doing. Regular inspections of these facilities are supposed to be taking place, but too often due to a shortage of inspectors, they do not happen. Wells do leak, and many times much of the well casing is unable to be examined anyway due to the specialized equipment and time needed to do a thorough inspection. Therefore, leaks may take years to be discovered, but in the meantime they are contaminating the surrounding soil and groundwater. Supposedly, this method is safe enough to assure the waste does not move for 10,000 years, but it is really another attempt to make it seem the waste has gone away.

The EPA estimates over four trillion gallons of industrial waste have been pumped into the Earth. In Ohio alone, three companies, Vickery Environmental, AK Steel, and INEOS USA, are running injection wells that pumped 247.5 million gallons of acids and cyanide into these wells. In fact, one of the wells in Vickery, Ohio, run by Vickery Environmental, is allowed to continue operating in spite of the EPA listing it as having major defects.

The facility in Vickery has had quite a history. Back in the mid-1970s, six injection wells were issued permits, and the company began injecting hazardous wastes. By the early 1980s, all of the wells had failed. Waste, which included dioxins, commercial waste, and landfill leachate, leaked into the rock above the 10,000 year containment zone. The leakage was estimated to be over 45 million gallons. Several of the wells were discovered to have holes in their casing, and others were corroded from contact with the waste. The company Waste Management was fined more than $12.5 million for repeatedly mismanaging the leaking wells and dumping over one million gallons of PCBs and dioxins into lagoons on the property.

Sometimes permits to operate hazardous waste injection wells are given out overzealously. In Pasadena, Texas, ExxonMobil wanted to have its own injection well to get rid of hazardous waste at its plant. The company convinced the EPA there was a reasonable degree of certainty the waste would stay where the company put them until they were no longer hazardous, that for the next several thousand years, thousands of gallons of toxic waste would remain in the exact same place and cause no problems. The EPA accepted this self-evidently false reasoning.

The rock and subsurface materials that the hazardous waste is injected into can be incompatible with the waste which can destroy confining formations. Corrosive chemicals, primarily acids, can eat away at the rock formations that protect the water-bearing zones, causing two major problems. First, the permeability of the receiving zone increases, allowing wastes to move into unprotected areas. Then, corrosive chemicals expand the receiving formation cavity by destroying the walls of the injection zone. The size of the cavity grows, causing instability in the overlying rock formations.

In at least one documented case the public has found out about, acid waste was pumped down an injection well in Florida to dispose of it. Unfortunately, the rock structure the acid was supposed to have been contained in for thousands of years reacted with the acid. The acid ate away the rock, which resulted in an underground cavern some estimated was at least 100 feet by 40 feet in size. This cavern may have led to collapse of the underground structure or release of the acid waste into other rock layers, potentially contaminating underground water supplies. Based on this example, the probability of this happening elsewhere with some frequency over the years is quite high.

3. Transportation of Wastes

Hauling hazardous wastes from the generator to the disposal company is also a large part of this process. Tank trucks filled with liquid waste roll over our highways all over the country. Truck trailers and rail cars filled with boxes and drums of wastes travel as well. Trucking companies haul hazardous waste like any other cargo. Trucks carrying hazardous wastes have registrations from the states they are licensed in on the sides of the tankers. People who know where to look can see the initials from the state environmental agency listed with registration numbers next to the initials. Other than that marking or a placard on the truck or trailer, the average driver could not tell what its contents are. Tony and I have seen hazardous waste trucks, along with every other type of hauler, routinely parked on the sides of the road.

There are two major concerns with toxic waste transportation: company truck yards being used for storage or transfer of hazardous wastes and the location that the trucks are cleaned and washed out in.

My family had an instance when I lived in upstate New York and a large hazardous waste trucking company wanted to build a terminal and transfer station in our town near a highway. Many people initially saw no problem with a trucking company, but when they realized there was going to be hazardous waste sitting at the location, sometimes for days, that was a different story. When the company was asked about how they were going to clean and wash out their trucks, initially, they said on the site. They quickly backtracked when the issue of the hazardous contents they would be carrying and residues needing they would need to clean out came up. Typical shipments would consist of PCB-tainted liquids, used and contaminated acid solutions, and paint sludges. Suddenly, the people in the area readily recognized the dangers from the waste and cleaning chemicals that potentially would be stored or used on the site. . These substances could possibly contaminate the wells, depending on the groundwater. Because the company was unable or unwilling to add safety measures to address these issues, their permit was denied.

In Vernon, California, a battery recycling center (which has since been closed down) and a trucking company worked together to illegally transport lead-contaminated materials. The recycler, Exide Battery, and the trucking firm, Wiley Sanders, moved over 128,000 pounds of this hazardous material from one location to another. There was no preventive packaging or liners to secure the lead contamination, and, at least twice, hazmat crews were called to respond to liquids dripping out from beneath the truck trailers. The county is suing the transport company, alleging that for almost 20 years hazardous materials leaked from the improperly secured trucks on roads, highways, and the ground. The storage yard where the trucks returned after hauling the toxic materials quite possibly may have been contaminated as well, spreading the lead and other toxic chemicals into the surrounding area over the years.

In Springfield, Oregon, a company involved in hazardous waste cleanup, testing, and training was fined for mishandling hazardous waste. Northwest Hazmat was cited for transporting waste to locations other than listed on official paperwork and storing waste onsite months longer than allowed. One memorable violation involved one of their employees being allowed to bring waste contaminated with cyanide home with him. The citation didn't specify why he brought the waste home, but Northwest Hazmat let him do it.

In Stockton, California, a Texas-based hazardous waste transport company was caught illegally storing hazardous waste at one of their facilities less than 500 feet from nearby homes. Nobody gave them a permit for doing so, and nobody knows how long these hazardous materials have been regularly sitting in the neighborhood.

In Salisbury, Massachusetts, Safety-Kleen was fined for violations at their hazardous waste storage facility. Numerous violations were noted, including improperly installed leak detection equipment, incomplete records, mismarked storage containers, and improperly trained staff. These were just some of their violations; several other issues were noted at the facility at the time as well.

In Cherokee County, Georgia, an alert driver spotted a tanker truck draining out the tank under an overpass along the interstate highway. Responding police and an environmental team found the hazardous liquids along the roadway. Analysis showed the substance was toxic leachate, most likely from a landfill. A warrant was issued for the truck driver who fled the scene prior to the police arriving.

4. Blending and Preparation of Wastes

Some companies decide to get into the game of hazardous waste disposal without having an incinerator themselves. They make themselves into an entirely problem solving company and negotiate disposal deals with the incinerator companies. These companies collect the hazardous waste and either directly bring it to the incinerator or, many times, take it to their own facility to sort, combine, and repack to maximize their own profit.

These sites can be almost as dangerous as the incinerator as many accidents and spills occur due to the increased handling and storage of this waste on the site.

There are many more of these facilities than most people think. In California alone there 78 such places. Many of them blend in with all the other companies, and people passing by would never know they are handling hazardous and toxic waste inside, unless, of course, there is a problem. Out of those 78, five failed for mishandling the hazardous waste they are supposed to be experts in. Nine other companies received only conditionally acceptable grades when inspected. There were violations including failure to store hazardous waste properly, accepting toxic materials they were not authorized to handle, and using trucking companies not licensed to carry hazardous waste. Thousands of these companies operate across the country and may be in your town.

Outside Chandler, Arizona, a company that does solvent recycling, a process that removes contaminants from solvents so they can be reused, was cited for a variety of violations. Investigations into the company practices revealed that hazardous materials were released to the air on numerous occasions. These releases went unreported, and no measures to prevent this were taken. If it weren't for employees and others who told about how some of the containers with aerosols were simply released into the air, we wouldn't even have known about this practice. In addition to solvent recycling, they also did blending of hazardous wastes, and repackaging which led to other violations as well.

Furthermore, safety measures to minimize risk of fire and explosions were not done, and hazardous waste was not handled properly at the site, which resulted in mixing incompatible wastes. In addition, there were ten open dump areas the company allegedly had used, which the EPA is now forcing the company to clean up. At another of their seven operations, a release of 4,000 gallons of a toxic solvent formed a toxic cloud that drifted away from the plant over wetlands areas near the facility.

Near Tacoma, Washington, in 2018, a hazardous waste facility run by Stericycle mismanaged handling and disposal of tetrazole, a highly hazardous material, resulting in a fire. The state environmental agency fined Stericycle $1.9 million, saying that during the investigation Stericycle gave misleading and incomplete information. The state department of ecology said people could have been killed or injured by the mishandling of the hazardous wastes.

The company had another fire in 2018 when they mixed chemicals together in direct violation of orders of compliance from the state. The agency also reported that employees were improperly trained and did not follow procedure.

Near Winona, Texas, Gibraltar Chemical resources operated two injection wells and a hazardous waste fuel blending operation. In addition, they also ran a solvent recovery and recycling business at the same location. Over 17 years, people in the area suffered a range of negative health consequences. Investigations by the EPA confirmed chemical releases at the facility caused nearby residents to continuously have dizziness, nausea, and respiratory problems tied to the releases. Documentation showed multiple explosions, fires, chemical releases, and spills occurred. At one point the operation was closed down when the EPA discovered one of the injection wells was missing 800 feet of required concrete casing.

In Detroit, Michigan, a hazardous waste processing plant has received approval from the state to expand its operations. In spite of U.S. Ecology racking up over 150 citations for discharging excessive amounts of cyanide, mercury, arsenic, and additional toxic chemicals into the wastewater, the company will now be allowed to store up to 677,000 gallons of toxic waste onsite at one time. Previously, they were limited to 76,000 gallons. Revenues for U.S. Ecology was over $167 million in 2019.

Every company who wants a piece of the hazardous waste business finds a way to get some action. Some even do it illegally. Down in Houston, Texas, mixed in with oil refineries, chemical companies, and numerous places producing hazardous waste is a company called Wright Containers. They offer customers used empty plastic containers that can hold several hundred gallons of hazardous liquids. Apparently, they were caught telling employees to empty any remaining hazardous liquids from old ones down a concealed storm drain in the warehouse.

The business of hazardous waste management is lucrative. Estimates put the 2020 market at over $13 billion. With that much money available to companies in the business, there are some prominent players and millions of dollars being spent to keep the rules of the game in favor of the industry. Billion dollar companies own and operate many of the disposal facilities and transport companies involved in hazardous waste disposal, and they're doing everything they can to keep it that way.

The operation of these facilities has always been questioned by people doing research into the risks associated with hazardous waste disposal, especially incineration. The chemicals being released regularly are not fully known and therefore unable to be fully monitored for. There are too many negative incidents related to these facilities of people experiencing adverse health effects to take a chance.

A hazardous waste incinerator or landfill is one type of facility that everyone should avoid living anywhere near. Long-term exposure to these facilities and their emissions are suspected of causing all types of cancer, diseases, and birth defects. These companies handle the most toxic substances known to man, and in many cases also dispose of the waste from Superfund sites. Hazardous waste is shipped to these facilities from all over the region, and, in some cases, the country, depending on who offers the cheapest disposal price.

Any industrial operation that deals with the treatment, storage, or disposal of hazardous wastes is a danger. Transportation companies who move the waste around also have hazards associated with their places of work. There is not one positive feature of the entire industry. If so much hazardous waste was not permitted to be made in the first place, there would be much less need for the industry. Ending the creation of hazardous waste is not a hopeless goal, but it does not appear that it will happen for the foreseeable future.

It would be prudent to leave a one mile buffer between you and any type of facility involved in transporting or repacking hazardous waste. Disposal facilities especially present a wider-reaching danger zone, and a minimum of four miles would be the minimum we would even suggest. In addition, regular testing of your water supply especially if using well water should be done annually.

Chapter 11. Factory Farms

A well-run and managed animal farm is one of the best neighbors that we could hope for. Beautiful to look at with its wide open spaces and well-kept animals. It seems to bring us a little closer to nature than most of our daily routine allows us to do. Farm fields with their animals lazily grazing on grass while they walk around is so calming. Many people move to the country precisely for this reason. Even with the smell and the noise from the farm, it is preferable to houses on top of houses and living elbow to elbow with other people.

Family-operated farms have been handed down for several generations, each new one proud to carry on the family farm. Someplace along the line over the past few generations, though, things started to break down. Family farms gave way to factory farms, literally farms run like factories. The official name the government has for these farms is CAFO, standing for concentrated animal feeding operation. As this is an industrial operation, we will use the term “factory farm.”

With the enormous number and concentrations of animals in one place, issues have arisen, creating significant consequences on the environment and surrounding communities, issues that also impact people beyond the community in ways nobody foresaw during the rise of these factory farms.

These factory farms are so tremendous that the amount of manure some produce far exceeds the combined amount from all the humans in many nearby towns. The CDC said, "Chemical and infectious compounds from swine (pigs) and poultry are able to migrate into soil and water." There are five main factors that can impact those who live by or one of these operations.

1. Manure

Overall, factory farms and the animals on them produce much more manure in a year than the entire population of our country. Human waste goes to sewage treatment plants to be processed and made less hazardous to human health. The same cannot be said for manure

The National Association of Boards of Health and General Accountability Office (GAO) have estimated a large factory farm can produce more manure in a year than some large cities in the U.S. According to the EPA, 2,000 cows produce 240,000 pounds of manure a day for a whopping 90 million pounds a year. Many factory farms have far more than 2,000 cows.

An older EPA report estimates the animals on factory farms produce anywhere from 3-20 times the manure of the population of this country. The big difference, and the big problem, is the manure from factory farms is not required to be handled in a way that minimizes the potential for dangerous pathogens and other issues. Nothing even close to the treatment of animal waste is done, nor are there any requirements to do so.

On large factory farms the movement of animals is usually restricted to a single area. The urine and manure drop beneath the animals and is only collected and removed when needed. Many farms simply allow the manure to be driven into the surface beneath the animals until their feet and legs are caked with dried manure and urine. When the manure is removed, it is then placed into giant storage piles or large storage lagoons. The factory farm has few options for the disposal of all this waste. They can choose to spread and spray it onto the fields they have or pay to have somebody truck it away to some other location for disposal.

With the livestock headcount of some factory farms numbering in the tens of thousands, the amount of waste is easily hundreds of thousands of gallons, even millions quite frequently. Returning the manure to the fields for nutrients for the next year's crop is a historically sound use for animal waste. But, proper farming management practices limit the amount of manure that can be applied to the land so that the land can use and absorb it. In most cases that would not get rid of a quarter of the manure produced by the animals on many of these factory farms. Sometimes factory farms pay other landowners to dispose of manure onto their property as well.

As a result, in the case of many factory farms, manure is applied to the land beyond the point of saturation. There are many instances of factory farms who have far exceeded the limits of absorption on the field and over-saturated the ground. Nearby neighbors can only watch as raw animal waste runs off of the farm property into nearby streams and creeks. Sometimes, it overflows onto nearby properties as well.

At times when the cesspools are full on the farm, liquid manure will be sprayed into the air on the fields to get rid of some. The liquids are taken away over the fields and to the areas beyond the farm boundary in the winds as liquid manure mist. Smells and airborne gas do not respect property lines.

Accidental spills because of both poorly maintained lagoons and weather events unleash hundreds of thousands of gallons of stinking manure and sludge into local waterways, polluting areas downstream for miles. Each time these spills occur, E. coli bacteria in the water rises to levels making the water unsafe for either human contact or use. Contaminated water used for irrigation on crops is responsible for many outbreaks of the disease sickening people each year.

In 2018 after Hurricane Florence, huge hog operations in Dublin and Sampson County, North Carolina, had two manure lagoons totally fail. According to the Environmental Working Group and reported by the *New Yorker*, 7.3 million gallons of liquid and sludge-concentrated hog manure spilled into the tributaries of the South River and the Northeast Cape Fear River.

At Lanier Farms in Jones County, North Carolina, the farm was suspected of releasing one million gallons of liquid manure from its lagoons. The release was traced to the farm by a 1.8 mile long trail of pinkish colored decaying waste. The waste poured into the Trent River at the end of its trail from the farm. Lanier Farms had been cited for releases and spills numerous times before.

In both of the above cases, the waterways were also used for irrigation of farmland, recreation, and a source of drinking water. Use of the water was deemed unsafe until the levels of bacteria returned to lower levels.

In a 2013 report about the Lower Yakima Valley in Washington state, the EPA estimated a dairy lagoon leaked million gallons of liquid manure per year into the surrounding soil.

Despite being pushed by the dairy industry to do another review, the EPA put out a statement that they were standing by their earlier findings. They confirmed that factory farms in the Lower Yakima Valley were responsible for extensive groundwater pollution, making the drinking water unsafe in many places.

In Michigan, legislation was introduced to ban manure application on frozen or snow- covered ground to prevent runoff of manure into nearby waterways and neighboring properties. The bill was bitterly opposed by the Michigan Farm Bureau.

2. The Smell and Odors

Factory farms emit tremendous amounts of dangerous airborne pollutants. Ammonia gases from extremely large quantities of urine and decomposing manure from hundreds of thousands of animals drift up and away from the farm and into the surrounding area and communities.

Hydrogen sulfide gas in the air is a constant pollutant that causes respiratory and digestive issues.

In the eastern part of North Carolina, approximately 2,000 factory farms have open air cesspools holding the urine and feces of 9.5 million pigs. Ammonia gases and bacteria seep into nearby communities, causing increases in asthma attacks, respiratory disease, and other illnesses.

Lethal levels of hydrogen sulfide gas are also possible in extreme conditions. In Minnesota one farm had a release with such a high level of hydrogen sulfide that the state declared a public health hazard situation and evacuated residents. The high levels of gas persisted for months, and, due to lack of real enforcement laws, the state was unable to do much.

Regulatory rules on emissions for factory farms are much lower than rules for human waste treatment in spite of the fact animal waste amounts to over triple the amount of human waste. If the industry was held to the same standard of waste management as sewage treatment plants, they most likely would go out of business. That is why lobbyists for the industry stay very active.

The watchdog group, American Promise, noted in 2018 that lobbyists working in favor of the new Farm Bill gave almost $1.7 million to House and Senate Agriculture Committee members. The committee member receiving the most money from the lobbyist was Mitch McConnell, who raked in $183,425.

Because of these lax regulations, large manure lagoons and storage tanks that are capable of holding over 600,000 gallons of waste are left uncovered. Basically, it is a huge cesspool of decomposing manure and urine heating up and sending the smell and airborne contaminants for miles.

Ammonia is one of the components in urine, and Poison Control has listed it as a substance to monitor. Neighbors of these factory farms and people downwind of the operation regularly experience irritated eyes, noses, and throats from breathing in high levels of ammonia. It especially affects older people and children.

Many residents living near factory farms have to live with their doors and windows closed all the time. Activity outside is often much less than in areas without the factory farms due to the continuous odor and exposure to ammonia.

Scientists have shown that when we smell feces, we actually are breathing in microscopic particles of bacteria from the waste.

Those who live near the factory farm and breathe that all the time are also breathing in particles of antibiotic resistant bacteria too. That comes from the tremendous amount of antibiotics fed to the animals to keep them alive in an environment they would normally be sick in. Breathing these, however, makes us resistant to many of the antibiotics that would otherwise make us healthy when we are sick.

3. Particulate Matter and Airborne Pathogens

Studies done by the EPA and other organizations have shown that large-scale livestock farms create huge amounts of dust particles. There are tremendous amounts of manure, feathers, straw, feed, bedding materials, and animal fur all together in one place. The movement of the animals churns everything up and sends all of this into the air surrounding the containment areas.

Several studies have linked an increase in illnesses to the particulate matter and dust generated by these farms. One study in particular stands out. John Hopkins University released results concluding living near a factory farm increased asthma in children. Another study also by Johns Hopkins concluded people living near high concentrations of livestock were at increased risk of developing MRSA, a type of staph infection resistant to not only penicillin but many other low level antibiotics. The study found as the concentration of livestock increases, so did the odds of people nearby having the bacteria.

In June 2019, the EPA changed the rules requiring some industrial polluters to report their toxic air emissions to state and federal agencies. Prior to this rule change, factory farms were required to send reports on the toxic emissions no different than industrial facilities, such as chemical factories and the like, have to. Despite the evidence that factory farms pour tons of toxic chemicals, gases, and harmful particulate matter into the air surrounding these huge farms, the agency has eliminated the requirement to report any information on pollutants released from the millions of tons of animal waste. The move is seen by communities affected by these factory farms as selling out protections for the people. The EPA is directed by many politicians who are taking thousands of dollars from lobby groups working for the meat, egg, milk and dairy industries and represent these factory farms who will be exempted under the new rule.

An article in *Scientific American* reported results of a study undertaken in 2016 in Pennsylvania. The findings showed after manure from hogs was applied to fields, people who lived near them had a 30% increased risk of contracting MRSA from exposure to pathogens in the manure. In 2006, an E. coli outbreak in spinach was traced to irrigation water that the investigators concluded was contaminated from pig and cow manure at a nearby farm.

Near West Bend, Wisconsin, neighbors of a large dairy farm discovered their wells had been contaminated with bacteria including E. coli and called the Department of Natural Resources to investigate. The department agreed the source was the manure from the farm, and so did the courts, who awarded the neighbors damages. One of the family's children contracted *bacterium avium* from consuming the contaminated water and needed to undergo surgery. This bacteria is found in the water, dust and bird droppings that frequently surround factory farms and people living nearby inhale in the air, drink in the water and eat in their food.

In Duplin County, eastern North Carolina, 2.3 million hogs produce twice the amount of manure as the people in New York City. The county is home to 32 times as many hogs as people. The health effects from all the manure, particulate matter, and other negative issues associated with huge factory farms has taken a toll on residents. University of North Carolina and Johns Hopkins showed elevated levels of fecal bacteria from hogs in waterways can cause hepatitis, typhoid, and dysentery. Higher infant mortality rate, asthma in children, and high blood pressure are also common in the residents.

4. Water Pollution

The huge lagoons and storage tanks totaling millions of gallons will have an impact on the groundwater. This can happen if there is a leak or spill, or it can also happen simply from continued application of tons of raw sewage to the land. The groundwater eventually becomes contaminated with nitrates, harmful bacteria, and antibiotics.

In Lincoln, Wisconsin, half the private wells already exceed safe levels of bacteria and nitrates in them. To make matters worse, the local dairy farm wanted to add another 2,000 cows to the already existing 4,000.

It's not just leaks from the lagoons and storage tanks affecting water. There are also illegal discharges of wastewater too. In Marseilles, Ohio, one of the largest egg producers pleaded guilty to discharging wastewater contaminated with manure from the egg washing operation into the local streams. They were charged with negligent violation of the Clean Water Act. At time this facility had over three million chickens in their operation.

In Valley Center and Ramona, California, another egg facility was given a violation for discharge of contaminated water. Previously, the county had issued six notices to the facility for allowing contaminated water to flow onto neighboring properties and into storm drains.

Between 2014-2016 in the Shenandoah Valley area of Virginia, a review of state environmental records showed more than 90% of the water quality monitoring stations detected E. Coli levels so high the water was unsafe for human contact. The manure comes from 159 million chickens, 16 million turkeys, and 528,000 cows in the valley. Most of the manure is spread onto the surrounding farmland and runs off into the waterways. Despite the dangerous levels of E. coli in the water, the state did not issue any warnings.

In the town of Casco, Wisconsin, one resident never knew her water had been contaminated with coliform bacteria from animal manure until she became ill with MRSA. At the time, she thought perhaps her well water was to blame. When a microbiologist with the U.S. Department of Agriculture later tested her well water, the levels of dangerous coliform bacteria were so high that the well was considered too dangerous to use. The location of the well was in the middle of two large dairy farms, each with over 1,000 cows. The biologist also determined the bacteria almost conclusively came from cow manure. Since cows receive a very high level of antibiotics, it made treating her MRSA that much more difficult. The well was considered unusable moving forward.

In 2018, an E. coli outbreak swept across the country, sickening hundreds and killing at least five people. Romaine lettuce from farm fields in Yuma, Arizona, was found to be the culprit. In a report by the FDA, their laboratory traced the strain of E. coli to a factory farm with 100,000 cattle nearby the irrigation canals and farm fields growing the lettuce. The FDA suspects manure containing the E coli either washed into the irrigation water used on the lettuce or blew off the factory farm onto the nearby fields.

5. Spills and Accidents with Manure Lagoons

Unfortunately, accidents involving huge amounts of manure leaking or spilling from storage lagoons are common. Because some above-ground storage tanks hold over a million gallons of manure, these accidents are considered by local and state health officials to be massive and cause extensive damage.

In Allegheny County, Michigan, a dairy had a failure with their manure storage lagoon. At the time of the accident, there were over a million gallons of waste that was then released and flowed into the nearby waterways. Area residents reported that a plume was visible for five miles in one of the main streams. Farms along the river that used the water for irrigation were forced to find other ways to water the fields until the bacteria levels returned to safe levels.

In Canton, Minnesota, a manure storage tank had a failure in one of the walls. The spill of manure was estimated to be about one million gallons. Two trout streams were filled with the manure as it was carried into the Root River contaminating the waterway with potential E. coli bacteria. Advisories to have no contact with the water were issued as bacteria surpassed dangerous levels.

Near Snohomish, Washington, an estimated 15 million gallons of manure and other matter spilled out of a waste lagoon holding over 21 million gallons. The liquid manure wound up going into the Snohomish River, causing a huge environmental impact in the area. The state health department issued a warning the river contained high levels of bacteria that could cause infections, fever, disarray, and vomiting.

In this case, there was actually something positive that came as a result of this accident. The spill prompted the state agency responsible for inspecting waste lagoons to do a preemptive inspection of the 500 other large dairy farms in the state. They wanted to ensure another accident of this magnitude was not about to occur. More government agencies should follow their example.

Factory farms number in the thousands. While many of them tend to gather in some of the same counties or areas of states with regulations favorable to them, they can be almost anyplace. Because the number of animals on a single farm can potentially be in the thousands for cows, tens of thousands for hogs, or in the millions for chickens, it doesn't take many to create all of these hazards. There are farms totaling dozens of times those figures, all concentrated into a small geographical area, such as the ones in Delaware and North Carolina. The residents have told Tony and me about the inescapable odor that permeates everything from their house, cars, and clothing--and with that odor comes a wide variety of health hazards.

I can empathize with these residents. I lived in Lancaster County, Pennsylvania, for seven years. Our house was directly behind a small factory farm that raised hogs. We never saw an animal outside the entire time we lived there. We sure did smell them, though. There was a giant above-ground manure tank behind the animal buildings. The farm would hook a hose up to the building on one end, and the other end would be attached to a giant sprayer that would shoot thick black manure on the field behind our house. It would spray until the manure would run off the field and into the ditch behind my house. We always had swarms of flies outside the house, and many inside as well. They would sit on the walls and ceilings and leave black spots from vomit and manure all over. Yes, flies really do this some of the time when they land. It was so disgusting. I would go to work some days, and my clothes would smell from the manure. Several times we would call the health department and complain, but were told there was nothing they could do. The only time we ever received any response at all was when the manure would run off the field so badly it threatened a local tributary of a river. Then, we were told they would send somebody out to look, but we never saw anyone.

This experience was very early in my life before my family and I realized farms were as much a hazard as factories. I know we experienced higher than normal amounts of colds and minor illnesses during those years. Our well water was contaminated with nitrates and several agricultural chemicals. We constantly monitored our water for bacteria and had a whole house reverse osmosis system in addition to other types of filters. A lesson learned the hard way, but now we know much more now.

Distances to be from factory farms varies along with several factors. To determine what is a safe distance you need to consider the concentration of the farms in the overall area along with the distance from the individual farm within the hazard area. The particulates and odors from factory farms travel distances of several miles and more depending on the weather conditions. The greater the number of farms in close proximity, the higher the concentration of pollutants in the air such as ammonia, hydrogen sulfide and other irritants. We suggest four miles if there is only a single factory farm, up to eight if there are multiple factory farms concentrated nearby. If the prospect for additional ones is high, you may wish to totally avoid living in the area at all.

Chapter 12. Nuclear Reactors

These giant concrete structures certainly are not hard to miss. Some of them, especially the water-cooled nuclear reactors, generate a plume of steam that can be seen from miles away. The EPA estimates we generate 19% of our country's electricity from nuclear power, as compared to 17% for renewable energies. For an industry that has been around for over 60 years, that is not a significant amount compared to the problems it created, especially considering the issue of disposing of nuclear waste, a problem that will haunt the Earth for thousands of years.

The nuclear power industry really started out a few years after the close of the Second World War in the early to mid-1950s. Scientists saw the enormous power of the atom and wanted to figure out how to harness the amazing heat and energy produced from nuclear fission. Researchers realized it could be used to power something directly like a submarine or harnessed and used to produce electricity.

In 1951 the first nuclear reactor to produce electricity started up. The reactor was a small experimental one located at Argonne National Laboratory in Idaho. It was not of commercial size but a functioning small scale unit of larger ones to come.

Looking into the future and seeing the potential of using nuclear energy to produce electricity, President Eisenhower focused significant government resources towards the goal of developing nuclear reactors capable of generating large amounts of electricity. In a way, the nuclear energy industry was born in 1953 with the government pushing private industry to get involved and backing research and development.

The Atomic Energy Act in 1954 was supposed to get private corporations to build nuclear power plants, but many prototypes and designs tested in research labs had partial meltdowns and other failures. Private industry did not want to take on the liability of a nuclear reactor they owned having a massive failure or accident. In 1957 the U.S. government passed the Price-Anderson Act, which essentially was a huge insurance policy issued by the government to take liability from corporations if something happened. It was only after the government was willing to take on the liability and insure nuclear power plants that the private industry really started.

Today, there are 450 operating nuclear reactors around the world, 96 here in the United States. A nuclear reactor is where the actual nuclear reaction occurs. A nuclear power plant is the overall facility that houses the reactor. Some power plants have multiple reactors on one site; several even have three nuclear reactors on one site, such as the Palo Verde nuclear power plant in Arizona. Incidentally, Palo Verde is the largest nuclear power plant in the country. All told there are 58 plant locations with a total of 96 reactors. Two more reactors are currently under construction in Georgia.

The nuclear reactors in the United States are aging, and many are already past the original planned life they were built to last. The cost to build new reactors runs in the billions of dollars, and public resistance to them is higher than ever. According to the power industry, the only solution is to keep the old ones running. That doesn't solve the problems that they are wearing out and many parts are difficult to replace or repair. The repairs at times are so vast that a plant may close the entire reactor for a year or more just to bring it back up to a safe operating standard.

The Nuclear Regulatory Agency does admit that accidents are possible. They have emergency planning procedures in place with local and state officials in case there is an emergency at one of the country's reactors. If there ever was an actual emergency and the response plan was triggered, it would start with a specific action plan for each of the two emergency planning zones. The first zone covers a ten mile radius of the nuclear reactor. Basically, that would be ground zero where people in that zone could have direct and immediate harm from exposure to radiation. The second zone is a radius of fifty miles within which all the water, supplies, farmland, and animals could be contaminated from fallout. Those living near nuclear power plants should keep this in mind.

Despite the dangers, nuclear reactors are currently a part of our energy generation. Other than some very basic knowledge that a nuclear reactor makes energy from atoms, most of us know very little about how they operate. We know there is a risk of a large scale accident such as Chernobyl in Russia or Fukushima in Japan, but the odds of that happening are extremely small. What most people are unaware of is that there are ongoing accidents of varying sizes each year that could mushroom into a catastrophe.

There are four main risks and dangers from nuclear reactors.

1. Ongoing Routine Releases in the Atmosphere

Venting gases from the reactor into the atmosphere is a regular part of daily operations. Inside the reactor there are many things going on at once. Every now and then something needs to be adjusted to keep the pressure, temperature, radiation levels, and humidity under control. It is similar to the release valve on a pressure cooker. To put it very simply, when it gets too hot inside, a nuclear reactor needs to vent off some steam to prevent overheating. During these normal events when releases of radioactive gases are made to the atmosphere, radioactivity levels are monitored, and the gases may pass through treatment controls before being vented. There is some level of radioactivity allowed in these releases. Also, very importantly, there are releases with higher than allowed levels frequently vented as well. Some of the radioactive elements released into the air according to the NRC include, tritium, iodine, bromine and cobalt.

The way the reactor and system is set up, there simply is no choice when the reactor needs to be vented. It is either vent or risk a blowout.

Routine venting of radioactive gases is a planned and regular part of the operation. Radioactive gases also build up inside the cooling water area, and when they get too high, they too are vented out when the radiation level becomes too hazardous.

A water-cooled reactor uses the cooler water to control the reactor temperature and then vents about 75% of the radioactive water out as steam and discharges the rest as liquid, creating the giant plumes of white we see coming out of the huge cooling towers miles away from the reactor.

The nuclear industry claims these releases are always very small and that they are monitored and reported by the operator according to regulations. They also claim the amounts released are almost always below what is allowed and does not threaten human health or the environment. That may be true with some of the radioactive gases they can filter and measure, but there are materials that there is no effective way to filter or measure for.

An example is xenon-135. This is emitted, but how much is anyone's guess. Xenon-135 breaks down into cesium-135. This radioactive isotope takes three million years to break down halfway. By their own admission, nuclear power plants do not know how much radioactive cesium-135 they are releasing.

Importantly, many scientists agree that continuous small doses of radioactivity are as bad as one large dose at once. The effects are cumulative, and the doses compound. Almost daily releases of radioactive particles into the atmosphere are taking place at nuclear reactors across the country. The industry downplays the significance of health risks to the exposure of the emissions and the effects on people living nearby.

However, the cumulative effects of being exposed to these radioactive materials is definitely something to be very concerned about, similar to the insidious way arsenic can affect health. Just a tiny bit of arsenic has little effect on the body, but when dosed over and over with tiny bits long enough, it can lead to serious health effects and even death. When you get an X-ray at the dentist once or twice a year, little effect or damage to cells is done. However, if you were exposed to that every week, week after week, the cumulative effect would be large amounts of cell damage, a highly increased risk of cancer, and possibly death.

2. Release of Radioactive Contaminated Water

According to the Nuclear Regulatory Commission, leaks of contaminated water in a nuclear reactor are a part of daily operation. Water to cool the reactor is brought inside from a large water source such as a river, lake, or even the ocean, and the water circulates through the cooling system of the reactor. As the levels of radiation in the cooling water in the reactor core become too high, some is removed and mixed with the rest of the water to lower concentrations.

In the same way radioactive isotopes are released into the air, radioactive isotopes in water are released from the plant too. Not all radioactive material present in the water is able to be detected or removed with the old equipment the nuclear plants use. Tritium, for example, is not removed and is released to the environment. This isotope, a radioactive form of hydrogen, is known to cause mutations and tumors in humans. Other dangerous materials that have detectable levels such as cobalt and manganese are usually released at the highest level regulations allow. These levels are not safe levels. There is no safe level for radioactive materials to be released. Nuclear plants generally claim that releasing no radioactive material is not technologically feasible. Older plants, which almost all operating nuclear plants are, do have options to upgrade and use new technology which would significantly reduce the amount of so called "accidental and routine releases".

The regulatory agency allows the plants in many cases to continue to use the old technology that is insufficient to reduce emissions simply because the industry claims it was the best technology available at the time and newer technology is not economically feasible to install.

According to the Nuclear Information and Resource Service, the nuclear power industry continues to claim there is no technology that is economically feasible to prevent tritium from being vented into the air and released into the water. Because of the industry claim, the NRC does not force the industry to do anything about the continuous releases.

Then, of course, there are leaks. All machines have leaks; it's inescapable. As a reactor gets older, it leaks more. The fuel rods inside the reactor, the tanks, valves, and pipes all leak.

Unfortunately, leaking pipes and tritium releases are more the norm than the exception. Thousands of feet of pipes are encased under the reactor and unable to be reached to check. Many of the other miles of pipe never get inspected for corrosion even though many reactors have been running for over 20 years.

An investigation done by the Associated Press found 75% of nuclear reactors are leaking tritium. When the AP examined records kept by the U.S. Nuclear Regulatory Commission, out of 65 sites reviewed, 48 of them were leaking tritium. The report went on to say 37 of the leaking sites contained tritium in concentrations higher than the federal standard for drinking water. Some levels measured were hundreds of times higher.

In addition, tritium leaks have been identified in 33 of the nation's 104 nuclear power plants and escaped into the groundwater at 27 of them, according to the Nuclear Regulatory Commission, the federal agency in charge of power plants. Exelon, the largest private operator of nuclear reactors, admitted that checking 100% of the pipes to ensure completely none are leaking "is not practical" as they put it. Practicality is directly related to cost. Exelon facilities have had their share of leaks in the past. One facility, known as Dresden Unit 2, is a reactor plant in Morris, Illinois, west of Chicago. Dresden Unit 2 has had leaks in the system supposed to cool the reactor core in case of an emergency. The same plant also has leaked tritium into the ground at levels 450 times the EPA limit.

According to records from the Nuclear Regulatory Commission, on the national level, there have been over 400 accidental leaks that have been reported. Some of the leaks involved millions of gallons of contaminated water. The scary part is some of these leaks were ongoing for years before they were finally detected. Despite all these leaks, the groundwater under the plant remains unchecked.

Ten miles away from Dresden is the Braidwood reactor site. The Braidwood facility is also an Exelon plant, and it too has leaked over six million gallons of tritium-contaminated water. No authorities were informed, and it was not made public until 2005, years after the multiple leaks occurred. A drinking water well offsite near Braidwood was found contaminated with tritium. While Exelon claims the contamination is harmless, the company bought out nine properties in the area.

In New Jersey, home of the oldest nuclear reactor in service, tritium-contaminated water has been found leaking from a pipe into two aquifers that supply water to over one million New Jersey residents. It was estimated that over 180,000 gallons of the tritium water already leaked at the time of the discovery. Exelon, the plan operator, said the contamination does not present any hazards to people using the water in spite of monitoring wells measuring tritium at 300 times the level the EPA sets as safe for drinking water. The plant solution is to pump contaminated water up from the aquifer, blend it with water used to cool the reactor, and discharge the diluted water into a canal leading out to Barnegat Bay.

At a different nuclear reactor in New Jersey run by the same company, there was another spill of tritium in 2002. At that time, the spill was measured at 15 million picocuries, which set a record for the highest radiation level recorded from any tritium spill in the United States. Exlon's proposed solution to clean up the spill was to dilute the water and pipe it back out into the environment. A much better and real solution would have been to remove the radioactive tritium from the contaminated water entirely. Simply diluting it only reduces the concentration, not the total amount still remaining in the water being piped back out to Barnegat Bay.

River Bend nuclear reactor near Baton Rouge, Louisiana, released over 25,600 gallons of radioactive waste water. A leak was first detected at the plant in October, 2012, but apparently nobody did anything about it for five months. Diluted or not, tritium was released into the environment surrounding the plant.

On the southwest shore of Lake Michigan, the Palisade Nuclear Reactor released what was called "very slightly radioactive water" into Lake Michigan. The lake supplies drinking water for 40 million people downstream from the plant. Entergy, the operator, was forced to shut down the plant while the event was investigated. This was the same nuclear reactor where in 2012, a leak from a 300,000 gallon storage tank directly above the control room flowed through the ceiling into the safety critical control room and into buckets placed to catch the leaks from short circuiting the electronics. Employees who feared the worst could happen as a result of the leak turned into whistleblowers and alerted authorities this was being done. In spite of this, the Nuclear Regulatory Agency gave Entergy an exemption to continue operating the plant even with the degraded storage tank.

3. Accidents

Accidents at our country's aging nuclear reactors happen much more frequently than the public probably realizes. The public never hears about most of these even though the potential is there for a catastrophe to happen. Except for Three Mile Island back in 1979, there may not be another nuclear plant accident anyone may remember happening within the United States. Sadly, though, accidents do not need to be of this magnitude to have an effect on humans and the environment.

Accidents occur at our nuclear plants every year. Usually, the situation is downplayed by the plant operator and the Nuclear Regulatory Agency. With many billions invested in nuclear power, the industry does not need bad publicity and an outcry to shut down.

Since the nuclear power industry began in the 1950s, accidents have been a part of it. The severity of these accidents ranges from an internal failure of systems in the nuclear reactor to the partial meltdown of a core in the reactor. The numbers of accidents also vary depending on the source and the definition of "accident" used, from as few as 25 to over 100. Most plants, instead, use the term "incident."

Many of these accidents had the potential to be catastrophes but were controlled. A perfect example is Three Mile Island in Pennsylvania. This is the worst nuclear accident ever in this country. The partial meltdown of the reactor core could have turned into a disaster, contaminating hundreds of square miles and tens of thousands of people if the full meltdown had occurred. Luckily, it did not, but roughly one billion dollars was spent cleaning it up over 12-14 years. The plant is still not fully cleaned but was put into storage instead, leaving it for a future generation to deal with.

Spills of radioactive materials during transfers, accidental releases of radioactive gases during routine maintenance, and leaks of radioactive water totaling thousands of gallons occur. With the age of the operating nuclear plants approaching an average of 39, the industry has successfully lobbied the government and the nuclear industry watchdog agencies to loosen regulations and allow more contamination to be released using the justification that doing so could prevent future accidents. Ironically, a gradual weakening of safeguards is making nuclear reactors even less safe and closer to a large failure than ever before.

Accidents also can happen after the nuclear reactor is closed. The radioactive waste remains on site, and unused radioactive material stays in the reactor until it is dealt with. Closing a plant in no way takes away all of the dangers.

In California at the San Onofre nuclear reactor, live radioactive material from the closed reactor was being transferred into onsite storage. One of the containers with radioactive fuel in it was mishandled during transfer and was almost dropped from the machine to the ground 18 feet below. If that wasn't bad enough, this nuclear reactor was already under scrutiny for poor planning. The temporary storage site they are still using for the waste is right at the edge of the Pacific Ocean just 100 feet back from the high tide line.

A Better Government Association review of federal and state records shows that since 2007, there have been at least 35 reported leaks, spills, or other accidental releases in Illinois of water contaminated with radioactive tritium. The most recent incident, which leaked 35,000 gallons, occurred over two weeks in May and June at Exelon's Braidwood plant, southwest of Chicago. The leak was ongoing even as state inspectors were at the plant, being assured by Exelon officials there were no problems.

A 2014 incident at Exelon's Dresden facility in Grundy County involved the release of about 500,000 gallons of highly radioactive water. Contamination was later found in the plant's sewer lines and miles away in the city sewage treatment plant at Morris. Most sewage treatment plants are not equipped to remove radioactive substances, so the contamination mostly passed through the treatment plant and into the wastewater discharged into nearby waterways.

At another facility also run by Exelon in Cordova, Illinois, the Quad Cities nuclear plant had a leak that some employees at the plant estimated had been leaking years prior to being discovered. The leak was so difficult to repair that it took eight months before it was stopped. Groundwater radiation levels were detected at levels up to 375 higher than allowed by federal standards to be in drinking water. When Exelon threatened the state of Illinois that it would close the facility, the state allowed Exelon to push through big rate hikes to its customers.

Several years ago, a disaster offsite created a real and scary threat that could have had horrible consequences for three nuclear reactors if it occurred. Boone Dam, located upstream from three of Tennessee Valley Authority's nuclear reactors, developed a sinkhole underneath it. There had been an earthquake registering 5.9 not too far away three years previously, and, according to seismologists, earthquakes in the area of the dam and reactors tend to have more underground movement than out in the western United States. Boone Dam is as high as Niagara Falls and holds back millions of gallons of water. Ultimately, this means that three reactors could potentially be hit by a wall of water if the dam had a breach and failure.

Workers noticed water and sediment running out from beneath the dam. Senior level inspectors were notified at the Nuclear Regulatory Commission, and crews from the TVA worked around the clock to try to figure out what was happening. They monitored the activity at the dam, and if no big storm occurred first, they hoped to solve the problem.

Six months later, the Tennessee Valley Authority was still trying to figure out a solution. If the dam was destroyed, not only would several towns including Knoxville be hit by the resulting floods, the nuclear reactors would also be hit with tsunami-like rushes of water, potentially leading to a disaster much like the Fukushima plant in Japan was. As of the writing of this chapter, it is now six years after the sinkhole was first discovered, and the construction to fix the sinkhole is still ongoing.

These issues are more serious and frequent than the government discloses. A reliability and risk engineer working with the risk analysis division of the NRC claimed portions of an investigatory report of potential dam disasters affecting downstream nuclear facilities had major portions redacted to conceal the true extent of vulnerabilities. One engineer specifically said one reactor operated by Duke Energy would almost certainly sustain core damage if the Jocassee dam upstream were to suffer a failure. That part of the NRC report was supposedly also redacted.

4. High Level Radioactive Waste

There is no place currently in the United States where nuclear waste from these reactors can be disposed of. The federal government spent over 30 years and $15 billion trying to establish a repository in Nevada at Yucca Mountain, but it failed. The repository, according to the government, would cost upwards of $96 billion to complete. However, despite still trying, there is still no place for the waste.

Even more disconcerting is the fact Yucca Mountain was planned to hold 70,000 tons of nuclear waste and already 79,000 tons await. Ultimately, this repository is over capacity decades before the facility opens--that is, if it ever does.

The new plan is to have a permanent location open and accepting waste in 2048. Until then, the government is talking about interim storage in the next eight years but leaves open the possibility of that not happening. Their back-up plan is to leave the high-level radioactive waste exactly where it is now: at currently operating and formerly running nuclear reactors. So, in addition to a nuclear reactor and its hazards, there is also the radioactive dump. At most nuclear reactors, almost all of the high-level wastes ever generated by them is still sitting inside the gates.

When the Rancho Seco nuclear reactor near Sacramento, California, shut down, scientists had a unique opportunity to study the relationship between cancer and the nuclear reactor. Based on reviews of 20 years of data tracking all cancers, the area near the now closed reactor experienced a decline in cancers, but what stood out was a steep decline the 10 years following the closing. Other studies looking at other reactors after closing also found declines in the population near the plant. These studies led scientists to conclude once an active nuclear reactor is closed, the cancer rates and health risks to people living within miles of the plant decreases. This is related to the elimination of atmospheric releases and ongoing leaks from daily operations. However, there remains the continued risks from wastes stored on the site and accidents which may occur during the decommissions of the facility which takes 5-10 years or longer to complete. Any contamination of the water supply will still be affecting the population as well. To state it bluntly, a closed reactor plant is not as dangerous as an operating one but, to use the word safe, is far from the reality of the continuing dangers.

Millions of people live within the 10 mile radius of a nuclear reactor considered ground zero. Tens of thousands of them live with a bottle of potassium iodide pills in the house in case of a nuclear accident spreading radiation their way. Accepting the fact that a meltdown is possible is just part of everyday life near the reactor. These people generally don't realize, though, that leaks and vented radioactive gases are an everyday part of their lives as well. By living near a nuclear power plant, you are gambling that your body will not be affected over the long term from cumulative exposure. At minimum, people deserve to have all the information about what they are breathing and being exposed to. Once they have the full truth, they, not the government nor the nuclear plant, can decide what counts as an acceptable risk.

Because of all the hazards associated with nuclear plants, living within 10 miles near one is always dangerous in our opinion. Tony and I have always maintained that a buffer of 20 miles between the plant and a residence was acceptable. Knowing the information about releases of radioactivity in the air and water made this distance far enough away except, of course, in the event of a major disaster and release. You can decide what distance works for you.

Chapter 13. Radioactive Wastes

When we think about radioactive waste, we usually think about nuclear reactors and the fuel they use to generate power. Probably, we also think about Three Mile Island or Chernobyl, depending on our age. Very few of us will ever need to be concerned about exposure to those. Much of our radioactive waste, however, is not high level, but still is dangerous. There are several different classifications of radioactive waste depending on the level of radioactivity it has. That determines how long the waste is expected to continue to be dangerous for and how its disposal needs to be handled.

Despite governmental oversight, there are exposure issues with the way certain radioactive waste is handled. That includes some of the high level waste at government sites. In addition, there is soon going to be a crisis from the lack of ability to control radioactive fracking waste.

Radioactive waste dumps supposedly closed years ago will present grave threats to people and the environment for thousands of years. The potential for accidents and uncontrollable events, especially due to sloppy handling of these deadly wastes, will haunt the surrounding communities unfortunate enough to be located nearby. One such disastrous event happened on October 18, 2015, at a closed radioactive waste dump outside Beatty, Nevada. The official story told by federal officials and contractors at the storage site say there was a huge explosion underground in one of the radioactive waste pits. The explosion created a crater 20' x 30' wide and sent 11 drums of waste flying through the air. Corroded 55 gallon drums were scattered around the crater, and two drums were actually found outside the fenceline. The town of Beatty had inspectors with radiation detectors checking levels in town. The radioactive waste dump had, according to federal records, 4.7 million cubic feet of waste buried in several trenches. Another source put the number at 7 million cubic feet. There was no mention of the estimated 47 pounds of highly radioactive plutonium and uranium disposed of at the site over up through the early 1990s.

The investigators looking into the cause of the explosion found that cracks in the cap over the storage pit had allowed water to penetrate the areas with the wastes triggering a reaction resulting in the massive explosion. The cause of the explosion was something that was never supposed to happen and undercuts the government's main argument that these radioactive waste storage and disposal sites are safe and will remain safe for thousands of years. In less than 50 years, not even a tiny fraction of the thousands, the storage site failed.

1. High Level Radioactive Wastes

High level waste is the absolute worst it can get. These are the spent fuel rods from nuclear reactors as well as leftover materials from military weapons systems and military installations where all types of nuclear and radioactive materials were used and disposed of. This waste will remain highly radioactive and deadly for thousands of years. Much of this waste remains onsite at the nuclear reactors that produced it because there never was any disposal site developed to take it. There have been many plans, and millions of dollars have been spent to find some place considered safe for disposal, but every attempt has fallen through. For now, the waste remains exactly where they have been for decades.

If anything ever happens to the onsite storage area at one of these nuclear reactors, the resulting damage and radiation exposure could be many times higher than a simple small meltdown would be.

The federal government also has enormous amounts of high level radioactive waste sitting around. There are places like Hanford in Washington where plutonium levels remain so high in places it is deadly to be exposed for long. Many other government installations across the country are also full of high level radioactive waste. Because they are the government-owned, though, they can create some of their own disposal sites where waste can be taken from one site and moved to another. This practice is an open secret. There are quite a few locations the government owns and operates where they dump their high level radioactive waste.

Many of these government disposal sites and places that are still extremely contaminated with high level radioactive waste are in communities and near populated areas. Because these are the most dangerous types, military installations and government property present an enormous hazard to people.

The new administration has been looking for ways to clean up some of the most deadly, dangerous, and highly radioactive sites created by the military and government over decades. With help from the Department of Defense, they have discovered a highly controversial process that will turn high level radioactive waste into much less harmful low level waste. The federal government will try this new method on high level nuclear waste that has sat for decades at three nuclear weapons facilities in Washington state, Idaho, and South Carolina.

The Department of Energy will attempt to change the current radioactive waste classification from high level deadly nuclear waste into much lower level radioactive waste using a simple change in language. There is absolutely nothing at all scientific about this. What the government is doing is impossible to do in reality, according to physicists and nuclear engineers familiar with nuclear waste. It takes tens of thousands of years for high level nuclear waste to degrade down to lower level waste. Once the waste has been reclassified, the government can dispose of this lower classification of waste at a low level disposal facility at a much lower cost even though the waste in reality has not become less hazardous.

If this works, it is possible the government will try this on all of the most deadly radioactive waste stored for years onsite at government nuclear weapons complexes across the country. Apparently, reclassifying something as less harmful makes it so. As an added bonus to the government, simply by changing the classification of the high level wastes to low level, the government anticipates saving upwards of $40 billion it would have had to spend to clean up the wastes before they changed the classification.

2. Low Level Radioactive Wastes (LLRW)

The vast amount of businesses who work with radioactive substances do not create much high level radioactive waste in their regular use of these materials. Almost all of the waste this group creates is classified in the LLRW category. By volume, LLRW is the majority of all radioactive waste produced annually and needing disposal. Hospitals, medical schools, universities, and some industrial manufacturing facilities produce LLRW. The EPA previously estimated there were about 20,000 places that were commercial generators of LLRW, which includes hospitals and universities. The Nuclear Regulatory Commission estimates 95% of radioactive wastes falls into the lowest level of this three tier category.

Items that simply come in contact with some radioactivity do not become very contaminated. These types of waste are usually only hazardous for short periods of time. This waste may be slightly contaminated papers, clothing, and tools that have been used in medicine and industry. Some facilities are allowed to store and keep this waste onsite until the level of radioactivity decreases to the point that the waste can be disposed of in regular landfills. Much of the clothing, paper, medicine cases, and other items used briefly have levels of radiation that quickly decreases. This is the low end of the low level classification.

The radioactivity level of items in this classification ranges from just over normal background levels found in nature up to much higher levels of radioactivity. This class could include, for example, parts taken from inside the reactor of a nuclear power plant. It also could include sludge removed from inside the reactor during routine cleaning.

LLRW is not always low level, nor is it safe. Some of the radioactive elements stay harmful for over 100 years as they continue decaying. Part of the requirements for the storage of this waste is that it stays in storage until it is harmless and its radioactivity level is almost nothing. However, there are currently only four low level radioactive waste disposal sites operating in the country.

Complicating the situation, this material stays hazardous for a longer period of time, possibly up to several hundred years. These LLRW dumps will be a storage facility where this radioactive waste will remain until it decays and becomes harmless. The structural integrity of any storage facility and ability to maintain this integrity presents the greatest danger. Leaks from these storage sites are possible, and radioactive tritium and other isotopes have turned up in groundwater of other facilities.

Add all of those other sites the waste from the Department of Energy facilities, nuclear power stations, fuel fabrication, and conversion plants. All together, over five million cubic feet of LLRW was disposed of at four operating disposal sites in the U.S.--five million cubic feet of waste that has to be kept safe and contained for over 100 years.

This wasn't always the case. There were other private facilities that operated back through the '70s. However, one by one, they closed down for several reasons. It was kept quiet, but some of the reasons included leakage at the sites. In Illinois near the town of Sheffield, a LLRW dump was closed after accepting over three million cubic feet of radioactive waste. It appeared that radioactive tritium began to migrate away from the burial trenches. Also, there were two other dumps, one located in Kentucky and the other in New York State, that were shut down in the late '70s. These facilities were also having problems keeping the radioactive waste where it was supposed to stay. At one site, in Beatty, Nevada, there were two incidents that involved trucks transporting the radioactive waste into the storage facility. These places will remain radioactive and dangerous for decades to come. Hopefully, the waste does not escape containment and contaminate millions of gallons of water.

In the meantime, there are quite a few companies who specialize in LLRW. Many of them consolidate this waste from many different generators and pack it for shipment to the licensed disposal sites. Some also may be permitted to hold the LLRW for a period of time, such as a year, until they are no longer a hazard and then dispose of them into a regular landfill.

The Nuclear Regulatory Commission allows a practice known as waste blending. It is where a higher level of radioactive waste is combined with a lower one to reduce the level of the new blended waste. Then, the cost involved in disposing the higher level one is reduced. The NRC points out this is not the same as diluting wastes where the waste is blended with something non-radioactive to reduce the level. Even if this sounds the same, according to the NRC, it is not.

In addition to the types of radioactive waste, any reasonable discussion of radioactive waste has to address a few more topics.

3. Radioactive Fracking Wastes

This category of wastes is a fairly new one and is currently presenting a significant problem for safe disposal. From around 2005 to 2015, the beginning of the fracking boom, the waste managed to stay under the radar in many places. The industry tried their best to ignore the fact that there were millions of gallons of radioactive waste that needed to be handled safely. There are now frequent stories in the news of truckers driving tanker trucks for fracking waste disposal companies without being told what they were handling was potentially dangerous to their health. Very little if any protection was used, and the results are serious issues with people having health problems and millions of gallons of water being contaminated.

The industry knows the wastes are radioactive and have tried to hide the danger from regulators and the public. Regulators knew about this but turned their heads and ignored the problem for years. The issue is about to become explosive because of the potential for so many people to be harmed by this, especially the workers handling these wastes simply trying to make a living and support their families.

Rolling Stone magazine ran an article written by a man writing a book about the radioactivity that is part of oil and gas production but almost never talked about. The article shared the story of a tank truck driver delivering a load of fracking wastes at the disposal facility and being told the level of radioactivity contained in the wastes in his truck was the most radioactive the facility had ever seen. The trucker had never been told by his supervisor about this nor been given any special precautions to follow or protective equipment to wear for protection. The driver said the liquids get all over their clothing and body, and at times nausea and headaches as well as numbness and joint pain rack his body. He said he knew of other drivers who got sores on their bodies like lesions that do not heal for months. When the trucker brought samples of the waste he had been transporting to a lab for testing, the levels of radium levels registered an average of an astounding 3,500 when 60 was considered safe. Thousands are exposed daily while little or no regulations are in place.

The fracking industry is currently having a really hard time trying to control the problem they created. There are many legal battles over various aspects of the fracking industry going on in the courts and political arena over radioactive waste again. One major issue is what is going to be done with all the radioactive waste the fracking industry has created. The waste, known as TENORM (technologically enhanced, naturally occurring radioactive material), is a result of fracking. Fracking concentrates the natural level of radioactive materials in the Earth into a much higher level of radioactivity. This material is then too radioactive to be disposed of into regular landfills or disposed of into nearby waterways.

Millions of gallons of this waste are being generated, and it needs to be disposed of somewhere safely. Incredible amounts of money are at stake for both the places where this waste can be disposed of and the industry that wants as low a cost as it can get away with paying for disposal.

The oil and gas industry in addition to the waste disposal industry are heavily lobbying to raise the legally accepted level of radiation in the current restrictions and regulations just high enough so they can dump radioactive waste in landfills.

Michigan was looking at regulations that would open up their landfills to accept this waste and even higher contamination levels of some radioactive waste with a special license. Fracking companies in Pennsylvania were already looking to Michigan to be their dumping grounds.

There are also some companies unwilling to wait until it can be legally done. In Arlington, Oregon, Waste Management has been accepting illegal shipments of radioactive materials at their hazardous waste landfill totaling at least two million pounds. The materials have been shipped from several states as well as Canada. The radioactive waste has been coming from Oilfield Logistics, which handles fracking wastes. Waste Management claimed to have never checked or tested any of the waste over the three years, a time period during which any responsible landfill operator performs such tests as a regular safety procedure. Since the year 2000, Waste Management has been fined a total of $500,719,123 for 208 recorded violations for everything from environmental problems to workplace safety and wage and hour violations. Considering they had a gross profit in 2018 or $5.9 billion, paying a half billion over 20 years to violate the law means very little.

The hazardous waste landfill operated by Waste Management now has two million pounds of radioactive waste illegally disposed of. If this landfill gets approved to now legally continue accepting radioactive wastes, Waste Management will continue endangering the people of Arlington with millions of pounds more. Once the landfill leaks, as it eventually will, underground water supplies to the community will be in danger of contamination not just from the hazardous materials already at the landfill but now the radioactive waste in addition.

4. Radioactive Waste Treatment Facilities

With the thousands of sites the government needs to have cleaned up all over the country, there are companies who package and treat some of the radioactive wastes at their facility.

Perma-Fix has such a facility in Richland, Washington in addition to others they operate across the country. Currently, they take high level radioactive wastes from the nearby Hanford Nuclear Reservation, prepare and package it for eventual disposal, and send it back to Hanford for storage at the government operated high level nuclear waste disposal site. The company is located in an industrial park in a rapidly developing area. While they have been operating there for about 20 years, their original permit expired back in 2009. A new one was submitted back in 2009, but it was rejected by the state environmental agency for deficiencies in both 2011 and 2015. It is still currently being worked on with the state Department of Ecology in 2019.

The interesting part is the company continued to handle deadly radioactive waste for the government even though their original permit to do so was no longer valid and many changes had taken place in areas the company was originally permitted for. The radioactivity of some waste the company now handles, such as plutonium has increased to the most dangerous levels. In addition, the company is handling waste from at least two other countries and seeking to import from additional ones.

Moreover, the area Perma-Fix is located in has developed significantly over the last decade. Transporting wastes in and out on local roadways and on-site storage is rapidly becoming a tremendous potential danger to the surrounding area and community as a whole.

5. Nuclear Fuel Facilities

There still are places in the country working to keep a supply of nuclear fuel available. One such plant is run by Westinghouse outside Columbia, South Carolina. Over the past 20 years, the plant has had a variety of issues with the Nuclear Regulatory Agency. Now, the plant is leaking potential contamination through a protective liner meant to contain any leaks or spills.

Inspectors found 13 small holes in the liner and are hoping pollution has not yet dripped into the soil and groundwater beneath the plant. There were also leaks in 2008 and 2011 that were not reported to the NRC as required. If the radioactive contamination gets into the underground water, it may cause kidney and liver damage as well as cancer in people who ingest the water.

The NRC in 2019 began to consider extending the existing permit for the plant to operate for another 40 years, and a Westinghouse spokesperson said the company had no plans to clean up any of the contamination under the building until their permit expires. Additionally, Westinghouse acknowledged the plant had problems over the years and said it would do a better job. The NRC is looking favorably at the plant, hoping many of the past problems are being addressed.

While there is evidence of many past issues and failings on the part of Westinghouse and the NRC report did conclude, "there could be noticeable impacts to the soil, surface water and groundwater" there is a larger reason the NRC is leaning towards approval of the plants continued operation. Simply put, the Westinghouse plant is one of only three nuclear fuel factories in the country and is important to the nuclear power industry as a source of fuel rods to power the nation's nuclear power plants.

While all of this was still ongoing, in May 2020, it was reported leaks in waste storage lagoons may have penetrated into the groundwater. A spokesperson from Westinghouse said, "It is expected that some contamination will exist in the soil underlying the east lagoon liner, given the long operating history of the lagoon and the potential for a liner system leak."

The admission of the high probability for the failure of the liner and contamination of the groundwater is little consolation for the people in the nearby town of Hopkins who depend on the groundwater for their water supplies. In addition, the contamination may move into the tributaries of the Congaree River and further spread radioactivity. Exposure to the radiative contamination may cause kidney damage, one of the specific elements leaked is Technetium 99, which collects in the thyroid and in the gastrointestinal tract, possibly causing cancer. It certainly sounds like the industry already knows these dumps all leak through the liners even though they never admit it in the beginning.

In Newhallville, Connecticut, an old nuclear fuel manufacturing facility is finally getting cleaned up after decades of sitting in the neighborhood. While environmental officials are downplaying the amount of radioactive waste and contaminated building parts, apparently there was more than enough to require special transportation routes and extra precautions. The radioactive waste being removed is going to be shipped to a LLRW dump in Utah. For 50 years, neighbors lived next to the contaminated site that also contained large amounts of asbestos and lead. While there are no documented illnesses directly tied to the radioactivity which sat there for decades, neighbors, aware of the threat to their health it poses, are more than happy to see it gone.

In Piketon, Ohio, a tremendous uranium enrichment facility built by the Department of Energy in the 1950s still sits with much of its contamination. In addition to enriching uranium, the plant was also used for development of the country's nuclear weapons. The DOE has big plans now for the 1,200 acre site. They want to create one of the largest nuclear waste dumps east of the Mississippi. The plans call for building a LLRW landfill able to hold two million tons of waste. They claim there will be a liner, a cover, and safety systems for any leaks of radioactive waste that could occur. The proposed site is not too far from the Scioto River, so any leaks would contaminate the drinking water of Piketon and the surrounding areas.

In Washington state, close to the tri-city area of Pasco, Richland, and Kennewick, the toxic winds are blowing. The Hanford Nuclear Reservation, one of the most contaminated places on Earth, is nearby. The massive multi-billion dollar clean up of high level radioactive waste has been ongoing for years and will continue for decades. Many mishaps have occurred at this site, but only a few ever made it out to the public. During one incident, for more than a year, dust flecks containing radioactive plutonium have been allowed to blow away in uncountable quantities. Workers have carried radioactive dust into their homes in levels detectable by equipment. After decades of residents who live nearby the reservation getting sick, the courts and the government are acknowledging and settling some of the lawsuits and claims of dozens of cases of thyroid cancers linked to the radiation at Hanford.

6. Known and Clandestine LLRW Disposal

To reiterate, there are only four licensed waste disposal facilities in the country. It is to these facilities that, supposedly, all the commercially generated LLRW is being shipped to and disposed of. Permission to dump at these sites is granted in one of two ways. Some states belong to a compact allowing only generators in those states able to use the disposal site as a group. If the site is not limited to only compact states, the facility is open to accepting any state's waste. If a company that generates some of this waste feels that they can hold onto it until the level decreases enough to be disposed of in a regular solid waste landfill, they certainly will do that. Disposal of any waste, especially this type, is very expensive and, therefore, avoided at all costs.

The LLRW dump in Richland, Washington actually sits in the middle of the federally owned Hanford Nuclear Reservation. Operated by U.S. Ecology, the site also had been a hazardous waste disposal site in past years. The groundwater is already contaminated by the chemicals from waste pits long ago and also from the Hanford nuclear waste. Because the dump is located on federal lands, there is no way to really determine the true extent of the contamination. The landfill for radioactive wastes is expected to continue being used until around 2050. Disposal of LLRW is limited to states in the Northwest Regional Compact, Alaska, Hawaii, Idaho, Montana, Oregon, Washington, Utah and Wyoming and states in the Rocky Mountain Compact, Colorado, Nevada and New Mexico.

At the radioactive waste site in Barnwell, South Carolina, the state Supreme Court ruled in 2019 that the waste dump has been clearly violating state rules in handling radioactive waste allowing contamination of the groundwater. Radioactive tritium in the groundwater comes from leaks in the trenches holding the waste, which are left uncovered until full, allowing rain to soak in and carry away the radioactive liquids.

In Clive, Utah, the LLRW dump was fined by state environmental authorities for failing to follow a procedure designed to keep radioactive and hazardous dust from blowing away from burial cells for waste. The repeated violation resulted in the more than $50,000 fine, but opponents of the waste dump question why it took authorities almost three years to take action. In 2018 as well, a tractor trailer on the way to the facility with a load of contaminated soil from a hazardous waste site caught on fire along the highway near the facility. People opposed to the dump have always raised concerns for accidents along the highways and transportation routes used.

Another currently operating LLRW dump operates near Andres, Texas. At the review and permit hearings, a group of scientists and engineers confirmed the location of the waste site was located too close to an underground aquifer and could result in contamination. In spite of the technical data showing the site was not appropriate and presented risks, the head of the Texas Commission on Environmental Quality approved the application for the company. One week later the director was hired as a lobbyist for the company. In the following several years, the dump received approvals to take waste from 36 other states and was allowed to triple its capacity. Between 2015 and 2020, the waste dump has been expanding again, this time looking to become a high level nuclear waste disposal site and home to the only mercury dump in the country. All this is located near the aquifer scientists and engineers said would be an issue.

But, these are the only known LLRW disposal facilities. There are radioactive waste disposal sites operated by the federal government at locations that nearby residents may not even know about. The government usually does not share information about many of their practices concerning hazardous and radioactive waste. There are millions of pounds of waste at contaminated facilities. In fact, using one or more sites to bring waste from other areas to dump is not out of the question. The storage and transportation of LLRW is putting oblivious citizens at risk of a variety of illnesses, especially cancers.

Near North Las Vegas, Nevada, the governor of the state accused the U.S. Department of Energy of disposing illegal shipments of LLRW at the Nevada National Security Site. The government acknowledged some waste may have been sent by mistake. The governor claimed that waste has been shipped there regularly for over 12 years and called it a secret plutonium smuggling operation. In court filings, the government admitted the

Department of Energy had been quietly sending highly radioactive waste from a storage site in South Carolina to a federal storage site north of Las Vegas. The court declined to issue an order to the Federal government to cease shipments as they claimed all the material needing to be shipped was already completed. The court also declined the state's request to have any material already in the state be removed.

As the federal government continues to crawl towards finding a permanent home for high level nuclear waste, there are two towns hoping to be able to host a temporary site and collect the money that comes with it. One is about 35 miles outside Carlsbad, New Mexico, which would put close to 30,000 citizens at risk in addition to the 460,000 plus annual visitors to the national park located there as well. The other is Andrews, Texas, which would put its 13,000 residents in danger. Andrews is already the host of the only disposal facility of all the mercury waste across the United States.

As I write this today, May 7, 2020, the federal government just announced the location in west Texas may move ahead with plans to become a temporary facility for high level nuclear waste. The initial permit sought is for a 40-year period in the hope a permanent location will be completed before then. There are many more hurdles for the west Texas facility to jump over before they start construction; however, it appears the project will be fast tracked to get it done before something happens to derail it.

The government isn't the only group disposing of LLRW without informing the local population. There are not enough disposal facilities for all of the LLRW needing a place, creating a high stakes scenario that sometimes leads to dishonest practices, such as the two million gallons of radioactive waste illegally disposed of in Oregon.

How many other instances of illegal disposal activists will be uncovered? Mixed in with thousands of tons of solid waste at landfills, how much of what is coming off trucks and quickly being covered over with dirt and trash is really radioactive waste? Unfortunately, once it is buried, that question may not be answered until the radioactive waste has long since polluted the groundwater and made people nearby sick.

While we all wait for something to be done about the radioactive wastes sitting at places across the country, the Department of Energy has managed to open one facility they are using for their own high level radioactive nuclear wastes.

Located 26 miles outside of Carlsbad, New Mexico, sits the federally operated Waste Isolation Pilot Plant known as WIPP. The only permanent long-term storage facility in the country where highly radioactive waste is brought to remain for thousands of years, safely decaying until they are no longer a threat. On February 14, 2014, one of the 55 gallon waste drums containing radioactive waste burst open, leaking radioactive isotopes that traveled through the plant tunnels and exhaust shaft, finally being detected almost 3,000 feet away from the source of the radiation. About 22 workers in the facility at the time received low level doses of radiation.

The accident, which shut the supposedly safe high level nuclear storage facility for almost two years at a price tag of $2 billion in clean up costs, was caused by simple human error. The investigation determined the nuclear waste had been packaged in the wrong kitty litter. It turns out kitty litter works in nuclear waste as an absorbent, but the brand workers used at Los Alamos National Laboratory was the wrong brand of kitty litter, which triggered a reaction. Unfortunately, there are close to 500 drums of high level radioactive wastes already stored underground at WIPP. The facility meant to safely store radioactive waste for thousands of years failed after only 15.

Tony and I believe safe distances to put between you and any type of radioactive disposal site would be at minimum, 10 miles, with 20 or more being best. Like a nuclear reactor, in the event of a major accident, there are going to be zones of impact. Within 10 miles will most likely be where the radiation levels will be the highest and may remain that way for quite some time after an accident.

Places that treat and reship radioactive waste such as Perma-Fix apparently also have the potential for accidents including spills, releases, and explosions. I could not see being closer than four miles from any facility where these quantities of radioactive wastes are regularly being handled.

Chapter 14. Landfills

The companies who operate the landfills in our country would like us to believe one thing: landfills are safe. That statement could not be further from the truth. Almost undeniably, landfills are definitely not safe, and there are issues all of them share. Whether we are looking at the integrity of the landfill itself, the operation of the landfill, or the trucks and trash, the problems are not pretty.

If we were to pick the most dangerous problem common to all landfills, it would be that all landfills eventually leak, some sooner than others, but all landfills leak. The EPA confirmed this fact years ago. When they do start to leak, it is a collection of liquids called leachate. Leachate is made up of all the liquids accumulating inside the landfill from rotting trash and food mixed with any industrial waste thrown in the landfill. Leachate can even include prescription medications, old pesticides. and paint--anything that was thrown away can be part of leachate. Frequently containing chemicals like PCBs and acids, leachate is considered toxic and, in the past, has contaminated thousands of wells and water supplies across the country near these leaking landfills.

The landfill industry has produced studies that claim not all landfills leak, but according to the EPA, at some point, the liner, if the landfill even has one, will allow leachate from the landfill to leak into the soil underneath. This applies to single liner landfills, double liner landfills, all landfills.

The majority of people simply see our trash as something we put out for the garbage truck to pick up once or twice a week. Later that day, the garbage is magically gone. It just disappears, so we don't think about it any longer. Just like flushing our toilet--we know it goes someplace, but we really don't care because we no longer see it or smell it. However, those who live near one of the thousands of garbage dumps where trash is disposed of think about it a bit differently.

Landfills have existed throughout all of human history. People historically disposed of their waste items and no longer needed possessions in a pile out back away from where they lived. Archeologists have dug up piles of cast away items dating back thousands of years. For the longest time there was never a negative impact because the items disposed of were made of natural materials. In time, much of the waste biodegraded on its own, and the remnants left were non-hazardous.

When the world started to industrialize, the types of waste in a landfill rapidly changed, and so did the impact on people. The waste stream quickly went from broken pottery shards and animal bones to petroleum products, industrial chemicals, and plastics. Health issues by trash not properly disposed of became the norm. Up until the 1970s, landfills were simply created. Old landfills from the '50s, '60s, and '70s were usually no more than a big hole in the ground, a canyon, or an abandoned quarry. The operator just started a dump on a piece of property.

Then, they began to fill it up with garbage and kept filling in the hole or piling the garbage higher and higher.

For many of these old landfills, it mattered not what was dumped in them, and they accepted everything. This included regular household trash, construction debris, asbestos, old industrial chemicals, toxic waste from factories, medical waste from hospitals, and dead animals. There was no liner underneath to prevent leaks or any system to prevent runoff from flowing away. Back then, landfills accepted anything and everything with almost zero restrictions. Industries were pleased as they had a very inexpensive way to dispose of highly toxic and hazardous waste.

There were many problems with this, however. As moisture in the form of rain and snow fell on the landfill, it mixed with all the rotting garbage and liquids inside to form a toxic leachate. The leachate and sludge would then soak into the ground underneath and around the landfill and run off into local streams and creeks, poisoning the areas. On the sides of large piles of garbage, the rain eroded away at the soils and trash, carrying it away to nearby properties and streams. Eventually, in many areas, this highly toxic liquid did wind up polluting the groundwater, local aquifers, and many wells supplying water to people nearby.

While Tony and I concede technology has made some improvements possible in landfill containment and leakage monitoring, these only exist in newer landfills constructed after stricter regulations went into effect. And, in our experience, even these new improvements are woefully inadequate protections. The liners are still known to leak, and even landfills using double liner systems have experienced the same issues. Also, this still did not address the basic fact that all landfills eventually will leak (and all the other, still unmentioned issues landfills have besides that). Furthermore, these safeguards only apply if the operators strictly adhere to them. So, not only will all landfills eventually leak, but in the meantime, people living nearby have to deal with many hazards that affect their health.

Up until the disposal of hazardous wastes started to become more regulated in the late 1970s and early 1980s, millions of gallons had already been disposed of in landfills across the country. Most, if not all, landfills were simple pits with absolutely no liners or protections from leakage of the contents dumped into it. At best, a layer of compacted clay was placed to slow down the inevitable leakage that would occur.The landfills still contained hazardous wastes and created hazards through its leachate, and emissions remained. Landfills that either reached their capacity or were ordered to close due to not meeting new environmental regulations were covered up with dirt and left to weather away, if they were covered up at all. Many were not even fenced off from the public and became fun areas for children nearby to play on.

These landfills are not tiny places. Many landfills were meant to operate for 20 or more years and be able to receive the trash generated from hundreds of thousands of people everyday. The quantity of rotting trash in some of the larger landfills is measured in the billions of pounds dumped into them each year.

Places like Orchard Hills, 50 miles west of Chicago, had two million tons, that is four billion pounds, of trash dumped there in 2008. For years neighbors would complain about the sulfuric rotten egg odor coming from the landfill. Puentes Hill in Los Angeles County took in about three million tons of trash, about six billion pounds of rotting, stinking garbage. Some areas of this landfill are over 500 feet deep of decomposing trash.

There are a host of environmental issues associated with a landfill, and studies that have monitored landfills and their effects on people living near them show negative health issues directly related to the landfills.

1. Water Contamination

Based on the fact that all landfills eventually leak, both groundwater and surface water are potentially affected. Residents who depend on their own wells for a water supply sometimes find themselves with water unsafe to drink. Usually, the discovery of toxic chemicals in a private well happens in one of two ways: when the homeowner is diligent about getting their water tested regularly or when people suffer cancers and other illnesses until somebody finally sees a connection to contaminated water.

Public water supplies usually depend on groundwater as well. Many towns and cities have large municipal wells pumping water up from the aquifer to the water treatment plant. There is no shortage of towns who had to close one or more of their municipal wells due to toxic contamination traced to a landfill near the well or aquifer recharge area.

In Ellington, Connecticut, the local landfill is contaminating residential water wells. A group of chemicals known as PFAS was found in local residents' well water. Even after the state found the contamination, it was six months until they started supplying them with bottled water. Until then, people either were supplying their own water or decided the risk of drinking contaminated water was one they accepted.

PFAS chemicals are also known as “Forever Chemicals” because they never degrade or go away. They will persist in humans forever. Environmentalists believe PFAS contamination is widespread, but the public is first finding out that many of them have been drinking contaminated water for years. Eventually, there will be enough evidence to strongly assert that millions of people will be affected by this.

The research done on the effects of PFAS being consumed in drinking water has shown a wide range of health problems. Some of them include reproductive and childhood developmental issues, kidney cancer, and diabetes.

In Bath Springs, Tennessee, the county landfill for years took in household trash and other waste from the residents in the region. In 1999, the county decided it made sense to privatize the landfill and allowed a private company to come in and operate the facility, and the county wastewater treatment facilities would take the leachate. The company that took over the landfill decided there was not enough money in taking household waste and decided to start accepting many types of highly toxic industrial waste. The small amount of leachate from the landfill grew exponentially. In fact, some of the waste was so toxic that the leachate the landfill produced killed off the bacteria at the treatment plant where the liquid was sent. After the landfill company found they no longer had any place nearby to send their leachate, they began spending hundreds of thousands to ship it out of state. Soon, they decided to walk away from the landfill. They simply decided to leave the thousands of tons of toxic materials they had brought in over the years behind and walk away, leaving the landfill and toxins for the county and the residents to deal with. There is much concern over groundwater contamination for the industrial chemicals disposed over the years.

In Camden, Tennessee only a short 35 miles from the Bath Springs landfill, another landfill also took in “special wastes” and allowed thousands of tons of hazardous aluminum waste and other highly toxic materials to be dumped at the landfill. When major environmental issues started occurring at the landfill, the operator decided it was cheaper to declare bankruptcy and walk away fro the landfill leaving the local area stuck with the problem.

In central Massachusetts, people living near the state's largest landfill have a solvent used in industrial applications and products showing up in their well water. The landfill nearby has been in operation since before liners were required to slow leakage from the landfill. According to the EPA, 1,4 dioxane, the solvent in their well water, is considered a likely carcinogen. The landfill is disputing they are the cause of the contamination in spite of monitoring wells near the unlined section also showing high levels of 1,4 dioxane as well. The people with contaminated wells used their water for years because nobody ever said there was a problem until it was too late.

2. Odors

All active and many closed landfills have a distinct odor to them. The smell of rotting garbage is a cross between a rotten egg and an old, decayed melon. Some days the odor is strong enough to make people's eyes water and not want to breathe.

Information on the New York State website health.ny.gov talks about studies done on communities near landfills to evaluate what health effects the landfill and exposure to the chemical emissions from the landfill is having on the residents. During the several months the studies were conducted, the health issues reported included eye, throat, and lung irritation, as well as nausea, sleeping issues, headaches, chest pains, and worsening asthma symptoms. All of these are consistent with exposure to hydrogen sulfide gas, one of the primary emissions from the nearby landfill.

Sulfur dioxide is another landfill gas, similar to hydrogen sulfide in some ways, but even more of a health threat and more damaging. The health issues associated with sulfur dioxide exposure range depending on the concentration in the air. Locations within a half mile of a landfill are affected more by these emissions than ones further away. Sulfur dioxide irritates the eyes and nose along with the throat and lungs. It also causes inflammation of the respiratory system. Doing any moderate physical activity increases the effects on the body. Symptoms may include chest pain when breathing, coughing, and breathing difficulty in general. Higher levels of sulfur dioxide can affect lung functioning, trigger asthma attacks, and can worsen heart disease in people who have heart issues by placing more stress on the cardiovascular system. Lastly, sulfur dioxide has the ability to react in the air with other chemicals and become small particles able to be inhaled deep into the lungs, which creates additional health issues.

In Northern New Jersey, a landfill in the town of Kearny has been subjected to enormous amounts of sulfur dioxide coming from a nearby landfill. The landfill is actually located right near a soccer field, and levels of the poisonous gas have been so high that the children have been ordered to stay off the field. In addition, many community events have needed to be canceled, and the residents are forced to live close to their houses due to the horrible odors. While the town waits for the landfill to have a cap placed on it to reduce odors, they all suffer with exposure to the hazardous hydrogen sulfide until and if that happens

The landfill industry claims they cover up the new trash every day with clean dirt to keep odors at a minimum. That may be true, but as the trash decomposes, it creates enormous amounts of methane gas. This smelly gas percolates up through the mountains of trash and into the air. Methane gas smells because of the large amount of sulfur dioxide in it. So much of this gas is produced by these mountains of garbage that it creates issues of its own.

Because of the need to handle such a high volume of methane, landfills have come up with a variety of systems to deal with it. Landfill operators needed to find a way to vent the methane gas out of the mound so it doesn't accumulate and cause a landfill fire or explosion. A nice green hill on the side of the highway that has tall white pipes sticking up from the ground all over it--that's a landfill trying to vent methane. Long lengths of perforated pipes are placed into the landfill to capture the methane gas as it rises from the mound of decomposing trash and is routed to the surface through those solid white pipes sticking out from the mound.

Some landfills have set up an igniter at the top of these pipes that burns off the gas when it gets to that point. They are often along highways and at night appear to be giant flares sticking out of a mound; they almost look like birthday candles.

These two methods are a bit outdated, though. Now, landfill operators more frequently capture the methane gas and use it as a fuel to power items at their facilities. These can be from items as simple as their electric lights to ones as complex as fleets of vehicles. That is great for people who happen to live nearby. Unfortunately, despite the growing number of landfills using their methane, most still simply vent off the stinky gas in the neighborhood air.

3. Particulate Matter

One of the more serious problems with living near an operating landfill is the large amount of dust created by the daily operation. A landfill is a giant pile of dirt with very little vegetation on it. Dirt roads crisscross all over the mountain of trash, carrying trucks and heavy machinery. Each day tons of dirt are dumped on top of thousands of tons of trash to try to cover up the fresh trash. Some landfills have water trucks spraying the roads in and around the landfills trying to hold the dust down, but they can never keep up.

The main purpose of the landfill is to hold solid waste being disposed of there. Significant amounts of the particulate matter come from the activity of the hauling, dumping, and spreading of the waste on top of the landfill. Other sources of particulate matter come from additional activities such as any composting done on the site, waste unloading and sorting, and transport into and off of the disposal site by trucks.

The particulate matter that comes from landfills is not limited to dirt from the ground and dust in the air. Whatever is also decaying in the landfill or was dumped there gives off particles that are picked up in the wind and blown into the surrounding areas. Significant debris from decomposing trash, rotting food waste, construction debris, and medical waste if disposed at the landfill all contribute to particulate matter. People who live downwind of these landfills are exposed not just to the dirt and dust particles in the air they breathe, but also to particles of contaminants and pathogens from the wastes.

According to a study published in Science Daily, people who live near landfills and other waste sites frequently have a higher incidence of respiratory issues. Links were made between inhaling microorganisms and endotoxins present near these waste sites and respiratory issues associated with exposure to the bacteria.

4. Truck Traffic

All day long, a constant flow of trucks heads into the landfill and back out. Operating hours usually run from 7-5, so during the time they are open, some landfills have upwards of 500 trucks daily dumping loads of trash and other assorted materials. The noise created by the clanging of dump trucks and the rumbling on the roadways is in the background for everyone nearby the landfill. Many times the roads used by these trucks were not meant to hold the weight of the vehicles now using them. Small neighboring roads where the children used to play and people had walked their pets became dangerous during the daytime.

A significant health hazard is the large amount of exhaust fumes the trucks and equipment put into the air all day long. The chemicals in diesel exhaust are known to be carcinogenic. The long-term exposure to high volumes of this exhaust over months and years may cause respiratory problems in nearby residents and certainly will negatively impact people with conditions such as asthma and other breathing issues.

Near Boyertown, Pennsylvania, residents already burdened with heavy truck traffic due to the nearby landfill have now found out the landfill operators want to expand the landfill's life. Originally expected to close in a few years, the landfill is requesting a ten year extension and permission to bring in another 16.8 billion pounds of trash. Truck traffic, already at around 250 trucks per day, could double to close to 500. In addition to the expansion, the landfill operators now want to expand operating from five days a week to seven.

5. Closed Landfills

Old leaking landfills are one of the biggest threats to human health and our water supply there is. Because many of these landfills have had industrial wastes dumped at them, they are toxic time bombs. Thousands of closed landfills are already designated as hazardous waste sites sitting on an official list, but waiting for action to be taken. Many hundreds more are on our nation's Superfund list already known to be some of the worst in the country. They too, in many cases, are still seeping poisons into the surrounding areas.

What makes old landfills an extra threat is that some of them are already polluting residential areas, but the risk remains unknown. Some of these landfills were too small to make official lists or were abandoned so long ago that they aren't on anyone's radar. And yet, they remain full of toxic, carcinogenic chemicals slowly poisoning the neighboring areas.

In Long Beach, California, a mobile home park was built on top of an old municipal landfill that closed back in the 1970s. The garbage beneath the park is slowly decomposing under the 182 homes built on the landfill, causing the land under some homes to buckle and shift. In addition, the county public health agency found explosive levels of methane emitted from the decomposing trash in some areas of the park.

Residents living near an unlined old municipal landfill in Wheatfield, New York, are fighting the state and the town for allowing the disposal of hazardous industrial wastes for years without a liner to prevent seepage into the groundwater or surrounding soil. The landfill, already listed as a hazardous waste site, has allegedly leaked the toxic chemicals into the surrounding neighborhood. Some residents have found high levels of these in their homes. Attorneys representing the residents said some people have been sickened by the leak with illnesses including neurological issues, heart problems, reproductive disorders, and cancer.

Up by Richland, Washington, the Hanford Nuclear waste site is locally famous. But, many people do not know about the closed Pasco landfill, which in some ways is even more dangerous to the people and groundwater. This landfill, which was finally closed in 2001, had accepted tens of thousands of barrels of all types of waste. For many years officials hoped the waste was not leaking from the unlined landfill, but several studies are showing there are leaks The state wants to begin removing toxic substances from one section of the landfill. While the rest of the toxic chemicals continue to leak into the groundwater, 35,000 barrels are planned to be excavated and disposed of at another facility. Groundwater contamination in a plume extending 1.5 miles from the landfill to the Columbia River has been detected. The City of Pasco is only 1.5 miles away in a different direction. There are restrictions in place on wells in the area, and vapors are known to be coming up through the soil in some places.

6. Landfills Accepting Illegal Substances

The practice of disposing of special waste at regular landfills instead of designated disposal facilities is a sizable problem. We can say with certainty that the number of times this has been discovered happening is only a small fraction of the times it actually did. The high costs of disposing of hazardous waste, radioactive waste, medical waste, and other such waste drives the black market for illegal disposal. Sometimes the waste is repackaged to appear to be regular solid waste and dumped at the landfill. Other times the operator of the landfill is in on it and sends the illegal waste to a spot in the landfill where it is buried and covered up quickly.

At the Hunter's Point Naval Shipyard in San Francisco, an investigation found soil contaminated with dangerously high levels of radioactive waste was routinely sent to conventional landfills in California handling regular waste and not radioactive materials. The contractor bypassed radioactivity detection equipment at the cleanup site and simply dumped the waste at the landfills unknown to the operators..

The practice of illegally disposing of hazardous materials in unpermitted landfills extends to some companies we would never suspect to be involved. In California, the attorney general's office concluded in an investigation that Target has improperly disposed of medical waste, pharmaceuticals, aerosols, batteries, and other regulated materials into landfills instead of proper disposal facilities. The state listed 2,038 items considered hazardous waste and 94 considered medical waste that the company unlawfully disposed of. Target Stores was fined $7.4 million for the violations.

A landfill near Williston, North Dakota, was ordered by the state health department to remove almost 950 tons of illegal material it had accepted. The material was radioactive and beyond the limits allowed for disposal at that facility. In addition, illegally disposed radioactive materials were discovered in at least two other landfills in the state.

In the beautiful town of Irvine, Kentucky, in the middle of town across the street from the school is the local landfill. The town found itself trying to figure out how to deal with almost 2,000 tons of radioactive waste that had been dumped at the landfill. The illegally dumped radioactive material came from a company in West Virginia, and radioactivity was measured at 340 times the legal limit for regular landfill disposal.

Illegal dumping seems to be getting more frequent with the thousands of tons of radioactive waste the fracking industry is now generating without enough disposal options. Since landfills are supposed to inspect and in some cases test the material coming in for disposal, either the landfill operators are derelict in their responsibilities, or they are knowingly accepting illegal shipments of materials that do not belong there. Fracking and the waste generated during the drilling operations create radioactive fluids and sludges that are banned from disposal at landfills in many states. Some states are trying to ship their waste to others to get around restrictions. Others, apparently, have resorted to breaking the law.

Medical waste is another type of waste illegally disposed of in landfills. According to the World Health Organization and other scientific agencies, medical waste contains viruses, parasites and pathogens that if not disposed of properly , can easily be transmitted to people. By improperly disposing these wastes in a landfill, the particles can be transported through the air off of the landfill and breathed in by people living nearby. Lung infections such as influenza and pneumonia are a couple examples.

Parasites thrive in waste, and parasitic infections from improperly disposed medical waste can be easily communicated to humans. This can happen both through skin contact and breathing in particles from waste that have been transported through the air off of the landfills. Lung infections such as tuberculosis, pneumonia, and influenza are caused by pathogens released into the air by medical waste.

In Ventura County, California, three hospitals were caught sending untreated medical and pharmaceutical waste to a landfill near Santa Paula. The waste contained items such as blood-soaked bandages and old intravenous tubes still partially filled with medications. The waste was disposed of as regular solid wastes instead of the regulated and controlled method that medical waste needs to be handled to avoid exposing people to potential deadly bacteria, viruses, and other pathogens. The California attorney general's office said they were unable to determine how long this had been going on and how much waste was disposed of prior to this discovery.

These violations by different hospitals occurred over June and July, which spawns concerns over exposure to illness, bacteria, and drugs. Questions linger over how long the materials were dumped.

In Pennsylvania, medical waste was discovered being disposed of at a local landfill. Twelve UPMC (University of Pittsburgh) hospitals were discovered sending untreated medical waste still packed in bright red infectious material bags to the landfill instead of a medical waste facility. The bags of waste included bloody bandages, used needles, and body fluids. Neighbors living near the landfill saw the bags and told landfill employees, who confirmed the illegal waste and called the state environmental agency. Ultimately, they blamed it on poor employee training at the hospital. The landfill operator, Waste Management, was the same company who didn't notice the 2.5

million gallons of radioactive fracking wastes illegally disposed of at their Oregon landfill either.

7. Fires and Explosions

More than just an isolated instance, landfills do catch on fire sometimes, and when they do, it is powerful. With an enormous supply of material to feed it, these fires tend to get going very quickly. The large amount of plastic we throw out in our trash makes an excellent fuel, causing thick, billowing clouds of toxic black smoke that can drift for miles from the fire site. Sometimes a fire actually can start inside the landfill itself due to the extremely high temperatures created by thick layers of decomposing matter.

When one of these fires gets started, it is almost impossible to reach it due to most of these fires starting deep under the surface of the landfill. Extinguishing the fire sometimes is not possible. Many of the underground fires have burned for months and, in several cases, years.

In Bridgeton, Missouri, a fire in the landfill there has been burning for years. The landfill company ingeniously calls it a "subsurface smoldering event." The bureaucratic language choices obscure how dangerous the situation actually is: Bridgeton has an uncontrollable fire burning in a landfill.

Because there is no real way of extinguishing the fire, it continues to smolder and move further through the landfill to new areas. Not too far away from this fire, another landfill is loaded with radioactive waste. It is a race to put out the fire before it reaches the radioactive waste. If the fire does spread that far, the results will be deadly. Officials from the landfill company expect the fire to burn until at least 2024. That is over a decade of a smoldering fire in a landfill near hundreds of residents exposed to the fumes everyday.

In Miami-Dade County, Florida the landfill caught on fire on a Sunday night and burned all week as firefighters struggled to extinguish it. The landfill again burst into flames days later, and the fire crews said it would be days until the winds died down when they hoped to get the fire under control. While no direct cause for the fire is known, spontaneous combustion from the heat of decomposing trash is a possibility. Health officials are warning people in the area, especially those with respiratory problems, to stay inside with their windows closed.

In Tucker, Georgia, part of the U.S. Highway 29 was shut down after an underground explosion damaged a section of the roadway. A HazMat team that responded to the scene found dangerously high levels of methane gas at the site of the explosion. The state environmental agency said the closed Crymes landfill located next to the highway was most likely the source of the methane gas that caused the explosion.

In Kentwood, Michigan, officials found methane gas outside the property of a closed landfill. There are about 150 homes within 1,500 feet of the landfill border where the leaking methane gas has the potential to build up in the residence and explode. The town has no idea how far from the landfill the explosive gas is leaking and is trying to prevent a major explosion while they figure out how to mitigate the problem. Residents who may be at risk were sent a notice in the mail advising them of the danger with no guarantee of a solution.

A landfill is an especially risky neighbor to have. Any one of a large number of issues can cause problems for your life and your health. I have never seen a landfill that did not have issues of one kind or another. Every one either had issues with odors and particulate matter or was leaking toxic waste into the aquifer underneath. At many of the landfills Tony and I visited, we saw evidence of documented disposal of hazardous wastes and other materials that create an imminent danger to the residents' health and the local environment.

Landfills can operate for many years. Once they reach their so-called capacity, some can apply to expand in one direction or another: either higher if a permit is issued, or further out if property is available. There is never any guarantee that a landfill will close on any projected date. We also have seen closed landfills reopen. People who rejoiced once the old landfill closed were shocked to find there was a new operator willing to reopen and expand an already permitted operation.

If there was a hazard that had the potential to not just make your everyday life miserable but also potentially ruin your health, that would be a landfill. Avoiding any type of these operations can only be to your benefit. There is no good reason to locate yourself in a position where harm may come so easily. At minimum, we would suggest placing at least four miles between you and any type of landfill, closed or operating.

Chapter 15. Airports

When we really think about it, there are many places we see every day but never really pay much attention to beyond a quick notice and perhaps a brief thought. We can't help but notice when we are near an airport due to the noise, but when you live near one, what else should you be noticing? The scientific community for a long time focused much of their research on the effects of airport noise on human health. The past decade or so, however, scientists and health officials have been looking at other potential risks to health. The results, which have been coming from a continuing investigation into other potential health risks of people living and working near an airport, are becoming increasingly concerning. Very little has been done over the years to attempt to manage the associated health risks. Some scientists even say living near airports is a greater risk to health than the impacts of living near a major highway.

The sheer size of some metropolitan airports and the enormous volume of planes, cars, trucks, and buses all combine to make an invisible toxic cloud. Every vehicle, whether an airplane or ground transport, runs on gasoline, diesel, or jet fuel. Recently, though, some effort to use electric vehicles where possible is beginning to occur. All of these combustible fuels create very large amounts of toxic particulate matter. This chemical cloud and its toxic gases have been measured at elevated levels up to 10 miles from the airport. Depending on weather conditions, the flow of fumes from the airport can surround the entire area. There are very few business operations that have the sprawl and intensity of activity to create such a dense and toxic cloud that almost never goes away or disappears. It hangs over the area like a shroud.

It is not only large international airports such as the ones in Chicago, New York, or Los Angeles that are the most hazardous. While in volume they may be on top of the list, there are also dozens of smaller regional airports with many flights coming and going all day too. As Tony and I discovered in our research, some of the hazards that the government regulates at big airports are not regulated with as close scrutiny at smaller airports. Thus, some smaller airports still present hazards.

A study published in *Environmental Science and Technology* addressed issues at these smaller airports. In Southern California, thousands of private planes and corporate jets use the small, regional Santa Monica Airport. Ultra fine particles, the most dangerous ones, were found in significantly elevated levels downwind of the airport. This is in addition to spewing lead from their engines because airplanes are still allowed to use leaded fuel due to a loophole in EPA regulations that tried to phase it out years ago. While the EPA has the power to regulate aircraft emissions, they must take into consideration the technology available, the cost, and potential negative impact to the airlines and aircraft safety.

Between the noise pollution and air pollution, the bodies of the people living nearby are being continuously assaulted in five major ways.

1. Pollution from Airplanes

The pollution caused by the aircraft, especially the ultra fine particles, are the biggest concern. Those particles because of their small size are able to pass through the lungs' filters and imbed themselves deep inside of the lungs. From there, they pass into the bloodstream and spread into the vital organs such as the brain, heart, and others. These particles also worsen conditions in people with any heat and lung conditions, aggravate bronchitis and emphysema, and may lead to blocked arteries in some cases.

Unfortunately for people living near airports, the EPA somewhat regulates larger size particles but the most dangerous, these ultra fine particles, are not. Because of the large number of airports throughout the country, that is a huge health concern and a big gap in regulations. Health reviews have shown this pollution increases asthma and chronic obstructive pulmonary disease by 17% and heart problems by 9%.

In Los Angeles, a study of the particulate matter in neighborhoods within several miles of the airport done by University of Southern California showed people living near the airport breathe higher levels of fine particles and toxic gases than people sitting on one of California's jammed freeways. In fact, the study said there was more pollution in the areas near the airport than almost all the pollution on 190 miles of L.A. freeways.

A Massachusetts study has found children living near Logan International Airport in Boston were up to four times more likely to exhibit signs of undiagnosed asthma than children living in communities further away. In addition, adults living in neighborhoods bordering the airport were almost twice as likely to have COPD than their farther-away peers.

While lead in jet fuel was phased out of use in large aircraft, small jets and planes still use fuel containing lead. The same lead that was phased out many years ago in automobiles due to its toxic effect on humans is being used in tens of thousands of small planes across America every day. Many small regional airports such as Teterboro Airport in New Jersey are still coating nearby neighbors with lead emissions long after large airports stopped doing so.

The pollution is not just airborne. The dust and particles of lead from years of exhaust live on in the soil and on the surfaces near the airport. Children still play in the yards and are exposed to more lead than they would be had there not been an airport. People living near airports where these planes fly are being continuously exposed to lead emissions in the exhaust. The EPA has stated without a doubt children exposed to lead may develop learning disabilities and brain damage.

A study done by Duke University found that people, especially children, who lived within a half mile of an airport where leaded fuel was still being used had lead levels in their blood higher than children further away. The EPA estimates show 16 million children live close to one of these 22,000 airports where leaded fuel is regularly used, and the lead in air near the airports can be inhaled by breathing or ingested after settling on soil and other places in the area.

Lead pollution is a serious problem in northern Sonoma County, California, where several local airports, including the Sea Ranch, Healdsburg Municipal, and Cloverdale Municipal, service small piston engine powered aircraft typically used for corporate and private travel. Such planes used avgas. According to the EPA, more than 167,000 piston engine planes in operation are responsible for nearly half of all lead emissions nationwide.

2. Pollution from Vehicle Traffic

Traffic converges at airports. Thousands of cars, trucks, and buses enter and exit all day. Many idle while they load or drop off passengers, adding to the toxic load already present at the airport. All of them move at slow speeds, exacerbating the tons of particulate matter emitted each day. Diesel engines from trucks and buses are especially high contributors of toxic and hazardous emissions. Computer modeled images showing the extent of pollution near airports look like a thick dense cloud completely enveloping the local area with several dense streams representing the roadways covered in fog of pollution as well. The fog density stays quite high for up to two miles from the airport in all directions.

3. Noise

Sound has been recognized as a pollutant since 1972 with the Noise Control Act . A study published by the National Institute of Health found a tie between health issues and the amount and volume of sound a person is exposed to. People who are exposed to constant high levels of sound, such as those caused by jet engines taking off and landing, experienced several effects. Constant lack of sleep or continually interrupted sleep can lead to irritability and depression. In addition, hearing loss, hypertension, and heart problems may develop.

Two studies in particular pointed to the already suspected conclusion that people living near airports, especially in the noisiest areas, are exposed to health risks as a result. Experts at the Massachusetts General Hospital agreed that exposure to excessive amounts or volume of environmental noise triggers a stress response in the brain that in turn causes the blood vessels to become inflamed. Health issues like hypertension and continuous inflammation can lead to elevated risk of cardiovascular issues like heart attacks and strokes. The body continues to be affected by elevated blood pressure, especially during the night from the continued noise, even

if not realized consciously. Conclusions were airport noise has negative health effects far beyond simply losing a good night sleep.

The National Institute of Health also published a paper on a group of studies showing a direct correlation between airport noise and a decrease in learning ability. They concluded that continuous aircraft noise at home or at school led to poorer reading and memory skills. There was growing evidence that the noise also was a cause for children to test lower on standardized tests compared to children not exposed to aircraft noise.

Aircraft noise can be so disruptive and unpleasant that, in Palm Beach, Florida, President Trump sued the FAA for $100 million in damages relating to the “unreasonable amount of noise, emissions and pollutants” from planes flying over his Mar-a-Lago Club near Palm Beach, Florida. If this is recognized as unacceptable to President Trump, it is unacceptable to the rest of us.

4. Potential for Expansion of an Airport

Like highway expansion, never say never. As air travel is becoming the preferred way of long distance traveling for more people, airports are finding themselves at capacity and running delays for flights more than ever before. This puts pressure on the airport to find a solution. Even those airports that are landlocked have been expanding.

Air travel and airports are regulated by the federal government. That means no matter what city or town the airport is located in, they have minimal say in final operations of the facility. Some airports have extra space to grow, but the new runways may impact previously unaffected neighborhoods. In cities, the government has even invoked eminent domain and taken property for airports. Sometimes, the airport gives residents the ultimatum of selling or simply dealing with the unwanted consequences.

In Atlanta, Hartsfield-Jackson Airport already handles 2,500 flights a day. During busy times there is a plane arriving or taking off every 45 seconds. The airport facility is looking at adding a third runway. While the expansion will not evict people, the runway directly impacts thousands of people in the areas the planes take off and land. The new runway and flight pattern run directly overhead existing housing.

New Haven, Connecticut, has a local activist group facing off against the airport authority seeking to expand the runways. The group says they will not be bought out and will fight the expansion. The fight against the expansion of the runway continued for a decade, and even the state eventually joined with the opposition. The entire issue was settled in March of 2020 when the United States Supreme Court refused to hear the appeal from a lower court ruling allowing the expansion to move forward. Airports come under the ruling of the federal government, and the court agrees with the expansion decision, allowing it to move forward. At the time of writing this, the airport authority is in discussions with the airlines to see how long a runway they need and any other requirements. While the people eventually lost this battle, it does not mean their efforts were not worth it. They gained the extra ten years they managed to delay the expansion.

In the past five years, the number of aircraft flying into Seattle-Tacoma International Airport—one of the fastest growing in the U.S.—increased almost 30 percent, and annual passenger traffic jumped from 35 million to 47 million people. Large airport facilities are under construction, and plans for further expansion, including a second passenger terminal, are taking shape.

Even smaller airports such as in San Luis Obispo, California, have the potential to not just double or triple the size of the facilities but add dozens of new flight routes and several new airline carriers in addition. Its location between Los Angeles and San Francisco makes it perfect for people seeking a smaller airport to use instead of dealing with the large cities. The city and the surrounding areas are growing with many new companies opening offices. Based on all those factors, airport traffic increased over 40% between 2015 and 2017 requiring the need for a large expansion. The new terminal is three and a half times the size of the old one. Runways are handling larger planes, and more carriers are already using much of the new space. The surrounding areas, which had been much quieter up until about 2016, are now feeling the effects of increased noise and pollution the much busier airport has. Since the expansion, noise complaints have tripled. Neighbors living under the busier flight paths and new landing patterns are being subjected to the noise and pollution from the planes. Flights are fairly common as early as 4:30 in the morning and run until 11 or so at night. The airport says it runs a 24/7 operation and will not restrict hours based on the complaints as the airlines are the ones setting the schedules.

5. Changes in Landing and Take-Off Patterns

Airlines highly value efficiency because it creates higher profits and drives the industry. As a result, changes in flight patterns to save fuel or maximize runways are quite common. This can come as a surprise to the people who suddenly find themselves under the bellies of hundreds of planes.

At Sky Harbor Airport in Phoenix, Arizona, flight patterns were suddenly switched to a new route. The new pattern took thousands of flights over several neighborhoods not heavily impacted with noise before. It took a lengthy four-year court battle, but the people won, and the pattern went back to the old one.

In Los Angeles, the FAA changed flight plans in and out of LAX. The planes now fly directly over downtown and the central district. The city is suing the FAA to change them back. Thousands of people have been affected by the noise and pollution.

In the city of Charlotte, North Carolina, construction is about to start during the night for work needed on a taxiway near one of the runways at the airport. The airport is shifting the flight patterns over the nearby neighborhoods during the seven months the construction is expected to take. While the people living near the airport are already used to the landings and takeoffs over their homes during the day, they will now be getting all these additional flights all night long. Officials at the airport feel badly about this, but they say the construction is needed, so there is nothing the residents of Charlotte can do but suffer through.

6. Groundwater Pollution

One new concern that must have been around for years but is starting to get some attention from forward thinking cities and towns is the prospect of groundwater pollution from airport activities, especially fuel storage and firefighting foam as PFOS chemicals have already infiltrated and contaminated groundwater across the country.

In Bellevue, Idaho, the city council realized much of the property the airport is in lies directly above the city and the groundwater flow. Potential sources of contamination are from eight underground storage tanks at the airport, of which three are known to be leaking. The airport also operates 21 injection wells close to the surface for storage or water disposal purposes. Chris Johnson, a member of the city council said most of the airport property lies within a six- to 10-year course of projected downhill groundwater flow to Bellevue, which could put the city's wells at risk.

All told, living near an airport can be a fairly miserable existence. People we spoke with said they lived near the airport because the housing there was the only place they could afford. Others added it was not quite so bad when they moved there 10-15 years ago. However, when the population of the area grew, the number of flights increased and the airport expanded, making the problem much worse. The only positive anyone has ever told us about being so close to an airport was that they did not need to drive far to get to their flight. They said that in a sarcastic manner and added they would very gladly drive a half hour rather than be so close.

Living near an airport can be quite damaging to a person's health. There are no reasons to place yourself in the situation of having your body assaulted everyday with the noise and pollution the airport creates. If at all possible, stay away from any airport for at least ten miles or more. While some air traffic patterns may still run over your neighborhoods, the noise and pollution levels are substantially less than if you were close.

SECTION TWO

This group of hazards depending on the specifics of the operation near you has potential to pose a health threat from medium level to dangerous. There are definitely going to be areas that you could be at risk and exposed to threats from any one of these.

- Chemical factories
- Industrial operations
- Fracking
- Mines
- Mine tailings
- Gas and oil wells
- Highways
- Agriculture
- Brownfields
- Sewage sludge
- Gas stations
- Sewage treatment plants
- Injection wells
- Sand and gravel pits
- Contaminated fill
- Junkyards
- Trash transfer stations
- Railroads

Chapter 16. Chemical Factories

The chemical industry has a special place in my heart. I grew up in New Jersey and spent almost thirty years living there. I like to say the chemical industry also has a special place in my lungs, liver, and the rest of my organs.

The chemical industry in the U.S. is huge. Their total estimated production in 2017, according to Statista, was $765.4 billion. They are one of the largest, best funded, and most powerful industries.. Because of this, almost all regulations are in their favor, especially any regulation that may limit the amounts of specific toxic substances the industry is allowed to emit.

As an industry, the chemical sector produces more pollution and hazardous waste than many other industries combined. The list of the top 100 polluters in the U.S. have 37 chemical industry companies making the list. Those 37 companies together produce more pollution than the next seven groups of industries on the list combined. The chemical industry is actually a group themselves made up of several manufacturing sectors, including petrochemicals, fertilizers, pharmaceuticals, and a few other players. Each of these groups by themselves release significant amounts of toxic chemicals into the environment each year.

In 2018, data from the EPA toxic emissions report showed just how much of these hazardous emissions really are dumped into our air, water, and ground. The chemical industry released 543 million pounds of toxins, and the petroleum industry 79 million pounds. According to the EPA, the number of chemical companies who reported emissions data was 3,455. On average, each chemical company was responsible for 157,000 pounds of toxic emissions. While that number is only the average of all of them, that alone is still huge. When we use the data to see emissions areas with concentrations of chemical factories, amounts of 10 million and 20 million pounds are common. Because of these concentrations, these places develop colorful nicknames to reflect their level of pollution: places like “Cancer Alley” in Louisiana and “The Armpit” up in northern New Jersey off the Turnpike.

There are many dangers and risks that come with living near a chemical plant.

1. Toxic Air Emissions

The incredible amount of chemicals poured into the air surrounding these plants is certainly the biggest danger. Many thousands of people have directly suffered as a result of exposure to toxic emissions from these plans. Entire communities suffer in the shadows underneath billowing clouds of chemicals pumped 24 hours a day into the air they have no choice but to breathe.

Miscarriages, birth defects, cancers of all types and illnesses of every part of the body are the direct result of long-term exposure to these emissions.

In Hopewell, Virginia, a former Honeywell chemical plant released over two million pounds of ammonia into the community in the year 2015 in addition to tens of thousands of other hazardous chemical emissions. Now owned by AdvanSix, the plant continues to make nylon and chemicals for fertilizer. It also continues to send toxic emissions out of the smokestack.

In LaPlace, Louisiana, DuPont operated a chemical plant that manufactures chloroprene, the main ingredient in making neoprene for wetsuits and other products. Since 1969, DuPont has produced chloroprene and neoprene at the plant, and since that time the residents in the area have reported the emissions from the plant have been making them sick. According to EPA data, the areas closest to the plant have the highest cancer risk in the entire country. In one specific area, EPA data shows rates 700 times the norm.

Even after the EPA finally issued an exposure standard that showed exceeding the threshold amount is harmful, the plant is still being measured with emissions of over 150 times that amount. Lawsuits are in court, and the toxic chemical is still in the air. DuPont sold the operation in 2015.

In St. James, Louisiana, a chemical company already in full operation wants to double the amount of toxic emissions it is allowed in order to create products like plastic bottles and bags. Manufacturing plastic products emits some of the most dangerous chemicals in large amounts. The plastic plant would put 800 tons of hazardous pollutants into the air each year.

In Riverside Gardens outside Louisville, Kentucky, there is a chemical company named Hexion. Their website claims they have a responsibility to make the world better, safer, and cleaner. Ironically, in just one 16-month period, the company had over 50 chemical emissions exceeding their permit levels. The cancer-causing chemical formaldehyde was one of them. Prior to that period of time, there were another 85 violations.

In Cancer Alley, the 80-mile long stretch of chemical plants between New Orleans and Baton Rouge, multiple plants owned by Exxon-Mobil were fined $2.5 million in a settlement with the EPA. Their plants in Texas were involved as well. Exxon-Mobil was violating air pollution standards at eight of their facilities, contributing to the area's dishonorable nickname.

2. Explosions and Accidents

Trying to find out about what materials are stored onsite at one of the facilities is made very difficult by the government and the industry itself. There are federal regulations that require companies who have more than a certain amount of especially dangerous and hazardous chemicals onsite to file a report with the EPA. The supposed purpose is for emergency planning in the case of an incident at the plant. In theory, as well, the company is supposed to work with the town or city it is located in so they are also aware of what potential materials the emergency crews may encounter in dealing with a fire or explosion at the plant.

In Port Neches, Texas, fireballs at a chemical factory just before Thanksgiving Day, 2019, forced the evacuation of 50,000 people. Dark plumes of thick toxic smoke were billowing into the sky at TPC Group chemical factory for two days after the initial explosions rocked the area. Earlier in 2019 the company was fined for excessive emissions and pollution. Failure to report these releases was another violation placed by the company as well. The plant manufactures chemicals such as butadiene and others used for synthetic rubber, fuel, and plastics. Butadiene is considered a known carcinogen by the EPA, who says high exposures to the chemical may likely cause damage to the central nervous system.

In Akron, Ohio, an explosion at Emerald Performance Materials forced an evacuation of the areas surrounding the plant. The chemical factory has a long record with state and federal environmental agencies for violations. One such serious violation was the failure to prevent non-sudden releases of hazardous materials. Records that were supposed to have names of hazardous waste managers and information for emergency coordination were incomplete and missing information. The location of fire suppression equipment was missing from the plan as well. Other violations included not completing required inspections of equipment and requiring workers to enter areas after chemical releases without having breathing protection.

In Crosby, Texas, there was an explosion and fire at the KMCO Chemical plant when a tank containing isobutylene, a flammable chemical, caught fire. A nearby highway was shut down, and all residents within one mile of the plant were told to stay indoors and shelter in place. One person was killed and two were injured in this incident. The plant already had a history of violations prior to the explosion in spite of the company claiming how compliant with regulations they are. The county sued KMCO in 2008 for releasing toxic chemicals into the air that caused severe headaches and other health issues in nearby residents. In 2010, after an accident at the plant, investigators issued 15 violations for serious issues including highly hazardous materials management, hazardous waste management, and emergency response management violations.

The company changed hands in 2012. Under the new ownership, despite claiming compliance with environmental regulations, in 2013, the company did not conduct emissions tests on its stacks as required and released much larger amounts of carbon monoxide into the air. In 2016, KMCO was criminally convicted on two charges of knowingly violating the Clean Air Act. A total of over $3 million in fines was issued. In that case, over a four year period the company failed to monitor leaks of several specific pollutants at the Crosby plant and when caught, admitted the company falsified records to secure permits.

In May of 2020, outside Boston, Massachusetts, in Newburyport, several explosions at a chemical plant blasted a hole through the roof. Firefighters responding were unable to enter the building due to thick black smoke. Records show the company, PCI Synthesis, has had 13 serious safety violations and five lesser grade ones since 2010. Federal violations in the past year included problems with safety procedures and chemical handling. The safety violations were related to preventing or minimizing the consequences of catastrophic releases of chemicals.

Again in Crosby, Texas, a chemical company called Arkema exploded after floodwaters from a recent hurricane knocked out power to some equipment, causing chemicals to erupt into a ball of fire with toxic black smoke pouring hundreds of feet into the air. The company, which worked with various types of peroxide chemicals, failed to do any emergency planning in the event of a flood caused by hurricanes, which the area frequently was subjected to. The chemicals that the company worked with had potential acute health effects that neither the community or the first responders were prepared for in an emergency situation. There were disputes over whether the first responders had the correct information on the toxicity of the chemical blaze, but, regardless, 21 first responders needed emergency medical treatment as a result of exposure to the fumes.

During the blaze, an evacuation zone was created a mile and a half around the plant. Toxic fumes pouring from the area caused eye and skin irritations as well as respiratory issues in residents. Some first responders totally unaware of the toxicity of the chemicals were immediately affected by the fumes and some of them literally became sick and overwhelmed in the streets. Police responders were doubled over vomiting in the streets, the lawsuit said, due to the toxic fumes they were breathing. Residents in the area said the effects from the chemical exposure were still being felt many months after the explosion. After the blaze, floodwaters containing levels of toxic chemicals released from the plant caused burns, rashes, and skin irritations on people who came in contact with the toxic water.

3. Water Pollution

Pollution of surface water and groundwater are frequently found onsite and nearby many chemical plants. Spills of chemicals and leaks from tanks and pipes all contribute to toxic chemicals used at the plant being washed into drainage ditches, nearby streams, or local roadways. In far too many cases, onsite storage ponds and lagoons have been used for hazardous waste storage or other purposes that then either moved off site into nearby waterways or seeped below into the groundwater under the plant.

In Pompton Lakes, New Jersey, a DuPont plant that made explosives detonated the defective explosives that did not pass quality control in a pond on their property. The pond, which has no protective liner, was also used to dispose of mercury, lead, and wastewater from manufacturing. A babbling brook actually named Acid Brook happened to meander over DuPont's property near the pond, traveled through a nearby neighborhood behind the houses, and emptied into Pompton Lake, the town's lake.

Along the way through the DuPont property, the chemicals from the unlined contaminated pond would flow over into Acid Brook. Everytime Acid Brook would overflow, the poisons would flood across people's backyards. Children played in their yards and splashed around in the water, totally unaware they were exposed to mercury and lead from the DuPont plant nearby.

When the backyards were finally tested, almost all the homes along the banks of Acid Brook showed high levels of mercury and lead--levels so high, in fact, residents were warned to not have any contact at all with the ground. It took many years of argument between DuPont and state environmental officials to agree on any type of cleanup. Because Acid Brook had so much contamination, 36 acres, over 2,700 truckloads, of sediment in the lake needed to be removed.

While not all the contamination will be removed, at least the worst will be, according to the EPA. Meanwhile, Pompton Lake is still used for recreation by boaters and fishermen. The contamination levels of mercury are still so high people are warned not to eat any fish. Frighteningly enough, the lake is actually a backup water source for filling another nearby reservoir used for drinking water. Recently, this water was actually used for that very purpose.

In Linden, New Jersey, a refinery now run by Exxon-Mobil has been operating for over 100 years. According to a report done for the state of New Jersey, the 1,300 acres the refinery sits on has leaked and spilled over seven million gallons of oil into the ground. Between two refineries there are over 600 different contaminants that have been found in the soil.

Some chemical companies and industry officials may argue that many of the places with contamination were caused by actions in the past. Even if that were entirely true, in many of those same places, contamination remains, still poisoning and haunting people for many years to come. Worse yet, the claim is not completely true. Contamination still happens today as well, and, unfortunately, it will still happen tomorrow.

Unfortunately for us, the U.S. actually allows some of the most toxic chemicals in the world, chemicals banned in other countries, to be made, sold, and used here.

In Orangeburg, South Carolina, a chemical company is set to manufacture pentachlorophenol, a highly toxic chemical banned by an agreement of 172 countries. The manufacturing plants in other countries are closing due to the ban. The EPA has no issue with US factories continuing to make the chemical. The president of the Treated Wood Council, Jeff Miller, went on record and said, "The EPA has previously certified pentachlorophenol for industrial use," so he's confident it is safe. Given that 172 countries have completely banned the chemical, it is hard to be convinced by his testimony..

If the US joined the ban, it would stop the sale and use of some of the worst chemicals that do the most damage, like DDT did years ago, and that persist in the environment for decades, continuing to poison people, animals, and everything else. Instead, we are not part of the agreement because it would hurt chemical companies.

The chemical industry is continually challenging any studies, reports, or research that show health hazards caused by any of these chemicals with studies of their own showing they do not. However, given the source of these studies, they are highly suspect. These companies also use tactics such as filing lawsuits against anyone or anything that can delay implementation of regulations limiting their emissions, a strategy that works because they can outspend local interest groups easily.

Over the years Tony and I have been doing research for information on specific companies, the chemical industry has always been the most difficult to get accurate information about. They usually claim trade secrets and do not share the information that the people who are most affected by the plants' operations should have the right to know. Living near a chemical plant can subject you to the highest levels of toxic chemicals and heavy metals in the country. Many chemical plants are known as "Super Polluters" because they rank among the top emitters of toxins anywhere in the U.S. Size is never an indication of potential for hazardous emissions. Many small chemical plants emit more hazardous chemicals than other industrial operations several times their size. When it comes to exposure to hazardous emissions and other dangers, chemical plants are one of the top places to stay away from.

Usually, the bulk of the heavier pollutants can be found in the immediate mile from the plant. However, the plumes of toxic materials from the smokestacks can be carried dozens of miles away. If you are looking for a zone to be beyond from the plant, we would suggest a buffer of two miles if you are able to. In the case of a large known toxic polluter or a large cluster of chemical operations, we suggest up to five miles.

Chapter 17. Industrial Operations

Sitting and talking with people about industrial operations and some of the hazards they can present is one of my favorite subjects. I have given dozens of talks over the years trying to share some of what I learned the hard way. There are thousands of different types of industrial operations and hundreds of thousands different products they make. To examine each individual industry one by one would be impossible, so this chapter does not deal with a very specific subject. Industrial operations are everywhere, including some unexpected places. This overview can provide some guidance to find out if the ones near you present an oversized risk.

Not all industrial operations are polluters and need to be avoided. However, most industrial operations have at least some part of their process that, if not handled correctly, does present a hazard. It may be the materials used in their manufacturing process itself that could present a danger. It also may be the waste products emitted during manufacturing or needing disposal afterwards that are hazardous and potentially dangerous.

Most people would agree that having some industry and business in their town is a good thing. And it is beneficial for many reasons. It brings jobs into the community for the local residents and tax revenues into the local economy to support the budget. Taxes paid by industrial and commercial properties help keep residential taxes lower. In many cases, local businesses may also donate money and support local organizations and community events. We have seen many Little League teams and other youth sports supported by wonderful businesses in towns.

As great as the benefits to having some manufacturing and light industrial business in town are, there also may be a negative side to some of them that we need to look at very carefully.

The first step is to determine what type of facility a given place is. Once we know what the operation does, there are places to get information so we have an educated idea about what we possibly may be exposed to, at least as much as we are able to find out based on information available and any conclusions we can draw from it. The first place to start would be finding out what specific businesses already exist in your town. There is a big difference between a company that makes wood trusses or manufactures windows for housing and a company that makes chemicals, paper products, or plastic pipes. The impact or environmental footprint the company has on the quality of life and the environment of the town can be drastically different. A single business making windows may have a very small negative impact or perhaps none at all. But, a single business such as a cement plant or solid waste incinerator can have a huge negative impact.

Determining what sorts of potentially hazardous materials the facility works with is usually the hardest part. Many businesses do not make that information easily available to the public. Sometimes they even withhold important information from the regulatory agencies trying to protect the public. We usually, though, are able to get enough information to make a well- educated decision. Even if the best we are able to find are generalities based on industry information, we can still make an intelligent decision. Usually, it is a safe bet that if the rest of the industry is using a certain group of chemicals and that the emissions from those types of facilities usually contain hazardous chemicals--for example, like high levels of mercury or dioxins--we can assume the plant in our town does as well.

We have to do the legwork to determine what exactly local industrial operations are releasing into the environment because the total amount of chemicals the EPA allows to be emitted each year is staggering. In 2018 industrial polluters in the state of New Jersey reported to the EPA they had emitted a total of six million pounds of toxic chemicals into the air, water, and soil. In just one year they emitted that total. Considering how densely populated the state is and how tightly packed many of the industrialized areas in the state have become, the people living anywhere near these facilities and sections are breathing these millions of pounds of chemicals.

This isn't a problem unique to New Jersey, either. It is shocking what the EPA will allow chemical companies to manufacture and emit into our air, water, and soil. Even more shocking is to find that chemicals that are so deadly they are banned in most of the world are allowed to still be manufactured right here in the United States at the company down the street from where you may live or work. Pentachlorophenol is a pesticide used as a wood preservative. Because it is so toxic and dangerous to the environment, 168 countries banned it from use. The last company in North America making it is located in Mexico, but it will be closing due to support of the ban. Sensing opportunity and profit, a chemical company in Orangeburg, South Carolina, decided they would go into the manufacturing of it due to the United States having not joined the other 168 countries banning it. After tremendous public opposition in Orangeburg and the surrounding area, the company backed down on its plans and decided not to.

Some places in the country have had one large company operating at the same location on sprawling complexes for decades. At some of these complexes, the amount of pollution is vast, affecting thousands of surrounding homes and contaminating groundwater for hundreds of thousands. Dupont, for example, who operated a facility in Pompton Lakes, New Jersey, since 1902, dumped millions of gallons of hazardous waste into unlined ponds and lagoons on their industrial campus.

When inspections from environmental officials back in the 1980s raised alarms due to what the inspectors saw, the company tried to downplay the impact the waste disposal had. The solvents they had dumped for decades migrated through the groundwater and entered hundreds of nearby homes, vaporizing up through the soil and into the buildings. People in the neighborhood were already dealing with mercury and lead contamination, and now there was evidence of the industrial solvents TCE and PCE in their homes and air. The people discovered their neighborhood had elevated levels of some cancers including kidney and non-Hodgkin's cancer. Several hundred homes needed to have special air venting systems to keep the toxic gases to a safe level in their homes.

Dupont resisted testing numerous homes and denied much of the pollution came from their site for years. In what once appeared to be a settlement between the state and the company, Dupont agreed to give the state 70 acres of woods for the public to use and enjoy. It turned out a large portion of that property was contaminated with arsenic and lead and, by giving it away, the company hoped to absolve itself of the responsibility and cost of clean up.

The Dupont story is only one of thousands we need to be aware of. Companies that have been polluting their lands for decades are in many places across the country. Dupont has other facilities just like this one. There are refineries, chemical companies, steel plants, and many other types of industrial operations that have a toxic past to be avoided at all costs. We highly recommend extreme caution before locating near any of these long-time manufacturing facilities. The disposal of hazardous wastes was unregulated for many of the years these businesses operated, and there are many secrets the property has yet to reveal.

There are also some individual companies that emit chemicals in quantities far greater than those allowed by their permits, even chemicals that are known to cause cancer. When self-reporting of these emissions is allowed to take place, underestimates and outright deceit is far more common than the EPA tells the public..

There are also companies that work with and emit large quantities of chemicals that are known to be extremely dangerous but because of weak or non existing regulations on these emissions, people living near these factories are subjected to the toxic chemicals. One example of this is the company Sterigenics.

Sterigenics is a known emitter of a highly dangerous and cancer causing chemical, ethylene oxide. Their emissions totaled as high as 169,000 pounds a year in the past. Known to cause genetic mutations as far back as 1948 and trigger cancers in animal testings in the '70s, ethylene oxide went basically unregulated until after 2016 when the EPA finally took limited action to regulate this chemical.

Sterigenics has multiple locations across the U.S. At their Willowbrook, Illinois, plant outside Chicago, the EPA confirmed toxic pollution from the plant was responsible for long-term cancer risks up to 10 times higher than considered acceptable by the EPA. After many years of exposing area residents to millions of pounds of toxic ethylene oxide the plant announced it would be closing instead of adding pollution control equipment as required. The company claimed the emissions of ethylene oxide were caused by people barbecuing and not their operation even though air monitors near the plant registered a 90% drop in the chemical once the plant closed.

In Smyrna, Georgia, another Sterigenics plant is operating. Concentrations of ethylene oxide are measured at 27 times the state level for safety. In most cases, the numbers used were reported by the company itself, and the state has not been active in collecting any air emission data on their own. The one air monitor the state did set up is not located near any facility emitting ethylene oxide. The state claims they want to measure the chemical in traffic emissions. The company temporarily shut down in 2019 to supposedly make modifications and add pollution control equipment. At the time of this writing, May 2020, the FDA is trying to have the plant reopened to assist with sterilization of medical equipment during the Coronavirus pandemic.

A powerful trade organization representing the chemical industry stepped in to do air testing near the plants. The Chemistry Council maintains that vegetation and cars are responsible for the ethylene oxide emissions.

Beyond the places in our own towns that may cause potential hazards, we also need to look at the adjoining communities as well. Usually where there are manufacturing facilities, there tends to be a group of them. An industrial park with multiple factories can have several dozen businesses of concern. Places like Petroleum Alley in Louisiana and Chemical Wasteland along the turnpike in New Jersey are extreme examples of the devastating effects of industrial development gone wild. Air pollution, water pollution, and toxic waste do not understand boundaries. Sprawling industrial zones spanning many miles have turned many urban areas into toxic wastelands. Along with those toxic industrial wastelands also go the residential housing areas and the people who live there every day.

1. Air Pollution

Pollutants emitted by local industrial operations can have a tremendous impact on health and quality of life. Even if there are only one or two factories in the area, depending on what they are, people living nearby could be in for a miserable experience. It is important to find out what the air emissions are for any factories located nearby where we live or work. Spending a large part of the day breathing whatever contaminants and pollutants the factories emit has a large effect on people's health.

Poisonous air emissions are legally allowed to be sent into our air by industry; they count in the millions of pounds each year. It is incredible how much mercury, lead, dioxin, and hundreds more known cancer-causing compounds are spewed into our air by companies looking to maximize profits and minimize expenses. Taking a look at an air emissions permit application submitted to the state or EPA from a plastic manufacturer or a cement plant would read like a list of chemicals found at a Superfund site. Many state environmental agencies have a searchable database of air permits from individual companies you are able to access and review.

The Toxic Release Inventory the EPA operates is also useful to search individual company data but, much is self-reported so always consider the numbers may be lower than actually emitted.

In the City of Industry, California a lead battery recycler was fined for emitting lead, arsenic, and other toxic chemicals into the air and surrounding neighborhoods. The lead and other chemicals were emitted over three years from 2017 to 2019 at minimum but may have been ongoing for much longer. The company was also cited for failure to minimize dust emissions, violating reporting requirements, and not properly maintaining pollution control equipment. In 2016 testing of 132 homes near the facility showed 100 had lead levels high enough to be of concern and may need remediation to prevent people from being further exposed. Exposure to lead may cause kidney disease, and it also limits brain development in children. 1,3-butadiene, one of the other chemicals emitted from the plant, can cause cancer.

In Pittsburgh, Pennsylvania a steel plant was fined for storing lead and zinc dust in a manner that the toxic particles were blowing around, contaminating areas outside of the storage site. The plant was fined for mismanagement of hazardous waste and illegally storing lead dust. The materials were found to have been dispersed through the area by precipitation and winds. The plant also was violating proper waste management practices.

In the southeast part of Los Angeles, California, lead from a factory that recycled car batteries for years contaminated the surrounding area. The Exide battery recycling plant processed 11 million batteries every year until it was forced to close in 2015 as part of a settlement for hazardous waste violations.

A study was done by the Keck School of Medicine at the University of Southern California. The researchers determined where the most impacted areas were from the lead emissions of the Exide plant and used soil contamination data from the California Department of Toxic Substance Control. They then gathered baby teeth from children who lived in areas where the concentrations of lead in the soil were above safe levels. The average level was twice that and ranged as high as five times. There was no doubt that the emissions in the air and the soil over the years showed up in the teeth of children. Lead causes damage to brain development in infants and is known to cause cancer in adults. Los Angeles County officials believe up to 2,500 properties near the plant are in need of remediation from dangerous levels of lead.

We can't emphasize enough trying to find out what the operations near where you live are putting into the air you are breathing. There are databases to locate this information listed in our resources section.

2. Water Pollution

The smaller cities and towns may not have giant clusters of industry stretching for miles, but the companies they do have may cause enough risk on their own. There are places like Hoosick Falls, New York, a small town about an hours drive north from the state capital of Albany. This small town was happy to have several large manufacturing companies call it home. The happiness ended when two companies caused almost the entire town's public water supply to be unsafe to use. The small village may have six Superfund sites as a result. One major contaminant was PFOA, one of a group of toxic chemicals linked to cancer and other diseases, such ailments of the liver and kidney. The level of the contaminant PFOA could well be the highest ever found in the nation according to a former EPA regional director.

In North Carolina a study from North Carolina State University found a DuPont plant discharged wastewater into the Cape Fear River with concentrations of PFAS as high as 990,000 parts per trillion. Near a water intake for drinking water supplied to four counties downstream from the plant, levels were 130,000 parts per trillion. The EPA has established a health advisory safety level of 70 parts per trillion.

Some remediation takes years and decades. In the meantime, residents are forced to find alternative water supplies. Sometimes the EPA declares the water safe enough to drink, but really the water is still contaminated with chemicals at some level. Once a town's water source is contaminated, sometimes cleaning and restoring it back to clean and safe may not always be possible. Certain chemicals, such as PFOS, TCE, some VOCs from petroleum, and Perc are next to impossible to fully remove from a water supply once they spread out in the groundwater aquifer.

There are very few places where the water has not been impacted and degraded by industrial activities. While fewer of our rivers, streams, and lakes are being used as cesspools, millions of gallons of hazardous wastes are still discharged into them every year. Groundwater contamination beneath some of these plants is an ongoing concern, and not a week goes by without another report of an aquifer being polluted by some chemical substance disposed of carelessly.

3. Concentration of Industrial Operations

Many times, once a certain industry has started up an operation, they attract other similar companies to move in and open up in the same area. Clusters of similar businesses such as car dealers all together in Auto Row are a low-impact example of this phenomenon. Oil refineries, scrap yards, and auto body shops are high-impact examples. Knowing the present and potential future zoning plans of an area makes predicting these clusters easier, but as many of these clusters already exist, it is important to know their unique risks.

Unfortunately, towns and cities look at each individual business as existing in a bubble. The government almost never talks about the cumulative effect of all the combined negative health impacts in an area. That is how an area that may have been a clean place to live starts to become a not-so-healthy place. To get the true number of potential health hazards, add up all the toxic emissions of all the businesses, not each individual business by itself like the government numbers.

Entire sections of cities are dedicated to heavy use industrial zones. For 34 years I lived in New Jersey. Many of those years were spent living and working in the area known to locals as The Armpit. If you have ever driven the New Jersey Turnpike headed into or out from New York, it's the section where the air some days starts to turn a strange shade of orange. We always knew when we got close because the smell was a cross between a rotting cantaloupe, a stink bomb, and burning plastic, a smell that permeated our nostrils. Some days it actually stung. All outside air to the car was turned off, the windows were closed, and we would hope to get through it as quickly as possible. The area was home to chemical plants, plastic manufacturers, an enormous Exxon refinery, power generation plants, and hundreds of other heavy industrial operations.

The tens of thousands of people who lived there, though, are not so lucky, The level of chemical compounds in the air they breathe is among some of the unhealthiest in the entire country. This type of heavily industrialized area is one to be avoided whenever possible.

There are many other places that have these huge clusters of some of the most giant, polluting, toxic, gas-spewing factories. From Texas to Maryland, they are everywhere. At least these places are visible and therefore easily avoided. Unfortunately, there are many thousands of people who live in these places. They seem to have no choice but to accept the conditions where they live and deal with it as best as possible. Unsurprisingly, some of the most heavily polluted places in the country are located in some of the poorest neighborhoods and towns.

Louisiana has its own Cancer Alley. An 80 mile stretch between Baton Rouge and New Orleans is home to over 100 refineries and petrochemical plants. One or two of these huge plants is enough to have an oversized impact on the air quality, but over 100 creates what the EPA said is "some of the most toxic air in America."

In a CBS news story, local residents from the town of Reserve shared some personal stories with the investigative reporter. Reserve sits smack in the middle of those plants and has a cancer risk calculated by the EPA at almost 50 times the national average. They are also the home of the Denka Performance Elastomer plant owned by Dupont, the same people who brought us non-stick pans that cause cancer. According to the CBS story, people in this town of 10,000 people for decades have had health problems ranging from dizziness and severe headaches to liver and lung cancer. Many believe a plant, hundreds of yards from some of their homes, is the source. The plant manufactures a chemical called chloroprene. The EPA calls it a likely human carcinogen. The Network for Human Rights also did a story on this town and the plant. It was called "Waiting To Die." When they did a survey of people living near the plant, they discovered an astounding cancer rate of 7%. To put that in perspective, data from the CDC shows most parts of the United States have a cancer rate of under 1%. Of course, the company denied they could be the cause, claiming the EPA's concern about the toxicity of their product chloroprene was based on faulty science. I will bet anything not a single one of their employees in middle management or higher lives within 10 miles of the plant. And other people should not have to either.

To help people see what is being released in specific areas or to avoid highly polluted areas, the EPA maintains a database called the Toxic Release Inventory. With the database, users are able to see what individual factories claim they are releasing into the air and water. It lists the specific chemical or pollutant and the quantity they are reporting to the EPA. This database sadly neglects the cumulative effect of all the industrial operations located in an area. Having 10 factories all belching levels of mercury and lead into the air just below acceptable air emission regulations means that the people living in the area are being subjected to levels 10x the acceptable limits for toxic chemicals.

4. Hidden Dangers

On top of all of the known dangers, there are the unknown ones lurking inside many of these industrial buildings, dangers people and authorities only find out about when there is a problem.

For example, there is the chemical company in Madison Heights, Wisconsin. Apparently, the owner of Electro-Plating Services decided he would store all the hazardous waste and old chemicals the business generated instead of having to pay to dispose of it. Over the years the company was cited for various violations; however, the actual situation apparently went unknown. The building contained 5,000 drums and containers of hazardous materials by the time environmental authorities acted. Some of the materials included chloride, cyanide, chromium, and acids. The Department of Justice said there was a pit dug in the lower level of the building to store the waste. The company was only 500 feet away from residential housing. How many of these illegal toxic dumps are inside of industrial buildings in any town are unknown--until something happens, that is.

The city of Ogden, Utah, purchased an old industrial building that they wanted to tear down and use for another purpose. The city, however, never knew what was stored inside of the building.

It turned out that the past owner, who used the building from the 1960s until he passed away in 2014, had accumulated an estimated 60,000 containers of hazardous waste inside the five story structure. The government is unsure where all of the waste came from, but many of the containers had markings from the U.S. Military. One theory is that back in the 1980s, when the government was allowed to sell surplus materials including old chemicals to the public, they were possibly purchased by the past owner and stored. The military would sell the chemicals at a cheap price to avoid the much higher disposal fees it would have had to pay if done properly. The old owner made his money from buying and selling military and government surplus items. The hazardous waste site is considered so dangerous that the EPA is using Superfund money to get the clean-up started. The clean-up of the building will take months, and if the structure catches on fire in the meantime, there would be a disaster in the entire area. Nobody had a clue what was sitting inside the building all those decades.

We are quite realistic that it is very difficult to find any place that is pristine any longer. However, with a little research and knowledge, you can avoid knowingly living in an area where the air, water, and land are so contaminated and unhealthy. Take the time to know what is in your neighborhood.

To decide how far to be from existing industrial operations, you need to determine what the plants are emitting into your air. Even before knowing exactly how much and what they are, you can draw basic assumptions to stay far away from a chemical factory or a refinery. However, a lumber mill may not present such a large threat if they also do not use wood preservatives. We always tell people to stay two miles away from any operation with smokestacks appearing to continuously emit smoke. Smokestack industries do require the largest buffer zone. Also check into the amount and volumes of noise and truck traffic these places generate. Remember, too, that multiple industrial operations multiply the quantity and number of potentially hazardous emissions you could be exposed to. It does take time to research these, but in the end you will have some comfort in knowing you are not at risk. In general, do leave at least two miles away from industry and five miles when there is a large group of factories.

Chapter 18. Fracking

First, there was the Gold Rush. Everyone was trying to get a piece of the action and find their fortune. After that, came the oil boom. A few lucky ones made huge fortunes. Now, we have the fracking boom. Fracking is being billed as the savior of everything from freeing countries from energy dependence on oil to putting almost anyone who wants a job to work. It seems, though, most of this hype is coming from the oil and gas industry and many of the politicians in their corner. What exactly is fracking and what is really happening?

In simple terms, fracking involves creating fractures in the earth, mostly in layers of shale rock, to free the gas and oil that is trapped between the layers. It is similar to oil drilling in that drilling a well into the earth is the first step. Then, the drill bit is turned to create horizontal wells shafts. However, in fracking, after the well is drilled, huge amounts of water, chemicals, and sand are pumped into the well at tremendous pressure to blow apart the shale rock and create fractures to free the trapped gas. Then, the oil and gas are pumped out of the well, and so are the millions of gallons of toxic and contaminated fracking fluids.

Fracking is not a new technology developed by some of the most brilliant scientists working for oil and gas companies. Similar operations have been around since back in the 1800s. Some actually used to send explosives down the wells to blow up and fracture the rocks to give up their oil. Fracking with regular wells combined with injecting fluids and sand to increase yields was used from the 1940s through the end of the century. George Mitchell was one of the pioneers of combining horizontal drilling with injecting fluids around 2005. The new methods of fracking involve drilling down to a certain depth and then turning the drill bit 90 degrees to drill a horizontal shaft thousands of feet long. The idea is, once the fracking fluids are sent down the well at high pressure, many fractures and rocks yielding gas and oil will be made all along the shafts. With the new technique, yields skyrocketed and shale gas drilling boomed.

For years once the boom started, it seemed everywhere a shale formation was mapped underground, people were racing to frack the area. It happened so fast that people were convinced the industry knew what it was doing and that fracking was as safe as they said. People who owned property in these shale deposit areas were told by drilling companies they could get rich by allowing drilling on their lands. What was never told to almost all of these unsuspecting people was the environmental consequences fracking was known to be causing elsewhere.

Many people, especially those in the industry, are declaring this the biggest oil boom ever. They may be correct about the tremendous increase in gas and oil this country has seen in many years. But, with it has come a tremendous number of people affected with health problems by the new toxic legacy currently being created. Where the fracking industry was once viewed as a positive for a community, now their promises of only good things to come are seen as the lies they always were.

The fracking industry has told people for years that they do not emit harmful pollutants. Back in the beginning of the boom in 2005, many people, seduced by promises of money, took the industry at its word. The evidence compiled now after over 10 years of almost uncontrolled fracking is indisputable proof against their claims

Despite the cash, resources, and political clout the industry has been using to keep the fracking boom going, it's time for it to end. Or, at minimum, it's time to enact some real regulations protecting people's health and the environment. There are states and communities banning fracking. Some weak regulations have been put in place along with a tiny bit of oversight, but these measures are not enough. Unfortunately, the current larger political climate is not in favor of the environment and puts industry's health over the people.

1. Air Pollution

Fracking wells emit various quantities of ozone. When we hear about ozone, we mostly hear about the ozone layer high in the sky; ground ozone is as much or more of a serious problem. Much of the smog hanging over some large cities is made of ozone. High levels of ground ozone increase the risks of asthma, strokes, and heart attacks. When research and testing was done in areas where fracking produced high levels of ground ozone, there were noticeably higher levels of asthma and heart-related issues. Some studies showed pregnant women in these areas having more premature births, birth defects, and babies born smaller than they should have been.

In areas of some rural states where fracking activity is very high, levels of ground ozone have been measured higher than in Los Angeles and other large metropolitan areas.

In addition to ozone, fracking creates volatile organic compound (VOC) emissions. Basically, these are the chemicals in the air we smell when we are in a gas station pumping our gas. Fracking releases VOCs including toluene, benzene, and n-hexane. Long-term effects of VOC exposure include damage to the kidneys, liver, and the nervous system. VOC exposure has also been known to cause cancer. In fact, benzene is known to cause cancer on its own. On top of all of this, VOC exposure is linked to birth defects, blood disorders, and neurological problems.

In one year, Uintah county, Utah, emitted VOCs estimated to be equal to the emissions of 100 million exhaust-belching automobiles. A study done in this area by the National Oceanic and Atmospheric Administration estimated between 6 and 12% of the gas produced in this heavy fracking region was actually escaping into the atmosphere.

In February of 2018, a fracking well owned by Exxon had a failure that resulted in one of the largest blowouts in the world. The well in southeast Ohio leaked methane for almost three weeks, totaling more methane than all the gas and oil industry in a country the size of France emits in an entire year. The cause of the blowout was attributed to degradation in the walls of the well, usually called casing failing. Many fracking wells, such as this one, only have casing for the first 20-30% of the well shaft. Because the cost of the casing may be as high as 30% of the well cost, the fracking industry tries to skimp as much as they can. There are no regulations requiring the entire well have casing so only the bare minimum is used.

Even when there are casings along the entire length of the well, that alone does not solve the problem. Oil and gas fracking wells use tremendous pressure to inject the toxic fracking fluids deep underground to create the stress on the rock structure to fracture itself. The pressure in the well cases are increasingly also fracturing the casings, allowing these blowouts and allowing fluids to escape into the upper layers of soil and rock the well passes through. More well failures have been occurring at an alarmingly increasing rate as the industry rushes to drill and pump and move on as quickly and profitably as possible.

2. Water Pollution

One of the biggest threats to any community is the loss of a safe, dependable supply of drinking water. The pollution in the air may eventually clear long after the drilling equipment is abandoned, but the contamination of the water resulting from fracking will remain for possibly generations.

The fracking industry is water-intensive. They pull millions of gallons of pure, clean,

irreplaceable water from aquifers that normally would be able to supply local towns and cities for many more years to come. They take our clean water and add their toxic fracking chemicals to it, then pump it down into the oil and gas wells. Once it shatters the rocks, it gets pumped up again from the well. This time, though, the originally fresh, clean water, the water they'd already poisoned with toxic fracking chemicals, is even more contaminated. Radioactive elements such as radium, strontium, uranium, and radon as well as heavy metals such as arsenic, lead, and iron get into the water.

A study done by Duke University reviewed test results from 60 different sites in New York and Pennsylvania. They looked at the relation between the distance from where gas wells were drilled and the level of methane gas found in residents' private water wells. They discovered that wells located closer than one half mile to the drilling activity contained 17 times higher methane levels than those further away.

In Dimock, Pennsylvania, back in 2008, residents saw their tap water turn a funny shade of brown, and some said it made them sick when they drank it. This happened at the same time fracking became popular in their area. After much back and forth with health officials involving the EPA, results were published in a report. Of the 67 wells tested, 27 of them had levels of lead, cadmium, arsenic, and copper high enough to be a health risk. In addition, in 17 of those homes, water with levels of flammable gases high enough to explode was found. The fracking industry usually claims the gas comes from natural sources and is not caused by fracking. To Tony and me, the correlation is too strong to be just a coincidence.

In Washington County, Pennsylvania, there were two separate instances of contamination from a fracking operation, and the operators were caught concealing knowledge of it from the residents in the area and the public. The first case involved a fracking waste storage tank that had leaked over 2,000 gallons of waste into a field that drained to a nearby creek. The second incident turned into a huge legal case because of the health impact and was actually made into a movie. In that incident the storage pond the company created for the toxic fracking waste had tears in the liner and was leaking into the ground. Two company employees testified at the trial that the company knew the leakage had contaminated the groundwater but chose to do nothing for almost three years. They continued to pour waste into the pond, further contaminating the groundwater, which the area residents used as a drinking water source. Again, the company knew they had contaminated the water supply but continued to do nothing.

Residents suffered a variety of negative health effects from the chemicals: headaches, dizziness, terrible fatigue, nosebleeds, and rashes. In addition, they claimed the company had conspired with two testing laboratories to falsify results about the extent of the contamination. The battle went on for six years, and finally in 2018 the company agreed to settle with the families. This case opened up other evidence there were other oil and gas companies in the state also involved in criminal activities, and the Attorney General's office began to pursue multiple investigations.

3. Fracking Fluid and Waste Disposal

Fracking companies claim the fluids they use are safe. The standard talking line is that the fluid is almost entirely water. This is only true if we look at what percent volume are the toxic chemicals. The average we have been told is 2% of the chemicals are mixed into the water. The quantity of fracking fluids used in any one well totals in the millions of gallons, so because 2% are these dangerous chemicals, that's thousands of gallons of toxic chemicals in total they admit to pumping into the ground.

As for what exactly these chemicals are, it's hard to verify. There have been many studies looking at exactly what chemicals fracking companies use. The companies in question claimed their exact mix of chemicals is a company secret and they are legally allowed not to disclose it. There have even been congressional investigations into determining what chemicals are used. One congressional investigation found 29 chemicals known to cause cancer and other serious health risks to people.

A report issued by a U.S. House of Representatives committee showed how difficult the problem is of identifying chemicals hidden behind the proprietary rules. The report found companies used 94 million gallons of 279 different products containing at least one secret ingredient. In many of these cases, when questioned, the companies themselves did not even know what the proprietary chemicals were.

A 2011 report to Congress estimated that from 2005 to 2009, 14 leading fracking companies used (before mixing with water) 780 million gallons of 750 different chemicals.

A study done by researchers at the University of Toledo identified 210 chemicals in the wastewater fracking creates, chemicals it therefore needs to dispose of. This is the same wastewater the federal government is seeking to loosen regulations on disposal of to allow it to simply be poured into the local streams and waterways. The researchers say there are still more chemicals to be discovered yet and the long-term effects of continuous exposure to them is nowhere yet known. Some of the worst chemicals found already included atrazine (a pesticide), toluene, and VOCs linked to bladder, liver, lung, and skin cancers. The government allows this wastewater to be used to irrigate agricultural crops, so these hazardous chemicals are getting into the foods we consume. On a local basis, the chemicals get into the groundwater used for drinking water, and any local produce the chemicals were used on goes on kitchen tables throughout the area.

Not only is the wastewater contaminated with chemicals, in many cases it is also contaminated with radioactive materials from the ground. Depending on the minerals encountered in the fracking well, the waste is subject to contain different materials at different levels. Many states have found their fracking waste has levels of radioactivity higher than can be disposed of in normal landfills and injection wells. The levels of radioactivity pose a high risk not only to people in the area but to the groundwater and environment.

It needs to be stated here that the waste is supposed to be disposed of in special landfills, but the industry is trying to get regular landfills to accept this along with everyday trash. Instead of ensuring this radioactive waste is disposed of safely, many states are raising the allowable radiation levels to have regular solid waste landfills be able to accept more of it. In effect, this allows more contaminated waste to be dumped. Industry groups are lobbying politicians to make these changes, and the landfill industry supports this as well.

On the federal level, these industry groups are fighting for changes that will allow millions of gallons of this waste to simply be piped and poured into local waterways near the wells. Industry groups claim they spend too much money transporting wastes for disposal. Being allowed to use more of the nation's waterways as toilets would be to their benefit and increase profits. So far it appears the government is leaning towards that direction. On the state level, they're already doing this in secret.

In Youngstown, Ohio, a trucking company saw its owner go to prison for telling his employees to drain truckloads of fracking waste into storm drains. The radioactive, contaminated waste eventually drained into the Mahoning River.

In Pennsylvania, a fracking waste transport company was caught dumping contaminated fracking waste into abandoned mines instead of disposing of it properly. The customers would pay, believing their waste was being legally disposed of. Illegal dumping had been going on for at least six years before the company was caught.

4. Depletion of Groundwater

Water is a critical component in order to extract gas and oil through fracking. The drilling process alone can use up to 300,000 gallons of water per day. According to the Susquehanna River Basin Commission in Pennsylvania, fracking a well in their area may use up to eight million of gallons of water in just one week.

Wells in the area, including many municipal wells supplying water to entire towns, are subject to lowering of the groundwater level, and available water resources drying up. In areas where water is not as plentiful and residents already have drilled deep wells, fracking depletes their available water even further, at times lowering the water table and drying out the local wells residents depend on.

There is a fierce fight going on between fracking opponents and everyone else. The fracking industry and many politicians see continued fracking and the huge amounts of oil and gas it produces as essential for our future. Opponents include not just environmental groups but thousands of people whose health has already been affected by the dangers of fracking. The scales are currently tipping in favor of opponents of fracking armed with hundreds of studies and piles of evidence showing the effects of fracking. Sooner rather than later, something big is going to change. Whether it will be a ban on fracking or much tighter regulations on where it is allowed to take place, it has to change.

The threats to human health and the environment from fracking are finally being seen as they should be: dangerous, imminent, and alarming. Fracking and the oil industry are throwing the full weight of their power and money to stop any further regulations that could affect their bottom line. At the present time of writing in early 2020, there is a very large push by the fracking industry and the Trump administration to loosen and rewrite regulations on how fracking waste must be disposed of. If they are successful, millions of gallons of what presently is considered hazardous materials will be able to simply be disposed of in our surface waters. Making a full circle back to the disposal practices of the 1960s and ‘70s will do irreparable damage to the environment and health of millions of people.

Many towns are metaphorically drowning in the millions of gallons of toxic fracking waste. If there is fracking activity near where you live, be very concerned for the continued degradation of your air quality and water safety.

Living near fracking operations can be quite hazardous overall to your health. We recommend not being any closer than two miles from any active or proposed drilling areas. In addition, use caution if you are near waterways such as streams or rivers that may be used to dispose of fracking waste in. If you do have fracking activity near you, check your drinking water at minimum once a year if you have a public water supply and twice if you are using well water.

Chapter 19. Mines

Without many of the resources that have been mined over the years, much of our industrial and technological progress would not have been possible. However, with that progress comes risks and damage to human health and the environment. This is compounded as a result of many years of irresponsible practices. While current mining practices are a bit better, they do still present many of the same hazards as in the past.

Mining for various resources found on the Earth has been ongoing for many thousands of years. The list of what materials are mined for and their uses numbers in the thousands. While the materials that mines extract may be vastly different, some of the negative impacts on the health of the people living near these mines are the same no matter what.

In many cases, closed mines present even greater levels of risks to communities than operating ones. Old mines fill up with water, and acid drainage has become a huge ongoing problem for decades. Giant piles of waste ore and mine tailings sit and blow around in the wind. Others leach their poisons into the water supply. Radioactivity continues to come from all the thousands of old uranium mines sitting open.

When the mining company decided it was no longer profitable to keep running, many operations simply took their machines and left everything else, including millions of tons of tailings, polluted lands and waters, open pits, and mine shafts. The toxic legacy, though, continued, and in many cases, still does today.

Counting mines in only four states (Arizona, Montana, Utah, and Colorado), there are over 145,000 abandoned and inactive mines. These are staggering numbers. These mines left behind thousands of miles of streams and rivers polluted. In addition to that, continued acid mine drainage may never be fully controlled or stopped.

Most of the different types of mines that exist could have their own chapter. Each has its own story to tell because there is so much to say. However, some of the more common dangers are presented by the more common mining operations. The impacts of copper mines, gold and silver mines, and coal and uranium mines are representative of the hazards presented to humans from living near a mining operation both active and closed. While other mining operations not discussed in this chapter do pose unique hazards, many also fit into the general descriptions of hazards here.

There are three main methods of mining. Surface mining involves large pits or strips of earth that are dug out, removing the ore. Underground mining entails shafts and tunnels being dug under the ground so the ore can be brought to the surface to process. In-situ, or leach pile, mines use chemicals either injected deep underground into the ore deposits or poured and soaked onto giant piles of ore on the surface so the chemicals can leach the desired mineral or metal into the solution, which is then processed to remove the target ore.

Mining operations like these create environmental issues which in turn create negative health issues. The mining industry as a whole causes more damage to the air, water, and soils in our country than many other industries combined, largely because regulations are weak

1. Particulate Matter

Depending on whether the mine is an open pit mine or an underground one, there may be a continuous cloud of dust surrounding the active mining area and processed ore storage areas. Coal mines and the towns they are near in particular frequently have levels of unhealthy particulate. On windy days, particles may even get into houses through any small crack to coat everything inside, including people's lungs. The outside may have so much particulate matter that there seems to be a haze in the air and outside activity should be avoided.

Dust from ore and waste rock is an unavoidable problem in mining. Depending on what is being mined, fine particulates, the type that our lungs can't always filter out, get into our bloodstream and our organs. With particles of copper, gold, or especially anything radioactive, damage to our body is a distinct possibility.

In Libby, Montana, a vermiculite mine that operated for decades had a very high concentration of asbestos as well. The dust and particulate matter from the mine released asbestos into the air, which all the residents in the town breathed for years. Vermiculite containing asbestos was used everywhere in the town, from baseball fields to insulation in buildings and homes. Over 400 people died from diseases directly tied to the mining in town. Thousands more were sick, including many children exposed to the dust. The town of Libby was declared a Superfund site by the federal government, and cleanup of hundreds of contaminated sites has stretched over 15-20 years.

In Westerly, Rhode Island, the beautiful beaches the visitors flock to in the summer hides a reality the residents live with year-round. Three granite quarries plague the town with dangerous levels of particulate matter. One of the mines puts eleven and a half tons of silica dust in the air each year. The quarries they mine the granite from consists of about 90% quartz. That just happens to be the most common form of crystalline silica, which is on the EPA's list of carcinogens. On top of that, the percentage of school age children with asthma is the highest in the state. According to 2016 Westerly Hospital patient data, about half the patients who are treated at the hospital have chronic asthma and/or chronic obstructive pulmonary disease.

A study done by West Virginia University looked at every county in the state. They reviewed the health data for communities living near mines, especially coal mines. After adjusting for other factors such as smoking and poor access to health care facilities, they determined that people who lived near mines have a 70% increased risk for developing kidney disease, have a 64% increased risk for developing chronic obstructive pulmonary disease (COPD) such as emphysema, and are 30% more likely to report high blood pressure (hypertension).

2. Water Contamination and Depletion of Water Supplies

Mines impact water in numerous ways. Some of them contribute to the pollution of surface water and underground drinking supplies. Others pump millions of gallons of water from rivers and aquifers, drying up waterways and depleting safe, available drinking water for towns and communities.

Mine drainage, however, is one of the largest threats to our water supply. Millions of gallons daily, containing arsenic, lead, and other toxic materials flow out of some of the most contaminated mining sites in the country. They drain directly into the streams, rivers, and ponds nearby without ever being treated to remove any of these contaminants. Metals such as aluminum, cadmium, thallium, and manganese are frequently found in toxic levels in the runoff from mines into rivers and streams.

Underground mining often uses vast pumps to prevent underground water from filling in the tunnels and shafts. The millions of gallons being pumped out goes into nearby streams and rivers, increasing sedimentation and blocking natural flows of the waters. This may cause flooding in previously dry areas along the waters, such as farms or developments.

Open pits used for mining can be giant. Many are up to 1,000 feet deep and several thousand feet wide. Pits dug this deep frequently dip below the water table. To keep the pit dry requires continuous pumping of water, which draws down the level of the water table as millions of gallons are pumped from the working pit. Once the mining is done in the pit, it usually is allowed to fill up again with water, and it quickly turns into a contaminated lake with acids, arsenic, and metals, reacting with the water and the rocky pit. This becomes toxic to wildlife, and any runoff from the pit may contaminate the groundwater and other sources.

In Salt Lake City, Utah, the open pit copper mine, known as the Bingham Canyon Mine, has the largest known plume of contaminated groundwater in the world underneath it. If the plume moves or gets released into the Great Salt Lake, unimaginable environmental disasters could occur.

Specific risks of water pollution occur in three major categories.

3. Acid Mine Drainage

Acid drainage is a natural process that occurs on a small scale in nature. Nature's process is called "acid rock drainage." It happens naturally when sulfide compounds in rocks are exposed to air and water. When that happens, the sulfide in the rock combines with the air and water to make sulfuric acid.

In mining operations, however, this process is multiplied many thousands of times over. Large quantities of rock are exposed to the air and water both in open mines on the surface and in underground mines along the tunnels and shafts. The dangerous chemical reaction happens not only when the mine is in operation but also afterwards. And it happens for hundreds of years, creating hundreds of thousands of gallons of sulfuric-acid-contaminated water.

According to a 2019 Associated Press story on acid drainage from mines, every day over 50 million gallons of untreated wastewater is released. AP reviewed the records of 43 mining sites on which hundreds of individual mines had operated for their report. The contaminated water drains into groundwater supplies, streams, rivers, and ponds. Unfortunately, this is not uncommon with abandoned mines and even currently operating ones.

The most acidic water ever found is mine drainage leaking out of an old mine at Iron Mountain near Redding, California. A researcher at the U.S. Geological Survey called it, "the world's worst water." The acid level of the water was measured off the charts literally. The PH level tested in some places at a -3.6. For comparison, battery acid measures a respectable .3 on the PH scale. The water set a record for the most acidic ever discovered outside of a laboratory.

Acid mine drainage has been an ongoing issue for over 100 years, and it is not going away anytime soon. If anything, it's getting worse.

A. Chemical Pollution

Some mining operations use chemicals such as cyanide or acids to process ores and remove the minerals or metals they are after. These chemicals frequently leak off site and find their way into the soil and water. In addition, much of the ore taken from the mine and its surface areas contains heavy metals, such as arsenic, lead, copper, cobalt, and others. These metals can leach into the water and contaminate it. When combined with acid mine drainage, the amount of heavy metals dissolved in the water drastically increases.

These materials end up in the same water that communities depend on. The toxicity of the water supply increases, and water treatment plants in many cases are unable to remove all the metals and toxins it contains.

In eastern Oklahoma's Tar Creek mining district, waterways are devoid of life, and elevated lead levels persist in the blood of children despite a two-decade effort to clean up lead and zinc mines.

B. Depletion of Water Supply

Mines use tremendous amounts of water in some of their operations. They take the clean water from water supplies and put back dirty and sometimes contaminated water. Many of these mines operate in places where water is a precious commodity with little rainfall to replace it.

In northeast Nevada, local tribes and environment groups are fighting a proposal from a mining company that would pull billions of gallons of water from the underground aquifers. The pumping of the water would change the natural flow of water, which currently supports a wetlands area containing 88 springs. The mining company says they will replenish groundwater by leaching the used dirty water back from the surface.

In addition, the mining company will need to lower the water table to be able to work the mine. A hydrologist studying the project said the actions by the mine would lower the water table hundreds of feet in the area of the mine and affect the water table possibly up to 30 miles away. As of May, 2020, the battle continues.

4. Radioactivity

In the western part of the U.S., there are about 4,000 abandoned uranium mines. While they may not be actively mined any longer, their toxic legacy and effects linger on. Many towns located near these mines have been tested for radioactivity and are contaminated to some level. It is not just the level in the air that is concerning. Many towns even used old mine tailings and waste ore to line their streets and build structures.

In the town of Monticello, Utah, uranium ore from mines was processed at a mill owned by the federal government. Residents used waste ore, which the residents were not told was radioactive, to build sections of the town. Monticello wound up with two Superfund sites due to all the radioactivity. Over the years the residents have had cancer clusters and increases in rates of birth defects.

5. Mercury

In gold operations particularly, mercury is leached when the ore is processed to remove the gold. Mercury is released into the air, especially when a mining method that requires the ore to be roasted is used.

In the big gold mining days in the late 18th and early 19th century, California produced almost 90% of the mercury in the United States. Over 220 million pounds were used around the world for gold mining and other purposes. Many tons of mercury were poured into streams and rivers in the California mountains used in gold mining to extract the gold from the ore. More than half of the waterways in the northern half of California are contaminated with mercury dug from mines and leftover from use in the gold mining operations long past but still contaminating the waters. There is a state park named for the metal, Quicksilver Park in the town of New Almaden. Signs warn visitors of the nearby toxic mercury mines throughout the area. There were more than 100 mercury mines feeding tons of mercury to the gold miners to use.

The dangers of mercury poisoning were not recognized until the early 1970s when it was way too late. The historic park was originally bought to be used as a recreation area back in the 1970s, but after mercury contamination was discovered, the park was placed on California's Superfund list. Many towns in California from the Sierra Nevada Mountains to San Francisco Bay remain hot spots of mercury contamination as do many rivers still under "do not eat the fish" advisories.

In California, there are over 150 reservoirs and lakes still polluted with mercury from the gold mining days. Clear Lake is considered the most mercury-polluted lake in the world. The nearby Sulfur Bank Mine still has pollution leaking from beneath a dam built by former mine operators. The EPA has spent almost $40 million over twenty years, and not much progress appears to have been made. The mine is a Superfund site, and the creeks in the area are still high in mercury, contaminating the fish, which people regularly eat, and sending mercury down waterways to San Francisco.

San Jose, California, has so many of the old mercury mines it is referred to as the largest source of mercury along San Francisco Bay. The state environmental regulators are overwhelmed by the sheer number of old mines throughout the state and do not even have a portion of the financial resources to clean up more than a few of the worst ones.

6. Smelters

Smelting has been used for about 5,000 years. Bronze, copper, and tin were some of the first metals to be made into their pure form through smelting. Now, the process requires an industrial-sized machine to heat up the ore and uses another compound to cause a chemical reaction to separate the pure metal from the ore.

Even though they are especially hazardous, many smelters operate across the country today in various industrial locations. The mining industry uses smelters extensively in the final processing of the ore. Many of these plants were built either near the mines or as part of a large operation. After the ore is processed, it is sent to the smelter to remove the metal from the ore.

Smelters are some of the dirtiest places and put out tremendous amounts of chemical pollution. At a lead smelter in Missouri, very high levels of lead have been detected in soil, dust, and air samples taken in the community nearby. Tests on children found lead levels above safe in 28% of them. Women in the community had lead levels in their blood twice the national average. High lead levels affect children's brains and learning, and high lead levels in pregnant women has a very negative effect on the development of unborn babies.

Mining communities are presented with unique hazards that many cities and towns luckily never have to deal with. From the giant piles of waste to the millions of gallons of hazardous water, the scale of these hazards are as large as the mines that create them. Sadly, the health risks presented do not go away when the mine closes down. They remain as a toxic legacy in the town and a reminder of their roots.

Many new mines are opening and exploiting resources in new ways, in some cases creating even larger environmental problems than before. Expansion of existing mines to get even more yield out of them is pushing areas to their limits in coping with dust, toxic water, and other hazards on a daily basis. Even with the jobs mines bring, in many cases they do not give the people working them nearly enough compensation for the hazards they face.

Any mining operation brings with it issues that are unable to be completely controlled. Particulate matter is impossible to prevent or in many cases even reduce. Breathing the metals, coal dust, and airborne chemicals is part of life for those who live nearby.

While Tony and I do not want to say moving to or living near a mine is entirely bad, we have to speak our minds and admit we would never consider this as an option if given a choice. Much of what is being discharged will never fully be admitted to and accounted for. That is a simple fact of doing business at the large scale these operations do.

Consider all the facts when you have a mine nearby, not just what the mine chooses to tell you. Many of these mines may cover hundreds or even thousands of acres. Adding in areas for tailings disposal, that makes for a huge footprint to stay away from. If you can be assured there are no closed, active, or proposed mines within four miles of your location, that may be as safe as you can be in an area where mining activity is commonplace. Even still, carefully monitor your water, having an annual test for metals and radioactivity in addition to checking its PH level regularly yourself.

Chapter 20. Mine Tailings

The past hundred years or so, mining has been done on such a monumental scale as to literally move mountains. Every kind of mineral and resource is being taken from the Earth. Large and small mining companies are looking for everything from metals such as gold and uranium to other resources such as diamonds and coal. Innovations over the years enabled hundreds of thousands of pounds of rock to be removed and ground up daily in search of the desired products. Many of the materials being mined for are present in such tiny amounts that tons of material are processed in search of a pound or two.

The mining industry uses both mechanical and chemical processes to extract what they want from the mined ore. Anything that is not the desired substances is cast off after the process. Mine tailings are the waste product that is generated from a mining operation processing ore. The National Institute of Health considers mine tailings one of the most dangerous environmental hazards we are facing today. Not only are the tailings themselves dangerous, but the extraction process is not anywhere close to 100% efficient either, so there is always some of the material being mined for leftover in the disposed tailings. This creates additional problems and health risks when the products being mined are radioactive such as uranium or poisonous such as arsenic.

Mine tailings are made up of the ground ore that is left and the chemicals or compounds used to extract the targeted metals. Tens of thousands of these piles exist from mining operations closed down years ago, creating a symbol of the industry's toxic legacy. Thousands more are created every year, and many present danger to the communities and environment. Because there is presently no value to the mining company in the tailings, they get disposed of as cheaply as possible.

In 1976 the Resource and Conservation and Recovery Act was passed. Its expressed purpose was to begin regulating hazardous substances and disposal. The mining industry managed in 1978 and 1980 to get much of mining waste excluded from regulations as a hazardous substance. Because of this, mines usually had little or no environmental controls on them, and waste was normally either dumped in giant piles around the mine or run into nearby streams and rivers to be taken away.

Mine operators saw the damage their operations were doing to the land and the streams being colored with the metals and acids from the mine drainage. Few people cared back then. It was all about the money. Years of mining without a thought to the environmental hazards to humans and the environment have come back to haunt the people and communities that once were boomtowns.

Some environmental rules and controls put in place during the past 40 years or so have helped limit the damage current mining operations have to some extent, but the tremendous increase in the scale of these operations created different and, in some cases, worse dangers and damage.

The mining operations for copper, gold, and uranium give an idea of how big a problem tailings are. The size of mines ranges drastically. Smaller mines produce a couple million pounds of tailing per year, and the largest mines such as the Morenci copper mine in Arizona generate billions of pounds of tailings per year.

A copper mine may produce ore containing copper at 0.6%. That is the average yield of copper ore. That would mean they would need to process 10,000 pounds of copper ore to be left with 60 pounds of copper. The other 9,940 pounds are considered mine tailings and useless to the mine company. They are usually disposed of either in giant piles onsite, used as backfill for filling in old sections of the mine, or placed into a containment pond. An actual amount of copper mined from the Morenci mine in Arizona in 2015 was stated at 902 million pounds. Based on an average of 0.56% copper, which was high for this mine, almost 160 billion pounds of waste rock and tailings was created in a single year. These are accurate numbers the industry group Mining Technology compiled. To put this number in perspective, it is equivalent to 987 Washington Monuments or 784 Navy Aircraft Carriers, not to mention over 13 million elephants.

A gold mine works with a metal content in the ore that is much less. There are a variety of different kinds of gold mines that produce varying amounts of gold and tailings. Underground mines usually have higher yields of gold content in ore. A high quality underground mine produces ore containing eight to ten grams per ton. A lower grade mine may contain only one to four grams per ton. That equates, even in the best scenarios, to a ton of ore having gold content of 0.000011%. To get an ounce of gold, the mine needs to process almost 6,000 pounds of gold ore.

Above ground mines have much lower yields of gold ore. An above ground pit mine has yields of under a gram per ton. The Fort Knox Gold Mine outside Fairbanks, Alaska, in 2018 had a yield of .43 grams of gold per ton of ore.

Furthermore, gold ore mining uses a chemical process to remove the gold from the ore once it is crushed. The main chemical in the process is cyanide. Gold mines remove ore by the thousands of tons. The ore is crushed and ground up, and then cyanide is added to react with the gold. After the gold has been separated, the remaining slurry is a waste product. The amounts of this toxic sludge are staggering. Some mines have millions of gallons of cyanide-laced slurry stored in lagoons that are known to leak.

Uranium mining has seen its better days. At its peak, in the 1950s and again in the 1980s, uranium mining had thousands of operations going in a wide variety of sizes. Today, according to the U.S. Energy Information Administration, there are only seven commercially operating mines in the U.S. Uranium has reached a historic low price, and imports meet 90% of the country's needs, so most mines are closed. The government is the largest producer at this time.

When uranium mining was in its most productive period, between the 1940s and early 1980s, many operations were pulling millions of pounds of ore out of the Earth every year to supply the U.S. Government with uranium for weapons systems and research. At the peak of uranium mining in the U.S., there was almost 43 million pounds of enriched uranium produced in 1980. Tailings were simply placed in huge piles as closely and as cheaply as possible. The Federation of American Scientists estimated over 230 million tons of radioactive mine tailings were produced by the uranium mines.

One of the many disastrous examples of this is the 16 million pound pile left on the banks of the Colorado River in Moab, Utah. The tailings were just piled and left when the uranium boom of the 1980s died. After close to 40 years and many years of fighting for its removal, there are now only six million pounds left sitting on the banks of the Colorado. But rather than actually solve the problem, they simply moved the tailings to another place away from the river and dumped it there.

The majority of mining operations use one of several methods to dispose of the vast amount of tailings and waste rock they produce. The easiest and least expensive disposal method was the most frequently used in the past. Much of the tailings and waste rock created by the mining companies was either disposed of into the nearest waterway or simply moved away from the active mining area and dumped, piling it high on any available land. Now, like in the past, small mountains have been created from tailings. These toxic mountains sit in the open, subjected to rain and snow, leaching their toxic wastes into nearby streams and rivers. The winds blow the fine particles into neighboring areas, covering everything with hazardous particulates for locals to breathe. In some states large mining companies have purchased large tracts of open land to use as dumping areas for tailings. This allows them to create and dispose of even more waste ore than ever before.

The main way the tailings are dealt today is the wet method, which creates a toxic, slurry-like material when the ore is processed. The leftover slurry still contains much of the chemicals used to extract the target ore in addition to the balance of the material mined for. The slurry like material is disposed of in enormous pits and lagoons. Many times, the mining company will create an entire toxic lake filled in with the slurry and mine tailings and build a dam to hold the disposal waste in place. Impoundments like this are known to leak, and at times, dams have broken, unleashing hundreds of thousands of gallons into streams, rivers, and the local environment. Millions of gallons of toxic tailing sludge sit near where the mining operation

There is a slightly better way to process and store the tailings, but it costs the companies more money and, if not required by their permits, they do not use this method. If the tailings are dewatered, they become more of a solid that creates less hazards from blowing away and does not require huge lagoons and unstable dams to hold. They are placed in rows called dry stacking and will remain in place better than dry tailings. Drainage still remains a big issue with the tailings exposed to the elements. Particulates blowing away and runoff contaminated with acid and toxic materials remains a large problem. They remain sitting wherever they were dumped for decades or centuries remaining hazardous.

If there are large open areas available to simply dump and dispose the tailings and waste rock onto, then the mining companies will use that method. In Baghdad, Arizona, a huge mining operation run by Freeport-McMoran bought 16,670 acres of land from the State Trust Land Department to use as a tailing dump. Interestingly, the mining company was the only bidder on the 25 square mile parcel of land at the auction.

When mining operations first started in the late 1800s through the mid-1900s, they were mostly located in more rural areas, but growth of towns and cities have brought many more people close to the piles of tailings. This is putting many more people at risk of exposure. Tailing piles are not a harmless pile of rocks. There is real danger in those tailing piles.

1. Inhaling the Particles

During processing, the ore is ground into very tiny, fine particles. Because of the small size of the particles, they are easily carried away by the wind. Many communities near mine tailing piles have a continuous coating over their houses and cars. The small size of the tailings allows them to invade the insides of houses, working their way through windows and doors. Due to the small size and light weight of the particles, tailings are able to be blown several miles and more from the dump. Asthma, other lung ailments, and cancer are associated with long-term inhalation of these particulates.

Breathing fine particulate matter is already dangerous, but breathing particulate from tailings comes with unique dangers. Depending on the type of mine the tailings are from, they may contain asbestos, lead, arsenic, and many other very harmful chemicals. Quite a few are known to cause cancer and respiratory ailments. High levels of metals contained in the tailings has been measured in humans and found in wildlife. Heavy metals usually found at many of these tailing sites cause adverse health effects such as memory problems, autoimmune issues, digestive problems, and poor kidney function. Metals such as arsenic can cause cancer and are associated with diabetes and heart disease as well.

In the town of Green Valley, Arizona, a mine operated by Freeport McMoRan was issued their fifth violation notice by the county for tailings dust. Tailings blew from the mine's property into nearby neighborhoods. The company had recently been fined $230,000 for similar violations.

2. Water Contamination

Runoff from tailing piles creates a toxic leachate similar to what seeps through and out of a landfill. The liquid leaking from the tailings is full of highly concentrated and toxic chemicals. This seeps into the ground underneath the tailings pile and contaminates underground water used by wells in the area. Much of this toxic liquid does run off the site, creating its own little streams looking for a place to flow to. Sometimes, it is the ditches along the road. Other times, it may be the pond next to a house where children like to play and swim.

In Rimini, Montana, the groundwater is so polluted with contamination left by the 150 mines in the area that residents are unable to use the tap water. The community is listed as a Superfund site, but the pollution continues to flow from the mines, making the groundwater even more contaminated.

In Kent, New York, tailings from an old arsenic mine were disposed throughout the areas near the mine. Back in the 1980s, a couple became seriously ill and almost died; it was discovered they had been poisoned by the arsenic tailings. Their well had been poisoned from arsenic leached from the tailings. The EPA investigated and trucked water in so the house had a safe supply of water to use. At that time the EPA closed the case and moved on. Fast forward to 2019, and now testing that should have been done decades ago revealed private properties in the area near the mine have levels of arsenic in the soil 1,600 times the EPA screening level for the substance. A health advisory the EPA put out in April 2019 told the residents that long-term exposure to arsenic in the soil creates significantly increased risk for cancer. For now, the government put down wood chips over the toxic soil and advises residents to not contact the soil and wash their hands.

In eastern Oklahoma in a mining area known as the Tar Creek district, the groundwater is so polluted that even the surface waters have nothing living in them. The children have elevated levels of lead in their blood, and that has not changed for almost 20 years as there is no real end in sight for a clean up to change this.

3. Exposure to Tailings Directly

In many of these old mines, the tailings piles are exposed, and the property is accessible to anyone who wants to go there. Many large mining areas near cities such as Butte, Montana, are home to hundreds of open and accessible piles of mine tailings. Chronic exposure to many of the metals in the tailings can lead to kidney damage, bone and lung disease, and some cancers. Many mine tailings contain radioactive substances and may have high enough levels to cause harm. Thousands of old uranium mines left tailings piled all around, still blowing radioactive dust into nearby communities for the residents to inhale. Kidney problems, liver issues, and cancer are a few of the potential health effects from exposure to these tailings.

South of St. Louis, Missouri, the Big River Mine Tailings waste site is contaminated with lead, cadmium, and zinc from old lead mining activities in the region. Dozens of homes are contaminated for the tailings that were used for driveways and fill projects. The large piles of tailings have eroded over the years, and lead has blown throughout the area. The EPA is testing the children in the nearby community for lead in their blood. There were at least 89 properties with unsafe lead levels in their soil.

In Dewey, Arizona, one old mining and tailings disposal area that is also a Superfund site has children playing and teens hanging out on the tailing piles. The EPA put nine signs up to keep people out. In the meantime, there isn't even a fence there after decades of exposure. The EPA found out the teens were disregarding the posted signs and issued a warning to the residents to stay out of the hazardous areas. There still is no fence around the area considered so dangerous it is a Superfund site. The locals continue to play near the area, possibly thinking if the EPA did not fence it off, it may not be as dangerous as they said.

4. Tailing Dam Failures and Leaks

For mines using the wet method of disposal, the tailings end up in large ponds that can cover hundreds of acres and hold millions of gallons of tailings. There have been many instances when the dams that hold back these ponds have had structural issues. Leaks are more common than most people are aware.

And it's not just ponds. Some mining operations fill old quarries or dam off canyons to use as places to dispose of their waste. Heavy metals, acids formed by the ore, and waste rock and other toxic elements eventually wind up in the streams and rivers nearby to the tailings' containment.

According to a review of data by the news agency Reuters, tailing dams in the United States are built using an architectural method that is not considered safe by engineers.

The challenge of mine tailings is not going to become any easier in the future. Through advances in technology, the mining industry is able to use lower grade ore than before. However, that will result in millions of more tons of rock being discarded as mine tailings and waste.

We cannot expect the mining industry to solve this problem themselves because there is no incentive for them to do so. The industry gets nothing at all in return for taking care of its waste problem. It costs them money, which they interpret as taking profit and throwing it away. The mining industry spends vast amounts of money on lobbying the government and politicians to keep regulations in their favor. Whenever possible, they try to weaken the few existing regulations to their advantage.

Currently, the Department of Interior is making it easier for the mining industry to profit and ravage our national lands. Newer regulations favor the industry, not the people or the environment.

There are no really great choices for a really sensible working solution. Covering up a pile of tailings with a dirt cap is not a real solution for future generations. The tailings are still underneath, poisoning the groundwater with leachate. The toxic leachate from the giant piles will continue to be a hazard for many decades no matter what we do. The only real partial solution is to remove the piles of tailings and place them into a more secure, double lined containment area (which will eventually leak too). The monumental amount of tailings needing to be handled makes any solution at this time unviable.

If the town you live in is an old mining town or currently is the site of operating mines, check to see what they do with the tailings and where they are located. Make sure your family is aware of the places to stay away from. If you live near a tailings pile, check with your state and local environmental agencies to see what you are being exposed to. There is no doubt you are being exposed to particles that may be dangerous.

Because of the airborne particles from many of the tailing piles and lagoons, distancing yourself 1-2 miles is advisable. If there is a large impoundment holding back millions of gallons of tailings, be certain that if there is a breach in the dam or other failure, you are not directly in the path the tailings will take. It is also imperative you have your water tested annually for heavy metals and any other contaminants which may be present in the aquifer you get your water from.

Chapter 21. Gas and Oil Wells

Oil wells cover the hills as far as the eye can see in many areas around the country. All day and all night, the pump goes up and down, bringing the oil and gas up to the surface. It was the dream of thousands of people to get rich drilling for oil or even having an oil company place wells on their property to send them a royalty check every month. Texas and Oklahoma have the reputation of being vast oil areas with thousands of oil wells pumping up that black crude oil. That may be our image of this, but almost all of the states do have a number of oil and gas wells. With the explosion in fracking for gas and oil the past decade, wells are now found many places they had never been before. Oil and gas wells are miniature industrial facilities like a small factory whose purpose is to produce oil and gas. They produce a product and along with that production, produce wastes and other by-products. In the case of oil and gas wells, many of the same emissions and pollutants found in factories exist here as well. Oil and gas wells release pollutants—including particulate matter, benzene, nitrogen oxides, ozone, volatile organic carbons, and carbon monoxide—into nearby air and water and have been linked to a host of health problems in people living nearby.

Studies and research coming out in the past decade show an association correlating the distance people live from oil and gas wells and certain health effects from the pollutants these wells give off. The oil and gas industry have dismissed the facts found in these studies and said there is no health risk, but more studies are coming out confirming that people living within a certain distance from oil and gas wells suffer adverse health effects.

In response, states are enacting new rules governing where drilling of any additional wells can take place. The distance that new wells are allowed to be from buildings and homes varies but are at least a good starting point. Many of these laws neglect to consider the concentration of pollution when multiple wells are placed onto the same drilling pad. In some cases, more than a half dozen wells occupy the same small area, making the potential for hazardous air pollutants many times higher than a single well. Additionally, the cumulative effects of multiple wells is ignored by environmental agencies responsible for monitoring emissions.

According to *Environmental Health News*, 17 million people live within a mile of an active oil or gas well. In fact, half a million or more people live near multiple wells in many high-producing areas. The closer people live to the well, the higher the negative health effects are.

In another devious loophole, the weaker regulations in the past apply to older, conventional wells that still fall under old regulations that allow them to emit far more pollution than even thought when the original rules were written. Pennsylvania has some outdated regulations controlling conventionally drilled wells. The operators use those regulations to their advantage to keep their conventional wells in an active status even though they no longer produce any product. Active wells covered under this old regulation have no pollution limits regulating how much they can vent. If a well switched status to inactive, it would be capped at 5,000 cubic feet of pollution gas per day. The way the regulation is worded, then, the operators have a choice to stay active and keep polluting in unlimited quantities.

1. Particulate Matter

There is no shortage of dust and particles from activity around the wells in addition to some of the smaller particulates emitted from the wells themselves. Fine particles clinging to the dust kicked up around the site contain many of the harmful chemicals emitted from the well such as benzene and hydrogen sulfide. Many, due to their size, are not screened out when inhaled, but the fine particles get deep into our lungs and attach to air sacs. Once in the body, they attack organs and deposit poisons in body tissues.

2. Emissions

The older a well is, the greater its likelihood that it will leak methane and other air pollutants into the atmosphere. Many of the operating wells date back over a decade; some date back even longer. While there has been improvement in some pollution control equipment on new wells, that does nothing for older ones. In fact, many of the older ones account for the majority of the emissions, and, as stated, they are not required to be retrofitted to reduce pollution even though they can be.

One study that collected measurements of benzene found concentrations of the cancer-causing gas up to 41 times higher than allowed within 500 feet of the well. The further away from the well, the lower the concentration of gas.

In the town of Pavilion, Wyoming, people living near these oil and gas wells were tested for chemicals. The scientists found eight of the chemicals found in the air near the oil and gas wells were also present in the bodies of the people tested, all VOCs not normally found in a human. These VOCs have short-term and long-term health consequences. These consequences varied from irritations to the eyes, nose, and throat, to headaches and lowered cognitive functions, to damage to the nervous system, kidneys, and liver. Several of the VOCs are considered to be cancer causing.

Researchers backed by the University of California conducted a study to see if there was a correlation between a trend they were noticing in infants being born smaller in areas where high concentrations of oil field chemical emissions were present. The resulting studies confirmed a link between low birth weight babies in the San Joaquin Valley, California and nearby gas and oil fields. Additional studies done in Pennsylvania, Colorado and Texas are also beginning to offer indisputable evidence of the effect of these chemicals on the infants.

3. Leaks and Spills

A report from American GeoSciences says there are about 55 spills for every 1,000 wells. The spills came from storage tanks, storage pits, and pipes near the wells and the drilling rigs. In the four states where data was compiled, the overall average spill was over 600 gallons. requirements based on the volume, type of material, location, and spread of the spill. The storage pits near these wells have a record of leaking hazardous substances, which is well documented. The organization Environment New Mexico Research and Policy Center totaled over 400 instances of water contamination caused by these waste pits used by gas and oil well operators.

Gas, oil, and condensation in the well is collected in tanks sitting near the well. Many of these tanks leak into the air or onto the ground. Some of the chemicals in the condensate are benzene and toluene, both of which are very harmful and known to cause cancer.

People living near properties with wells have been reporting a much higher frequency of headaches. Some of them are so severe as to be debilitating. The older the well is, the more likely it is to leak. As the seals and parts on the well pumps age and wear from use, they increase the amount of toxins that are able to escape into the air. People who have been living near these wells and already exposed long-term are now being exposed to even more hazardous emissions. Children living in these areas experience frequent nosebleeds, nausea, and also loss of coordination. In Karnes County, Texas, many residents are suffering with these and other strange symptoms they say came along with the rapid increase in oil and gas production. They are not alone. Communities from California to Pennsylvania all have similar stories about the health effects from living close to these oil and gas fields.

On a hot, dusty day in August last year, a group of regulators from the New Mexico Environment Department and the Energy,

Minerals and Natural Resources Department traveled to Counselor, New Mexico, to tour the oil and gas sites that dot the landscape of the Greater Chaco region.

The group included NMED's Air Quality Bureau chief Elizabeth Bisbey-Kuehn, Environmental Protection Division director Sandra Ely, NMED Secretary James Kenney and EMNRD Secretary Sarah Cottrell Propst — all key regulatory figures in the state's Methane Advisory Panel, tasked with developing new regulations around oil and gas emissions.

Teresa Seamster, Navajo Nation Counselor Chapter Health Committee member, used the opportunity to present the findings of a recently completed health impact assessment (HIA), which found periodic spikes of formaldehyde and other pollutants associated with oil and gas development, recorded at unsafe levels for short periods of time near homes.

"Formaldehyde is probably one of the most carcinogenic chemicals in air that you can have," Seamster told NM Political Report. "It will cause irritation of the respiratory tract, it can lead to throat and nose cancer, chronic respiratory inflammation and bronchitis, it's definitely something you do not want in the environment, and we were getting it in the open air at levels that require mitigation."

"Formaldehyde was detected at all sites at unhealthy levels," she added

4. Burning Off Gas

Many oil wells bring up large amounts of natural gas along with the oil. Depending on factors such as the operator's ability to capture the natural gas for use and the gas's profitability, the gas may be allowed to vent off or burn off at the site.

In the state of Texas, a new well is allowed to burn, or flare, as much natural gas as they want to. While a permit is needed after the initial ten day period, almost 7,000 of these permits were issued in 2019. Because of the expense the well operator would need to spend in order to collect the gas and move it through a pipeline to the processing facility, they simply are choosing to shoot it into the air.

The burning of the gas emits huge amounts of methane and carbon dioxide in addition to other pollutants. When there are many wells in a small area, the amount of methane and other pollutants is very large. Besides the noise and odor from flaring the gas, there are health issues related to the fumes it gives off. The Natural Institute of Health notes these releases cause headaches, dizziness, vomiting, and lack of coordination. There are other chemicals burned off as well such as volatile organic compounds which are known to cause respiratory and cardiovascular health issues.

In Karnes County, Texas, people are sick from all the health issues they have from all the flaming gas wells and emissions. The constant exposure has given many residents sore throats, headaches, and respiratory tract irritation. The area has so many flaring gas wells that one man sarcastically refers to it as Flare City, USA.

5. Orphan Wells

This word choice is an interesting term for a really big issue. There is a significant problem with old oil and gas wells that have simply been abandoned by their owners. The majority of these stopped producing enough to make them profitable, so the operators moved elsewhere. Drilling has been going on for well over 100 years, and according to the figures the EPA compiled, there are between 2.3 to as many as 3 million abandoned wells across the country. Sometimes they pop up in the most unlikely places, but back when the wells were active, these areas were still undeveloped.

Each abandoned well is a risk, although the danger depends on age, decay and proximity to people. They can channel gas and oil to the surface, pollute streams and drinking water, create explosion hazards when gas seeps into homes and emit climate-changing gases.

In many states, the current 2020 drop in oil and gas prices is causing many wells to be abandoned. In Pennsylvania, the state already has 200,000 abandoned wells from the last century. Past laws never required capping or sealing these wells, so many are simply exposed and still venting gases. Each individual well presents a risk for gas and oil to pollute the surface water and aquifers and create explosion hazards from released gas building up underground or seeping into nearby buildings and exploding.

In Los Angeles, California, in the Vista Hermosillo neighborhood, an overwhelming stench had been driving some of the residents crazy. Unable to locate the source and lacking much help from the city, one man decided to use a jackhammer to see what was under the ground causing the odors. He discovered an abandoned oil well that had never been properly capped was leaking potentially toxic gases. The area and the corresponding four mile stretch between downtown and Hollywood was an old old field known as the Los Angeles Oil Field. According to 2018 records stored by a division of the Department of Conservation, there are 854 orphan wells from the early 20th Century in this area. About 19 wells are located in the neighborhood where they were first uncovered. Explosive and potentially toxic fumes have been coming from these wells. Some of the leaked toxins are linked to respiratory issues and even cancer. The city knew about the 854 wells while all of these residents breathed in these fumes and waited to get sick.

Also in Los Angeles, there are 18 wells sitting unused and unplugged. City inspectors found one of them was leaking hazardous emissions into the neighborhood for two weeks. One family who lives in an apartment complex close to the wells said the family has frequent nosebleeds and headaches in addition to nausea caused by the leaking gas fumes. Over 800 people live within 600 feet of these abandoned open wells and experience the health effect of exposure to the fumes on a daily basis..

6. Transmission Compressors

Depending on what the well is producing--oil, gas, or both--small compressors are needed to compress the gas at the site and push both gas and oil into the transmission pipelines. Usually, these compressors are powered with diesel generators. The burning diesel emits particulate matter that is a known carcinogen. Since these generators need to run almost continuously, they are not only constantly creating noise, they are emitting large amounts of toxic fumes.

Unfortunately, all the gas and oil from these individual wells need to connect someplace and then be moved into the national pipeline system. Their compressor stations are very large, with equally large compressors. The lights will sometimes be on all night, the noise is a perpetual nuisance, and there are continuous emissions of fumes from diesel exhaust.

In Dish, Texas, residents who were tired of the noise, odors, chemical exposure, and lights from large compressor stations near them sued the companies running the compressors. They suffered from health effects tied to exposure to the noise, benzene emissions, diesel exhaust, and other chemicals emitted. The residents had respiratory issues from the emissions, nervous system issues from the constant noise and lack of sleep, and other health issues. While they won their lawsuit in a lower court, the Texas Supreme Court overturned their win on a technicality relating to a filing error with their lawsuit. The industry chalked it up as a big win, and the compressor pumps on.

Colorado requires a new oil and gas well to be 500 feet from a residence and 1,000 feet from high occupancy buildings serving more than 50 people like schools and hospitals. However, that may not be enough.

The Colorado Department of Health and Environment had a study done to see what some of the health issues may be for people living near oil and gas wells. The study determined there were several short-term effects on people who live closer than 2,000 feet from one of these wells, ranging from headaches and nosebleeds to occasional dizziness.

Another study conducted at the medical campus of the University of Colorado showed a correlation between the types of emissions and particulate matter at oil well sites and an increase in cardiovascular issues. The people doing the research looked at over 50 other studies done by well-respected people and peer reviewed. The oil and gas industry dismissed the evidence saying the researchers cherry picked the information.

This is unsurprising as decades-old studies on conventional oil and gas wells have been dismissed by the industry, which claims that newer technology is different and safer. However, nothing has changed but the time.

New wells or old, there are multiple issues. Many of the problems that affected people living near the oil and gas wells before the fracking boom are the same. Not much has changed in regulations requiring pollutants to be controlled at the source. The new oil boom is actually creating even more of the same problem with the addition of the millions of gallons of hazardous wastes it produces.

Entire communities who thought the new boom in drilling was going to lead to a town renaissance have been deceived. The only ones benefiting from the oil and gas wells are the operators making the vast majority of any profits. The pollution is so bad now that, in some places, the gas that normally would be captured and sold is being treated as waste, and the property owner who should be collecting a royalty gets nothing but pollution and an unhealthy environment to live in instead.

The old robber barons riding the wave of oil and gas production are laying waste to any place they are allowed to sink their drill into. The promise of money is seducing many people to sell out only to later realize it was a pipe dream.

Distances to these oil and gas wells that may be considered safer, but certainly not safe enough, is at least a mile. If you are able to put two miles between you and the wells and compressors, do so. The noise from all the compressors will be most likely still heard especially in the night, but hopefully the distance will be enough to not be constantly affected.

Chapter 22. Highways

There are over four million miles of roads in the United States according to the Department of Transportation. That is a whole lot of pavement. While a large percentage of these roads are in rural areas, there is a sizable chunk in our urbanized cities. Sprawling development has opened up the borders of cities to encompass miles and miles of suburbs surrounding them and stretching far from the city center. Large highways with multiple lanes clogged with traffic bring tens of thousands of exhaust-belching vehicles into and through the areas surrounding these highways.

Sitting on one of these busy highways in a traffic jam with the window open feels like clearly a bad idea. Shutting the window and not breathing all the fumes seems an obviously better idea. Unfortunately, too many people live or work next to a constant, never-ending flow of traffic spewing out exhaust fumes 24 hours a day. Emissions from cars, trucks, and buses contain a long list of harmful chemicals: carbon monoxide, sulfur dioxide, benzene, formaldehyde, particulate matter (especially ultra-fine in size), and dozens more. Many of these have been linked to various types of cancer. If all these invisible exhaust fumes were suddenly visible, we would see a thick cloud many feet high and wide on all of our highways.

Almost all the other hazards in this book impact a few hundred or even several thousand people at a time. Unlike those other hazards, the impact of negative health effects from highway pollution impacts millions of people every day of the year. The American Lung Association estimates between 96 and 130 million people, 30-45% of the population, in this country live near busy roads. Other estimates include one from the EPA, which puts the number of people living within 300 feet of roadways and terminals at 45 million individuals.

Living near a roadway could put people's health at risk. Elevated levels of chemicals from emissions from chemicals from cars and trucks contribute to tens of thousands of deaths each year. Even though we cough and wheeze, we tell ourselves the fumes we breathe are more annoying than hazardous. That is one big mistake we are making.

In a study Tony and I did of people living near highways, when they were asked about their concerns, less than half talked about health concerns from the roadway. An equal number were surprised to hear just how badly the pollution affects them. In spite of the widespread scientific knowledge about the dangers of exposure to highway pollution, millions of children, adults, and elderly people live, work, and attend school in the toxic zones near these roadways. Conclusive evidence shows links to decreased lung function, increased asthma, low birth weight and premature births, and cardiovascular disease. Making matters worse, cancer risk from the chemicals contained in auto emissions has not received the attention it deserves from the government and environmental agencies charged with protecting the health of the public.

Brown smog consisting of ozone pollution and clouds of fine and ultra fine particles present the greatest hazards to our health. Many of our large cities, and even some smaller ones, now have this haze of smog covering the denser areas. In Phoenix, Arizona, where I live, the air frequently is so heavy with particle emissions that the government asks people to carpool or not drive at all. We regularly exceed the ozone limits due to the pollution that hundreds of thousands of vehicles spew into the air.

The worst areas where the pollutants are the highest and most dangerous are in the actual travel zones, about 50-100 feet from the roadway. Heavily trafficked city roads with a lot of stop-and-go traffic can be as bad for pollutants as major highways. Roads with tall buildings on either side trap the pollutants and make the concentration many times higher than an open highway. Studies from the University of Southern California already show these hot zones have pollution levels at five to ten times higher than other places in the city.

Pedestrians along roads like we described are walking in the worst zone. In fact, at street intersections, the levels of pollution are even higher. People who live along these streets and have their doors and windows open are breathing the same amount of pollution as down on the roadway. Bicycle riders, joggers, and walkers are all subjected. People who exercise along these roads are putting their lung function at risk, breathing in many times more fine and ultra fine particulate matter.

There's even a difference in the heaviness and thickness in the smog between a weekday and a weekend. When thousands of vehicles are clogging the road, crawling along to wherever it is they go each day, they are causing the pollution to be the worst. Then, on the weekend when traffic is much lighter and mostly moves at a steady pace, the smog lifts a bit, and it becomes easier to breathe.

Time of day can even affect the level of smog. In April 2013 a study done by UCLA and the California Air Resources Board found that heavy traffic in the morning rush hour created pollution levels so high that the effects were seen up to a mile from the highway. The automobile exhaust created visible smog clouds of pollution. Billions of tiny microscopic particles are needed to create a cloak of smog that thick.

Smog is made up of billions of tiny particles, pieces floating in the air from all sorts of chemicals, toxic compounds, dust, etc. Particulate matter especially causes lung cancer, asthma, leukemia, and heart disease in adults. Children who are more active and also tend to be outside more of the time are getting more of a share of exposure. Their small bodies with still-developing lungs are subjected to more of the particulates than older, more sedentary people. Women who live in these high-traffic, high-pollution zones are at a statistically higher risk of having children born with some type of lung and respiratory issue. In fact, studies have shown evidence that pregnant women living within 500 feet of these roadways have a higher chance of having a premature birth, babies with low weight, and children with more medical problems overall.

Smog and exhaust pollution is literally sucking the life out of these people bit by bit. Pollution from smog makes the body work harder just to do its regular functions, putting continuous extra strain on all of the vital organs.The fumes that are breathed in take the place of oxygen needed to refresh the blood, brain, and body. They rob brains of well-oxygenated blood and cause disruption in normally well-functioning brain activity.

The exhaust that creates all the pollution that is putting so many people at risk is made of a variety of chemicals and residue from the combustion of gas in the vehicle engines. To get more technical, the exhaust and pollution caused by vehicles burning petroleum send hundreds of different chemicals into the air. Many of these chemicals have the capacity to combine with others to create even more hazardous substances for us to breathe. The chemical substances found in automobile exhaust reads like a shopping list for a chemistry lab.

To keep things a bit simpler for this chapter, we will focus on a group of four that are responsible for the majority of diseases, illnesses, and other health issues.

1. Particulate Matter

Particulate matter is a primary danger. Our lungs act as filters for our body, keeping harmful substances out. Our lungs are only able to filter out particles larger than a certain size. Others get through our bodies' defenses and either need to be caught by our secondary ones or get into our bloodstreams and stay deep inside our lungs.

Incredible amounts of particulate matter are emitted by every combustion engine there is. These are constantly emitted in automobile exhaust, pouring into the air, and mixing and attaching to whatever other particles happen to be there. The size is critical to what happens next.

Particles in our air are categorized into different sizes. The largest, called coarse particles, range in size from 2.5 to 10 micrometers. In comparison, a human hair is about 75 micrometers thick. Our bodies are able to trap most coarse particles before they get deep into our body.

The next size are particles between 2.5 and 1 micrometer. It would take 30-75 of the particles next to each to equal the thickness of a human hair. These particles do not always get trapped by the body and sometimes make their way into our lungs. Some may be caught in the lungs and expelled, but many will make it through the air sacs and into our bloodstream where they get sent to our organs and other parts of our body. These smaller particles can pass deep into our lungs, causing respiratory complaints and contributing to the risk of developing cardiovascular diseases.

The last category is particles smaller than 1 micrometer, ultra fine particles. These are the most dangerous, the ones that cause the most harm. Particles of this size pass completely through any defenses our body has. They are breathed into the lungs and head on through into the bloodstream. These tiny particles wreak havoc on our internal organs, similar to fine particles, but are able to pass into even more vital areas of our bodies.

Some of the particulate matter emitted by vehicles is a product specifically called soot. This is partly burned fuel emitted mostly by diesel fuel burning vehicles but is also created by reactions between other pollutants as well. While most newer diesel cars have special particulate filters to reduce the emission of soot into the air, the large number of very busy roadways packed with these vehicles still creates very large quantities of soot, which contains chemicals specifically linked to cancer. The amount of these emissions is so concerning and dangerous that some countries are trying to phase out diesel engines in an attempt to reduce these risks.

According to research done at Cambridge University, another source of this harmful particulate matter is also adding to the levels of fine particulate matter. Brakes on automobiles and trucks have been found to emit metal particles as the brake are used. The pads are composed of different types of metal, especially iron. Dust is created by the friction of the brake grinding on the rotor when the vehicle slows down. The dust goes into the air and, when inhaled, causes respiratory tract inflammation, especially in the lung cells. This important finding may show brake dust to be as severe as fine particles from diesel fumes.

2. Polycyclic Hydrocarbon

These chemicals are emitted from the exhaust system as unburnt components from fuel. They hang in the air around the highways and roadways and react with nitrogen oxides. The reaction, usually caused by sunlight, creates different chemicals including ozone, which is known to cause breathing problems, irritate respiratory systems, and increase symptoms and attacks in people with asthma.

3. Benzene

Benzene is classified as a cancer-causing pollutant. It is one of the most dangerous emissions related to vehicle exhaust. Busy roads and highways have levels of benzene that are extremely hazardous and affect the respiratory system, immune system, and central nervous system, as well as causing developmental problems. Long-term exposure to benzene in vehicle exhaust is directly tied to cases of birth defects, childhood leukemia, and other types of cancer. It has also been known to lower the count of red blood cells in the body, which can lead to anemia. Even relatively low levels of exposure are known to cause many of these illnesses. Besides occupational exposure to high levels of benzene in the oil and gas industry, the main route of exposure is through highway pollution, gas stations, and oil and gas wells. The chemicals may be inhaled directly, or they can be carried along with particles and inhaled as dust.

4. Carbon Monoxide

Another dangerous chemical found in large quantities in vehicle exhaust is carbon monoxide, created from incomplete combustion of the gas in the engine. On busy roadways the carbon monoxide levels are high enough much of the time to adversely affect the health of people living nearby. Carbon monoxide is inhaled and travels through the lungs into the bloodstream. Once in the blood, it reduces the amount of oxygen the body is able to deliver to the organs and tissues. The shortage of oxygen causes severe headaches and problems with the proper functioning of the respiratory system. People with heart issues and cardiovascular diseases are especially affected as the reduced level of oxygen available to the body makes the cardiovascular system strain and work harder. Long-term exposure to higher levels of carbon monoxide may also lead to poor learning ability, reduced dexterity, visual impairment, and more difficulty performing complex tasks.

5. Diesel Exhaust

Exhaust from diesel engines is a large part of the pollution on the highways. These engines burn particularly dirty, sending much more hazardous materials into the air from their exhausts than regular gasoline engines. Highways in particular have a larger number of vehicles using diesel engines on them. The exhaust from diesel engines contains at least 38 known chemicals listed by the EPA as known carcinogens.

All of these chemicals, and dozens more, are present to varying degrees in air polluted by roadway traffic. Inside houses within the high pollution zone by the highway, the air can be almost 70% as polluted as outside the building. Depending on the structure, there may not even be much difference since an average building completely exchanges the air with outdoor air every hour or so.

It is fairly easy to figure out if you are living near this type of a hazard. Do not be fooled into thinking very busy urban streets are safe just because they are not highways. The relevant factor is an elevated level of chemicals that stays high above normal levels for long periods of time.

An article in the respected journal *Scientific American* discussed studies in the U.S. that show babies born in areas with high levels of diesel exhaust and other airborne toxins were more likely to have autism than areas with low pollution levels. In addition, many of these chemicals have been linked to various types of cancer and other diseases.

Some states have gone so far as to have their toll collectors in booths containing special ventilation equipment. They know how unsafe it may be to breathe these fumes continuously during their shifts. People living along the roadways are not able to wear breathing masks or have air purifiers running all the time but have to breathe these harmful fumes. Too bad they are not afforded the same protection.

Because so many of our cities and urban areas are covered with high volume roadways, the general view is that nothing can be done; people seem to have accepted this is the way it needs to be. However, this idea comes from industries and institutions that benefit from our collective defeatism. The automobile industry wants to sell more cars because people driving less or owning fewer cars hurts their industry. Gasoline companies want us to drive as much as possible as well, encouraging us to burn those petroleum products and keep supporting the gas and oil industry.

For over 20 years there has been knowledge in the scientific community about the negative effects of vehicle pollution on people nearby. Unfortunately, the government simply has ignored it and pushed it aside instead of pushing laws and regulations with real teeth in them. The lack of urgency surrounding the problem kicks finding a solution further down the proverbial road. Our current laws and regulations do almost nothing to protect the population from the daily assault of toxic pollutants. The government claims they monitor the pollution and air quality, but their methods of monitoring are set up to measure pollution over a wide area, not the levels in the hotspots.

Tony and I know that people simply can't pick up and move at any time they wish. For the people who are living within the danger zone of a highway, we wish you could, but that simply is not realistic. This chapter is to help you see what the impacts of this could be on you and your families and the steps you may be able to take to lower the exposure and the risk.

While you can't do much about the outside air if you live in one of these places, you can try to take some steps to limit the most harmful exposure. Short of moving, you can try to limit your time outside doing any type of exercise or physical activity, such as jogging or playing basketball. If possible, try to find another place further away from the traffic for those activities. Also, if possible, stay indoors during peak traffic times when pollution levels are at their highest.

The inside air is something you do have a little more control over. If you have a central heating and cooling system in your house, you could consider adding an air filter system to what is already there. A high-efficiency house filter would reduce the indoor pollutants in your home, including particulate matter. However, filters are only effective at removing some of the pollutants, and different filters work better on certain chemicals than others. Hepa filters are particularly good at reducing the smaller particles that could get into your lungs. Carbon filters are better at reducing benzene and some other volatile organic compounds found in the exhaust vapors. Comparison shop, but be careful about the information and claims individual companies make about their products. In addition, they will only work the most effectively when all the doors and windows are closed. At the very least, it is a good starting point and better than nothing at all.

If you do not have a central heating and cooling system, you can at least get several room- sized air filters to run in the areas you and the family spend most of your time in when inside.

One of the best ways of lowering your risks to pollution from highways is to oppose highways near residential areas before they are built. With enough pressure from citizens, the government has made certain highway expansions illegal. For example, in 2003 California created a law mandating new schools must be at least 500 feet away from busy roadways. On the federal level, new EPA requirements put into place would force over 100 cities to place air quality monitors in high pollution zones close to highways and heavily-trafficked roads.

When there aren't laws in place to protect people from encroaching roadways, there are still actions to take. Highways take many years to build. The plans may stretch out five, ten, or even twenty years on the actual construction. Take the extra time to triple check there are no plans on the drawing board of any agency to either build a future roadway or greatly increase the volume a current roadway carries.

Do not trust anyone who says the proposal is years away from happening or that the road will never be built. There is a reason to fear. Consider, for example, Route 303 in Surprise, Arizona, and Route 208 in Oakland, New Jersey. For many years, these roads were nothing more than lines of dots on a map. Then, one day, the construction signs went up, and the bulldozers came in. Now, hundreds of people have highways carrying thousands of vehicles a day just yards from their homes. It will never happen, people said. Then, it did. You always need to use your own judgment and gut feeling. You are the only one who will be living with the consequences.

Living near a highway or busy roadways is almost inevitable in many cities and towns. Many times there is little choice based on how many roads there are. The amount of the traffic is really what you need to consider. A highway with heavy traffic all day long especially stop-and-go traffic would likely need a distance of a half mile just to reduce the levels of pollutants you would be exposed to. Other roadways that are just fairly busy but not jam packed all day long, you may want a 1,000 foot buffer between you and the road. Obviously you will be living off a roadway wherever you are. The less exposure you have to high volumes of exhaust pollution, the better off you will be.

Chapter 23. Agriculture

The agriculture industry of today looks nothing like it used to decades ago. As late as the 1970s and '80s, family farms were seen everywhere and made good livings selling high quality, good-tasting produce in a localized area. As the costs of land got more expensive in what had once been rural farming areas (but now were becoming the suburbs), a larger number of smaller farms started to close down. Real estate prices had skyrocketed, and small local farms were selling their land and cashing out. In the larger farming areas of the country, family farms, which had always been the backbone of the industry, started to get squeezed out by corporate farms that were able to grow much more product in the same size space. Farms in the suburbs became housing developments. In addition, corporate farms started buying up neighboring farms to increase the size of their operations even more. These large corporate farms were able to ship produce further out from the farming areas, putting even more strain on local and regional family farms. Efficiency became more important than anything as the size of corporate farms grew larger still. All the different varieties that once were the most popular and tasty were dropped in exchange for varieties that travelled better in trucks and stayed longer on grocery shelves.

The dawn of the agricultural chemical age at the end of World War II brought the use of chemicals into agriculture and it experienced explosive growth. Small farmers in the past used as little of these chemicals as possible and used real farming methods instead of simply dumping and spraying chemicals. These included rotating crops, planting cover crops to add nitrogen, using integrated pest management, and combining livestock and crops. They knew the land was their most important asset and did not want to destroy and contaminate it or the food it produced. Corporations, however, went the opposite direction. Using chemicals allowed them to grow more products at a cheaper cost, maximizing profits. When the land became depleted, they simply threw down even more fertilizer to stimulate it again. Routinely, corporate farms squeezed every bit of production they could from every acre. Through the years that followed, due to the enormous use of fertilizers, pesticides, herbicides, insecticides, and fungicides, bigger farms have been able to produce the perfectly round and shiny food products that we have come to know and expect even though, according to sciencemag.org, most have become totally tasteless, such as the corporate farm tomato. At some point, we allowed the agricultural industry to trade in good-tasting produce for large quantities of cheap food.

Today we apply 40 times more nitrogen fertilizer than we used 75 years ago while crop yields have only doubled according to publicintegrity.org. These millions of pounds of chemicals are going into the groundwater we drink, the air we breathe, and the nearby waterways we depend on. Plants can only absorb so many nutrients at a given time. The excess fertilizer that is not utilized by the plant raises the nitrate levels of the drinking water to an amount that is a danger to children and pregnant women. The excess chemicals in the fertilizer that are not utilized or absorbed by the soil and crops are also washed off into the groundwater and waterways. They will eventually wind up in our water supply and consumed by us in our contaminated drinking water.

Many large farms now also use manure and sewage sludge as a cheap fertilizer for their crops. Land-based disposal of sewage sludge from municipal sewage treatment plants has become an especially preferred method for disposing of the sludge. It is much less expensive for the farm operators than spending money on fertilizer.

Special caution must be taken when sludge from public sewerage treatment plants are routinely applied. While there are pretreatment standards in place to limit industrial hazardous material being discharged into public sewage treatment plants, they are neither always followed nor adequate.

This problem extends to any operation that grows a crop and uses pesticides and chemicals on the land. These crops can range from field crops such as corn or wheat to orange trees and almond groves. Even when the land is no longer used for agriculture, it still may present a danger. Unfortunately, too many people did not realize in time the danger lurking decades after waiting. While many of these are lovely to look at, there are several significant areas that may prove to be dangerous and a risk for our health.

1. Pesticide Drift

One of the most dangerous risks from any farming operation is direct exposure to the full-strength chemicals being applied. American farmers use more than 300 million pounds of pesticides alone on farmland each year. Out of the top 27 they use, the EPA considers 15 of them to be carcinogens or cancer-causing compounds.

These airborne materials have been documented to have traveled distances up to 1,000 miles from where they were originally applied. The areas closest to these farms are, of course, exposed to the highest concentrations of these airborne chemicals. If you are living in or moving into an area that has a lot of farming, it is wise to investigate what crops are grown and if possible what chemicals are applied. Some farmers, if asked, will notify surrounding landowners of what chemicals will be applied and when they will be doing it. In some states when restricted-use chemicals are being used by chemical application companies, they must place an ad in the local newspaper when the intended application will be taking place, and concerned individuals can contact the company to find out the exact chemicals that will be used and the location of lands that are going to be sprayed. It would be advisable to investigate this for your protection.

The government in 2019 proposed making some changes in the rules governing pesticide drift. The new rules lift protections to people living nearby and allow toxic pesticides to drift further and wider away from application areas. The EPA says the new rules will make the problem of pesticide drift easier to control. Groups such as the Pesticide Action Network and Beyond Pesticides, among many others, say the new rules will needlessly expose thousands more people to harmful chemicals.

All kinds of hazardous chemicals are applied to the fields at a staggering rate. Herbicides, pesticides, fungicides, and insecticides are routinely applied to fields. Much of this material never makes it onto the plants or the ground. The winds pick up much of the material and transport it off the farm and into the surrounding areas. Crops that are heavily sprayed with chemicals, such as apples and other types of fruit trees, are especially prone to have large amounts of materials drifting away with the wind. Since the chemical has to be applied directly to the trees and the fruit of the trees, large sprayers and misters must be used for this operation. Of course, it makes sense for the spraying to take place when the wind is as still as possible, but even so, the materials still drift. The larger the operation, the more chemicals need to be applied to cover the area. Crop dusters are frequently seen spraying cargo loads of chemicals onto the fields below. Clouds of drifting pesticides are also frequently seen moving off the fields and into communities and housing areas nearby.

In Tulare County, California, a group of homeowners told stories about how pesticides applied on nearby crops would drift onto their property. Sometimes, the cloud was so thick it would leave a film on the children's playset in the backyard. Many of the people talked about being sick, vomiting, and having skin irritations. Others talked about the children suffering from burning eyes, coughing, and having asthma attacks from the pesticides.

In Indiana, the office that enforces state pesticide laws logged 97 incidents between 2010 and 2012 in which the pesticides being applied on farms violated drift regulations. Most of the damage is usually to other property crops; however, at least a dozen times, exposure made someone sick.

The Northwest Center for Alternatives to Pesticides reported a substantial increase in the number of people experiencing pesticide drift health issues. Data from the Washington State Department of Health reported 22 occasions when pesticide drift events sickened a total of 129 people in 2014. Some of the effects of exposure to concentrated amounts of pesticides include trouble breathing, nausea, vomiting, burning eyes, and fainting. People who experience drift exposure sometimes have sore throats years after the initial exposure occurred.

In October, 2015, about two dozen students and staff at Coachella Valley High School in Riverside County, California, reported feeling sick after a nearby farm grower applied a pesticide near the school. When the wind shifted, the chemicals drifted towards the school and over the students and staff. Medical teams need to respond as 28 students and staff were affected by the chemical. Dizziness, nausea, irritated throats, burning eyes, and vomiting were all symptoms related to exposure from the pesticide.

In Grant County, Washington, an apple orchard applied pesticides using an air-blast method in March, 2015. A school campus was located adjacent and the chemicals drifted onto the campus causing an incident requiring emergency services to respond as people on campus experienced burning in the eyes and throats as well as respiratory problems.

2. Groundwater Contamination

An aspect of agricultural operation that may present a potential hazard is chemical residue. A large amount of residue remains in the ground on the farm and leaches into the groundwater. Some of these materials are long-lasting in the environment. They do not break down for many years and may accumulate to very high levels over time. DDT, chlordane, and 2,4,5-T, are several materials that were routinely applied for many years before the associated health risks became publicly known, causing the use of these materials to be prohibited. In spite of the fact that it has been quite some time since the actual application has taken place, they remain in the soils and waters, still wreaking havoc today.

There are hundreds of chemicals being applied to the agricultural land, and many have not been extensively tested and studied to determine their effects on humans. Herein lies the problem.

Because of the lack of studies, the safety of these chemicals cannot be determined. A few have an EPA maximum contaminant limit in drinking water, but many more do not and are totally unregulated. In addition, the synergistic effects of multiple chemicals consumed from drinking water may even be more dangerous than if the chemicals were separate. A very sadly little amount of research is ongoing on more than a few chemicals.

The Environmental Working Groups Tap Water Database shows that Atrazine, a toxic pesticide that is a known hormone disruptor is in water being supplied to about 7.6 million people. In 2015, their data showed over 800 public water systems exceeded a public health standard set by scientists in California. Again, the standard is non-enforceable as the federal government does not take action to make it so.

In Illinois, Nebraska, Iowa, and Minnesota, the levels of atrazine found in drinking water are high enough to be linked to birth defects. The average level found was more than 5 times the allowable level set in Europe where significant increases in birth defects occurred. Links to cancer have been found as well.

The potential for groundwater contamination is not just limited to pesticides and fertilizers; there are other areas that may cause an issue as well. When thinking about farms and farming, the subject of gasoline usually does not come up. Like any industrial operation, which is what a sizable farm is, they use lots of gas to operate the machines. Much of the farm equipment and all of the tractors run on gasoline. Almost all large agricultural operations need to maintain an onsite supply of gasoline for their equipment. Most of the tanks on these farms are metal, and over the years the metal has corroded. It is not uncommon for a storage tank on a farm to be over 20 years old. Similar to the issues gas stations faced with old rusty tanks, there is a distinct possibility of leakage and contamination here as a result of these tanks.

On top of those hazards, the application of animal manure and sometimes sewage sludge as supplemental fertilizer is common. While there is nothing wrong with using animal manures or treated human sewage, there are definite issues that occur if precautions are not taken. When human sewage that has originated from public sewage treatment plants is used, there is a distinct possibility that the sewage sludge may have high levels of heavy metals or chemical contaminants that are routinely discharged by industrial customers. A spokesperson for the University of California Extension Service advised caution about heavy metals when sewage sludge is used on agricultural lands. While businesses are supposed to pretreat their wastes, all metals are not prevented from entering the waste stream to the sewage plant. Heavy metals in the soil may leach through the surface. These heavy metals accumulate in the soils of the farmlands on which they were applied as do the chemicals contained in the sludge, which reach dangerous levels that can quite possibly contaminate the groundwater. Evaluations of the amounts of heavy metals that were available to leach through the soil and into the groundwater varied depending on two factors. Where the application of sewage sludge on the land was done primarily for waste disposal and the quantity greater than the amount the soils could absorb, higher levels of metals were found deeper in the soil closer to the water table. In addition, in areas where there is regular precipitation, the metals were found in even higher levels.

In Marinette, Wisconsin, treated sewage sludge from the Marinette wastewater treatment was spread on the farm fields. The state Department of Natural Resources found the waste contained unacceptable levels of PFAS, a chemical used in firefighting foam and now considered a possible carcinogen by the EPA. Nearby homeowners are very concerned about their well water after samples taken showed high levels of PFAS as it is a forever chemical: it never leaves your body. One of the wells tested showed over 57 times the level set by the EPA as the highest before concerns over toxicity are warranted. Unable to remediate the private wells, the manufacturer of the firefighting foam is providing bottled water for homeowners and not discharging the chemical to the wastewater treatment plant. In the meantime, the farm fields are contaminated and the groundwater will most likely be for a long time to come. The town is looking into extending public water to the affected area.

Extreme cautions of this chemical contaminating water supplies across the country is leading to concerns about continued use of sewage on agricultural lands, or any lands for that matter. In Vermont, the Department of Environmental Conservation found seven out of 34 sites with sludge applications had levels of PFAS over state groundwater standards. A coalition of groups is calling for testing of all sewage sludge prior to application on fields where it may lead to additional groundwater contamination.

In addition to sewage sludge, manure from animals, used for centuries, works great as a fertilizer. It is especially good at helping to replenish the nutrients in the soils that the crops took away. When too much manure is applied, however, runoff from the fields gets into the local waterways and also contaminates the groundwater that is needed as a water supply by nearby homes.

In Wisconsin, the USDA tracked manure runoff from a field to the tap water of a nearby home after the water ran brown. Three days later, genetic testing showed the same bacteria traveled 1,500 feet through groundwater to another home's well.

In San Joaquin Valley, California, almost one quarter of the nation's food is grown. As a result, the valley has been overfarmed. Hoping to get some of the nutrients back in the soil, the operations added huge amounts of fertilizers and manure, which are hazardous in high levels. The manure has been piled and spread all over depleted farm fields and orchards so much so that testing of communities' wells in the valley were detected at almost 500 times unsafe levels.

3. Soil Contamination

Fruit orchards were routinely sprayed with lead-based arsenic pesticides back in the 1990s. Many of the toxic compounds continue to persist in the soils where orchards remain, but particularly where they have been developed into housing communities.

The industry especially liked a group of pesticides called organochlorine pesticides because they didn't wash though the soil but actually attached to the organic matter in the soil. Like the banned toxic pesticide DDT, it does not break down and, when ingested, also stays in the fat cells of any animal, such as it did with humans and many birds. Unfortunately for humans, when chemicals such as Aldrin is used in addition to DDT, the health effects on the human body are even greater causing the DDT to remain much longer. Many thousands of pounds of these chemicals remain in agricultural areas long after they were banned.

In Washington state, the government has tested and mapped out a total of 187,000 acres that are contaminated with arsenic-laced soil. A subdivision outside Yakima was constructed on one of those old orchards. Sadly, both the developer and the government knew the property had tested for lead and arsenic at more than double the level the state recommends cleanup. Unfortunately, nothing was done, and none of the homeowners were informed about the high levels of contamination. As a result, the soil at hundreds of properties contains levels of lead and arsenic that, through long-term exposure, can lower children's IQs, cause behavioral problems, or increase cancer risks later in life. No one knows for certain yet what the long term consequences of this exposure will be.

In Saunook, North Carolina, a small town 30 miles from Asheville, a housing development built on an old orchard has been declared a Superfund site. For decades, pesticides containing lead and arsenic had been used there. In addition, DDT, DDE, and benzene hydrochlorides, had been applied to the orchard to control pests. When the state Department of Agriculture tested wells for contamination, 32 out of 80 wells had these chemicals above acceptable state levels. Recommendations to avoid contact with soil and water were issued.

A farming community or a community near an agricultural area in the suburbs can be a great place to raise a family. Using some caution and knowing there are chemicals and other potential areas of risk is the best course of action. My children loved when we lived across the street from a small dairy farm in upstate New York, especially seeing the cows at the fence every morning as they waited for the school bus. We knew the people who owned the farm and saw how they handled the chemicals and manure from the animals they used on the fields. Yes, we needed special filters on our well water, but we did live in a rural community with quite a bit of farming.

That is the main takeaway: know what is there and see it with your own eyes.

There is a giant difference between farmers who use responsible farming practices and farms whose sole interest is squeezing every last bit out of the soil they can and disposing of waste as cheaply as possible. Responsible farmers do not spread manure on their fields until it is running off the lands. Responsible farmers do not spray their fields and trees when the weather can carry pesticides and other chemicals to nearby homes and neighborhoods. Asking around the area is one way to see what you may be getting into. You may find out people's water is already contaminated and the farm is a bad neighbor. Using caution could help you from becoming another one of the many people poisoned by the farm next door. There are not too many options for people who are already living in places like these. Frequent testing of your water supply and staying inside when you see pesticides or manure being applied will limit your exposure. A distance of a half mile at least is advisable.

Chapter 24. Brownfields

A term you need to become familiar with when checking for environmental hazards is "brownfields." To most of us, a brownfield sounds like a description for a field that is not green, like the hills in California in the summer or in upstate New York in the winter. From what some people have told us, the term came from the real estate industry to distinguish developed property (brown fields) and undeveloped property (green fields). For our purposes, however, it has a totally different meaning.

"Brownfield" is an environmental term used to describe the current state of a specific piece of property. According to the EPA, "A brownfield is a property, the expansion, redevelopment, or reuse of which may be complicated by the presence or potential presence of a hazardous substance, pollutant, or contaminant. It is estimated that there are more than 450,000 brownfields in the U.S."

Hazardous waste sites and brownfields are basically the same thing. Both of these are likely contaminated with chemicals or other substances considered a hazard to humans and the environment. However, once a property is designated as a brownfield, it gets treated a bit differently than if it was still a regular hazardous waste site. Somebody decided that if the contamination of a brownfield were cleaned up, the property could again be used for some limited purpose. What those limits are depends on the extent of actual contamination and how clean the agencies responsible for the cleanup decide the end result needs to be. The big difference is that when a hazardous waste site gets designated as a brownfield, funds then become available to the prospective developers to evaluate the extent of the pollution and to clean up the site and return it to some type of productive use again. It also takes away any future liability from the developers if more hazardous waste is found in the future or if people become sick from exposure to the site.

The full extent of the contamination at any hazardous waste site, brownfield or not, is never really known until the cleanup is started. Assessment which is the first step may only show part of the contamination. However, most times there are significant environmental issues needing to be addressed and rectified. Sometimes, the extent found at a brownfield site is so bad it is reclassified as a hazardous waste site and development plans are canceled.

There are examples of these contaminated brownfields everywhere. Many brownfields are old manufacturing facilities with toxic waste left onsite or in the soil. They are closed chemical plants with old drums left behind, sitting and leaking in long-closed warehouses. Thousands are dry cleaners who dumped toxic cleaning fluids down their drains and leaked fluids into the soil for decades. There does not need to be enormous amounts of toxic waste to make a place hazardous and make it eligible for a brownfield program. Sometimes, it may only be a half dozen drums of a hazardous chemical or a hole behind a factory which was used to dispose of old transformers with PCBs. The EPA estimates there are over 450,000 of these brownfield sites across the country.

An entire sub-category of these are called petroleum brownfields. There are thousands of these gas stations, DPW yards, and similar places where rusty old underground storage tanks contaminated the soil and groundwater with oil, gas and other petroleum based products and chemicals. The EPA estimates out of the 450,000 of these brownfield sites across the country, almost 225,000 are considered these petroleum brownfields.

The majority of these sites have been sitting vacant for years, contaminated and unused. Some of these sites are still owned, but the owners are unable to develop them because the extent of potential contamination makes it too expensive to do so. The owners can't sell these properties either because nobody wants the responsibility that the hazardous waste present. If these sites are able to receive a brownfield designation, that would help in several ways.

Owners of properties designated as brownfields may benefit from the cleanup of the site being partially funded by money available from the federal and state government. In addition, the property is more valuable to the owner if they are able to reuse it or at least sell the property with a clean bill of health. Many times the owners are simply unwilling to clean up the site as the cost may be far more than the value of the property. Then, the contaminated land simply sits in the neighborhood with its toxic chemicals still creating potential hazards for the people living nearby.

In many cases these brownfield sites are located in urban areas that are run-down and blighted. The mass migration of people from the cities to the suburbs was instrumental in the desertion of many of these properties. During the period of this migration, the urban areas started to become run-down. Over time the businesses there began to move out to the suburbs too, leaving behind old factories and buildings that in many cases also had large amounts of pollutants from years of use.

In an effort to revitalize the area, city leaders frequently look to work with developers in building on these sites. The benefit to the community is to get these properties back on the tax rolls and producing revenues. The cleanup and redevelopment of these sites encourages other businesses into the area. Property values and job creation often increase in areas where large brownfield sites have been redeveloped into sprawling multi-use projects.

Even though this is an improvement in many ways, it doesn't come without consequences. One negative effect we especially would like to mention. Along with higher property values come higher taxes and, frequently, gentrification of the area, pushing out the people who lived there during the bad times, people who are now unable to stay during the good times.

Legally speaking, the people primarily responsible for the costs of the cleanup are the polluters. Many times, however, those companies have long ago gone out of business. Sometimes, multiple companies have operated on the same property over the years, making it virtually impossible to figure out how much each company is responsible. In the end, it usually falls back on the community to try to arrange funding through the developer and the federal and state agencies with clean-up funds.

The costs of a cleanup can range from thousands of dollars to tens of millions. As a result, many of these sites that will be cleaned up are located in prime urban areas where the property values will substantially increase once clean and developed. Unfortunately, many of the over 400,000 brownfields, especially in rural areas, will stay polluted for some time to come.

Many residential areas are in very close proximity to brownfields. It is important to know where there are brownfields near us and what hazards they present. However, it can be very complicated to determine the extent of the risk of living near a brownfield for a number of reasons.

1. Future Discoveries

Sites designated as a brownfield first were listed as hazardous waste sites. There is knowledge of known substances on the site that make it hazardous to people and the environment. The quantity and exactly everything that may be at the site is still unknown. When the site gets the new designation as a brownfield, a site assessment study is done to get a better idea of what could be on the site. The assessment does do some onsite investigations and testing, but the actual excavation or removal of hazardous substances has not yet begun. Any site listed as a brownfield is on the list for a reason. It means the chances of something being discovered during the mandatory environmental assessment stage is quite high. It also may mean there is a greater chance that substances not discovered during assessment may turn up and in far greater numbers once remediation begins.

In brief, part of what makes brownfields dangerous are the unknown hazards yet to be discovered, and the known hazards that are allowed to be left at the site after meeting the lower brownfield cleanup standard.

In Tonawanda, New York, a portion of the old Tonawanda Coke manufacturing facility already designated a Superfund site is being moved to the state brownfield program. A portion of the site is being subdivided for a developer. They want to construct a project on it and if the property remains a Superfund site the cleanup process would be lengthy and much more costly. Moving the heavily contaminated property to the brownfields program also would take Honeywell, the company responsible for cleanup costs off the hook for more money. Critics say changing the designation of the property doesn't change the chemicals that already are known to exist and more likely will be discovered. The cleanup should not be shifted to brownfield status to allow a private company to leverage the system for their advantage. So eager are city officials to return a toxic property to tax paying status, they are willing to allow a developer to bypass normal procedures for ensuring the public is protected from these hazardous waste sites before it becomes too late.

It is of extreme importance that you are not lulled into feeling the term "brownfield" means the site is safe. Part of what makes a brownfield dangerous is that quite the opposite may turn out to be true. A brownfield during the active investigation phase may yield the finding of substances equal to the worst Superfund sites out there. Do not allow anyone to tell you a brownfield is safe or low risk until all remediation is completed. Even then, it may still present threats of some level. Very frequently, asbestos, lead, and PCBs, as well as petroleum products like benzene, cadmium, and chromium are found in these brownfields. Just because the hazard is unknown doesn't make it non-existent.

2. Exposure to Hazardous Materials

Living near an uncleaned brownfield site may expose you to hazardous materials such as industrial solvents, mercury, asbestos, and paint sludges. As a designated brownfield, the level of contamination would be high enough to create a risk hazardous to human health. In many cases, piles of disposed chemicals may be blown through the area into the neighborhood. Sometimes the site is not protected, so the neighborhood children use it as a play area. In many cases, materials may have leaked into the soil and may be entering the air in your home through the foundation and floors. Sometimes, there are exposed rotting drums of hazardous waste left sitting on the property.

In Washington County, Pennsylvania, in the middle of the county seat there is a contaminated brownfield sitting right in the middle of the residential neighborhood. An old car dealership that became a body shop has been sitting open to the neighborhood for years. The previous owner walked away from it, turning it over to the county instead of paying back taxes on the polluted land. Petroleum washed off the site during a rainstorm, and an old underground tank out back needed to be removed and the contamination remediated. The brownfield is completely surrounded by residential homes. Authorities have concerns the contaminants have migrated off site and beyond the immediate homes but, offers to test properties nearby have been rejected by the owners. It is unknown if any residents are having any health issues yet from the contaminants. In the meantime, the land still sits, contaminated and open to the neighborhood.

Sometimes brownfields are reclaimed from old contaminated places to be used for recreation and parks. Just because there were some cosmetic changes does not mean anything underneath has changed. In Washington, D.C. an area used as dumping grounds for old landfill holding as much as four million tons of waste has been undergoing a transformation for years into a supposed park for the mostly poor Anacostia section of town. Back through the 1960s, in addition to use as a landfill in the past, toxic dredgings from the nearby Anacostia River were deposited on the land, incinerator ash was dumped, and open burning of huge piles of garbage was allowed. When the landfill closed, more toxic dredgings were piled on the top, and then a layer that people say may have been sewer sludge. People used the park in spite of this. In the 1990s, between two and twenty feet of construction waste started to be dumped on the area, supposedly for the purpose of building up the land to develop athletic fields for the community to use. The waste contained nails, pipes, and rebar. In 2008, a report revealed asbestos and PCBs had been dumped at the park with the construction debris, and the records detailing the origin of where the materials came from are unable to be found. In 2016, environmental testing showed arsenic, lead, PCBs, pesticides, and chemicals found in petroleum wastes in samples taken at the park.

It is far from uncommon to have these brownfield sites unprotected and accessible to anyone wanting to explore.

3. Extent of Contamination

Sometimes brownfield sites are old gas stations or dry cleaners. In other cases they are large manufacturing facilities such as steel mills and chemical companies. Many old industrial buildings illegally dumped solvents on the grounds and buried drums of waste. Sometimes the amount of cleanup could be minimal and contained only to the site. In many cases, we have seen the contamination turn out to be very large and affect many properties in the area. The hazardous materials also may have contaminated wells that serve both private and public water supplies.

In Salt Lake City, a dry cleaner contaminated the soil and the runoff from their property, which spread to several neighboring parcels as well. The city wants to approve a redevelopment project for that area, but the current state of the pollution makes it questionable. Instead of removing the contaminated soils, a proposal to contain the toxins is being looked at. Even so, vapor intrusion into the buildings remains a potential hazard as the contamination in the soil rises to the surface.

4. How Clean is Clean?

When the brownfield is cleaned up, how clean is the standard for that specific site? This is how many property owners, polluters, and companies bend around the regulations. Many of these brownfields have been abandoned by their owners due to the enormous costs of cleaning up the toxins. Many businesses interested in using the property want to spend as little as possible to restore the land for industrial use. The standard of clean is much less than if people were living directly on the site.

For people living in a neighborhood mixed with industry, a brownfield site may only need minimal cleanup to be reopened again for industrial purposes, regardless of the people living nearby. The residential houses directly nearby the site are now living with a cleanup only passing the level needed for industrial use. Simply pouring a concrete slab or placing a parking lot barrier over the site is all that is required sometimes, especially in poorer areas.

The state and cities saw all this valuable property that was contaminated and looked for some way to turn these lands back into business opportunistic for developers. There were new programs created to help speed this process. Tax credits and special treatment of investments into these properties was offered by the federal government. The creation of the Brownfields Redevelopment Program offered thousands of dollars to do environmental assessments on the properties to determine the extent of contamination. Creation of the program also gave developers a break by allowing them to do less than complete cleanups depending on the intended use. It went one step further too. It covered any liability in the future for the developer, financial institutions and others involved in the project if there was additional hazardous materials discovered or liability for health risks if people became sick or died as a result. In essence it gave developers, states and cities the opportunity to redevelop contaminated property that otherwise nobody would have wanted to take a chance on. The downside is, there are less incentives to protect the health of the people working or living in these places.

Many states created their own State Voluntary Cleanup Program to assist and speed up the cleanup of brownfields to return them to becoming tax generating properties. In most cases the state programs work hand in hand with the EPA to move these cleanups along. Once the cleanup is completed according to agency standards, liability protection is offered to the developer and other parties involved against future issues.

Tony and I have seen mixed-use developments built on brownfield sites that included residential, commercial, and industrial uses. The developers are allowed to use lower clean-up standards where they locate the commercial and industrial portions of the project and can be allowed to construct the residential property on an area needing a higher level of cleanup.

Future discoveries of additional hazardous waste on a previously polluted industrial site are more common than the developers like to talk about.

Crestview Station in Austin, Texas is an example of a mixed use brownfields development. The property originally was an industrial site for a chemical research facility. During those fifty years, industrial wastes were land filled and dumped on the site. The Texas Voluntary Cleanup Program allowed the property to be developed into over 1,000 housing units and 150,000 square feet of office and retail space. In spite of the cleanup, a plume of toxic benzene remains onsite and potentially unknown quantities of industrial waste in areas that were not completely excavated or fully tested.

5. New Developments

The new development may potentially be even worse for the neighborhood and the environment than anything that could be imagined. A cleaned up brownfield that becomes a small strip mall or perhaps a new warehouse is not going to have a negative impact on the neighborhood; perhaps a new office building would even be nice. But, sometimes the development depends on what the town decides. That is the case with residents in Millsboro, Delaware.

In Millsboro, Delaware, a former Vlasic Pickle plant designated as a brownfield was redeveloped into a chicken processing plant. Despite finding levels of nitrates, PCE, TCE, lead, and chloroform higher than standards in monitoring wells, the plan was approved for the redevelopment with no remedial action needed to address the pollutants. Monitoring wells was all that was specified. The new manufacturing facility has turned into a nightmare for several of the residents we spoke with, discussed at length in the chapter on factory farms.

6. Residual Hazardous Materials

Many brownfields are not totally cleaned up of all the hazardous wastes located there. The government agencies in charge of the cleanup believe restoring a polluted site back to its original state is usually too costly. Depending on the new use the property is being cleaned up for, they want to clean up as little as possible. A site that will be a new factory will cost much less of a cleanup than a site that may be a shopping center.

People who are employed at the new business operating at the brownfield site may be exposed to volatile toxic chemicals that were never addressed in the cleanup. Because the site was only cleaned up enough to meet the new use, chemicals and pollution allowed to be left behind may seep into the buildings from the soil and foundation. The process is called “soil vapor intrusion” and presents a real threat in many cases.

Frequently, any site has had several different businesses operating at different times. There may be a wide assortment of hazardous substances accumulated over the years. You need to do some research on what information is available on the property and what stage of the brownfield restoration process it is in, if any at all. There may be nothing at all to worry about, or you may find out the cost and time needed to remediate the property for your intended use is not going to be a workable situation.

If you find out you are living near a brownfield or believe you may be having health issues related to potentially contaminated sites, you need to contact your state environmental agency. In many cases, they will come out and hear your concerns and possibly do a site assessment. Remember, brownfields are hazardous waste sites, many of which contain unknown quantities of unknown wastes, sitting there leaking into the soil and groundwater as well as into the air.

In spite of a brownfield being redeveloped and supposedly cleaned up, many sites still have chemical contamination. Whether the cleanup consisted of a parking lot to cover it up or another method, the continued existence of the waste presents an ongoing threat to the neighbors. We always recommend a half mile buffer between you and any brownfield, redeveloped or not.

Chapter 25. Sewage Sludge

The human waste we produce and the way we handle it has consequences on our health and the environment. Some of our waste is more dangerous than others; it can cause disease and illness in addition to poisoning our drinking water. While special attention needs to be given to how we handle those types of waste, the vast amounts we produce sometimes leads to less than ideal methods as we simply just try to manage the flow.

This brings us back again to the subject of human waste. As a nation, we are full of sewage sludge, which is the semi-solid product left at the wastewater treatment plant after it processes the sewage we send down the drain. According to information from the United States National Sewage Sludge Repository, we produce an astounding 7.18 million tons of sewage sludge in a year. It needs to be disposed of someplace.

Once the treatment plant has processed the sludge as far as it can, the next step is disposal. Up until the end of 1991, sewage sludge would be hauled out to the ocean and dumped. For states not near the ocean, any large waterway or body of water was used. This method of waste disposal is sometimes described with the phrase "the solution to pollution is dilution." Fortunately, the practice of ocean dumping was finally halted, mostly due to public outcry, especially over continuous beach and lake closings due to high coliform bacteria counts. The end to this method of disposal forced the industry to find more eco-friendly (and therefore more expensive) methods to dispose of human waste.

Currently, the most common methods of sludge disposal are landfills, land application, incineration, and deep well injection. Depending on where you live, your local sewage treatment plant is doing one or more of these, and each comes with its own consequences and drawbacks. You would want to check with the water district to find out how they dispose of the sludge. Otherwise, you may find yourself in a rather unpleasant smelling situation.

1. Landfills

Disposing of sewage sludge at the landfill along with regular solid waste presents a few issues, especially for the people living nearby. When the sewage sludge is trucked into the landfill, it first needs to be unloaded from the transport vehicle. Depending on the sludge thickness, it will usually be transported in either a vacuum pumper truck if it's wetter or in an open top cargo trailer if it's drier. State regulations usually require the sludge in the transport vehicle to be completely covered with tarps if the truck has no top, such as with dump trucks and open trailers. This is to prevent sludge from blowing away, getting rained on and creating leachate, or creating overly powerful odors. When the trucks arrive at the landfill, they usually are directed to an active area on the landfill to dump their loads. While the landfill is required to cover the sludge as quickly as possible to minimize odors and keep down the flies, this sometimes may not happen until the end of the day and all the trucks of sludge have completed their deliveries to the landfill. Many times the landfill dumps trash on top of the sludge to save money on using cover dirt, so hours go by while the sludge sits exposed.

While landfilling sludge is not the preferred method, it is widely used because it is the easiest disposal method for the sewage plant. Landfills were not made to take the enormous quantities of sludge that most plants need to dispose of, so the overabundance of waste being dumped causes the landfill to give off a much stronger odor. When the weather is warm and especially humid, the smell lingers in the air for days and permeates into everything nearby.

Sewage sludge decomposes more rapidly than solid waste due to the processes of bacteria in it. This gives off a much larger amount of gas. Unfortunately, the gas smells like rotten eggs, and the landfill may not have the ability to control the excess gas. As a result, anyone living near the landfill isn't just breathing in a bad smell; they're breathing in sulfur, pathogens, and other harmful substances.

In Parrish, Alabama, the residents wound up with a train nicknamed "The Poop Train." It seems that New York City had been sending their sewage sludge to a landfill 20 miles away for the past year or so. While the courts battled over whether or not the train could proceed to the landfill as planned, the waste sat and sat, all the time the stench growing more rank.

In Coventry, Vermont, the state Department of Environmental Conservation has been receiving dozens of complaints from its neighbors about strong odors, some of which would linger. The landfill accepts sewage sludge from the treatment plant for disposal, so locals expect the situation to only get worse. In addition, due to pathogens and bacteria in human sewage sludge, monitoring the water in the area is of critical concern. Many residents and towns get their water supply from wells in the area.

2. Land Application

The application of animal manure to the land is an ancient practice. Before the birth of chemical fertilizer, farmers routinely returned the waste from their animals back to the land to add nutrients. Using manure with its natural nutrients and organic matter reduces the need for other fertilizers to be used. The nutrients contained in the manure help build the soil up to help the crop grow. Unfortunately, over the past 20 years, the effective use of that practice has been compromised by the creation of factory farms (as was discussed in an earlier chapter) and millions of pounds of animal manure recklessly overapplied to the fields. While there are some small family farms using the practice responsibly, most farms are not.

Sewage sludge from human waste is different because it contains more materials that are harmful to us. Sewage treatment plants process waste sent through the sewage system from many more places than residential homes. Offices, stores, restaurants, hospitals, and industrial operations all share and dispose of their wastes through the same system. Spreading human sewage on agricultural land is usually only done with sludge that has gone through additional treatment to remove as many of the toxins and heavy metals as possible. That treatment process, however, is woefully inadequate to deal with the range and volume of chemicals, metals, pathogens, and other waste which enter the treatment plant. The sludge that comes out the other end of a sewage treatment plant is dangerous.

The term the industry likes to use is “biosolids.” No matter what we call it, biosolids or sewage sludge, it still smells and attracts flies. It still contains toxic chemicals and materials which pollute the land and groundwater and affect the people living near where it is applied. For example, the government recently discovered high levels of PFOS, a likely carcinogen, in sewage sludge. The likely carcinogen has been there all along, but finally now is openly being discussed as millions of pounds of contaminated sludge has been spread over farmland for years all over the country.

Even with these known dangers, many states use land application as the preferred method of disposal. In some, about 60% of the volume of sewage sludge is disposed of this way. In addition to agricultural land, sludge is also used for mine reclamation areas to build the soil back up. It’s also used In forestry projects and public works projects such as parks.

There are strict federal regulations and guidelines that are supposed to be followed when anyone disposes of sewage sludge on the land. On top of that, many states that allow this method also have their own additional requirements as to how the sewage sludge is to be applied. Many of the restrictions deal with what uses the land can have and what is allowed to be grown on that land. Even though the land it is applied to is regulated, the sludge has a large impact on the surrounding area as well.

Nitrates are found in large amounts in sewage sludge. While they are good for the soil and add needed elements, nitrates also have the potential to contaminate the groundwater with levels harmful to infants and may cause blue baby syndrome. Some studies are now suggesting there are some negative health effects on adults as well, but more research is being done.

As if that weren't enough, the sludge still contains many pollutants even after treatment. That leaves open the possibility for pathogens to be present in the sludge that were not destroyed during the treatment process. The sewage sludge contains high amounts of bacteria from the feces and urine that make it through the treatment process, including Listeria, salmonella, E. coli, and clostridium. All of these harmful bacteria have been found in sewage sludge tested after treatment.

The problem is widespread. In 2019 almost 60% of the sewage sludge from over 16,000 treatment plants was disposed of on the land. Used on farm fields, parks, schoolyards, lawns, and many other places, millions of tons were spread. The use of sewage sludge on land is allowed by the EPA based on the claim it is a beneficial use of a waste product because the sludge contains phosphorus, nitrates, and some heavy metals considered good for the soil even though the agency has no testing requirements for other chemicals in the sludge before being applied. The waste may contain thousands of dangerous chemicals and other heavy metals. Pesticides, lead, mercury, radioactive wastes, microplastics, dioxins, PFAS, pharmaceutical wastes, and many other substances are found in the sewage sludge being applied to lands across the country.

Even with all of these known dangers, because there is an almost limitless supply of sewage sludge available to farmers and other places, at times they put down too much. This presents several new problems. If over-applied, the sludge will not be absorbed by the ground but will instead run off into nearby properties and waterways. Along with the organic matter and nutrients, however, these nearby locations also receive whatever pollutants and pathogens were in it. If not properly monitored and managed, these could adversely affect human and animal health, soil quality, plant growth, and water quality.

In Lancaster County, Pennsylvania, for years my family found ourselves living in a farming community that not only spread the manure from the pigs and cows they raised on the fields, but also brought in sewage sludge from the local wastewater treatment plant to apply to the fields as well. The stench of the entire area during spreading time was unimaginable. In addition to the stench, we had constant sore throats, headaches, and much more frequent colds due to the airborne pathogens in the sludge.

In Franklin County, Alabama, farmers are suing Synagro, a sludge spreading company, as well as several industrial operations who sent toxic, carcinogenic chemicals through the sewers to the treatment plant. The chemicals PFAS and PDOAs were in the sewage sludge applied to the farm fields in the area. Five thousand acres of farmland and water were contaminated. The company said it was unaware of the hazardous chemicals and didn't test before spreading.

The town of Lapeer, Michigan, had disposed of much of their sewage sludge the past 20 years by offering it to the local farmers as free fertilizer. It was good for the farmers as it cost nothing and contained nutrients they could use to enrich their soil. It was also very good for the town as it saved them a large amount of money on other disposal methods for the sludge. However, the newly-suspected hazardous chemicals PFOAs and PFOS started turning up in tests. Concerns are now growing that the fertilizer that was applied to the lands all these years, has contaminated not just the soil, but possibly the water as well.

In Edmond, Oklahoma, according to a new report, over a million gallons of sewage sludge has been applied to a 44 acre site of local farmland in a seven day period. The sludge was injected into the ground. Putting 28,000 gallons of sludge in one acre of land in a single week isn't a good decision. Authorities are assuring people the sludge has been tested for safety; the Director of Water Resources for the town told the people and the news reporter, "I wouldn't do anything that would not be safe to do at my own house." Given the gravity of the situation, the director's words ring hollow with the locals who still have concerns.As a bare minimum, better testing for certain chemicals such as PFAS is required. The way the situation is now,there may well be some unpleasant surprises in store.

3. Incineration

At the time of this book's writing, there are 217 sewage sludge incinerators operating in various communities in the United States. Because sewage sludge contains a variety of toxic chemicals and heavy metals, the EPA regulates the emissions of these incinerators. There are actually two sets of regulations, depending on the age of the incinerator. There was a change made by the EPA in 2016 to lower the quantities of certain very hazardous emissions. However, due to how expensive it would be to upgrade the older incinerators built before 2014, the EPA allows a different amount of pollution to continue. Very unfortunately, though, the amount of hazardous materials incinerators were allowed to emit prior to tighter regulations was extremely concerning to communities affected by the operation of the incinerators. Some already have quite a few residents whose health was impacted as a result of delay.

Incineration of sludge reduces its volume to about 15% of the original amount. This creates a new problem for the 15% remainder, which needs to be landfilled. Because of the chemistry behind waste incineration, the incineration process increases the percentage of most of the toxic chemicals and metals to five to ten times their original concentration: many of these are not destroyed during the combustion process. In other words, the remaining sludge has even higher hazardous levels of heavy metals and toxic compounds. While some people such as Tony and I say this should qualify the incinerated sludge as hazardous waste and disposed of as such, it gets dumped in a landfill. There is no financial incentive to dispose of it otherwise, and there are no governmental restrictions against doing so.

Mercury is one of the most dangerous chemicals found in sewage sludge. When the sludge is incinerated, much of the mercury is emitted from the smokestack. There are guidelines for the amount of mercury pollution allowed, but many times the incinerator and its pollution control equipment are not capable of meeting the standards. Unhealthy and dangerous amounts of mercury are released into the neighborhood and community.

Sludge incinerators also emit dioxins, furans, cadmium, and lead into the surrounding area as well.

In Manchester, New Hampshire, the sludge incinerator emitted so much mercury that the EPA needed to step in to force the city-owned incinerator to comply with The Clean Air Act. The city settled with the EPA by agreeing to spend more than $6 million to install pollution control equipment, which will meet the standards needed. The amount of mercury the plant had been emitting was exposing residents to dangerous levels of mercury emissions. Inhalation of mercury vapors in emissions can harm the nervous and digestive systems. It also causes damage to the lungs and kidneys. Too much exposure to mercury can even be fatal in extreme cases.

In Bridgewater, New Jersey, the sewer authority was fined over $200,000 for multiple violations of the Clean Air Act involving the incineration of sludge at its facility. Among the violations was non-compliance with emission standards, not having a plan to comply with emissions, especially the highly toxic chemical mercury. The sewer authority operates two incinerators at the site. They agreed to bring one of them up to standards to meet EPA requirements. They told the EPA that the second incinerator had a catastrophic failure and would no longer be using it at the current time. While there were no details about what this failure was, it certainly was obvious it was not running in compliance at the time of the catastrophe.

4. Well Injection

Sewage sludge is one of the many waste products we like to dispose of in injection wells. Simply put, the sewage sludge is injected into the ground using a reverse-powered well. While many companies and municipalities like to talk about how scientific this all is, it is a method that will come back to haunt us at some future time. For years these injection wells have been sending millions of gallons of human waste down into cavities and porous rocks below the surface. Most of these wells extend down a thousand feet or more. Some go down several thousands. However, scientists and engineers cannot possibly predict when something below the surface may shift or rupture, sending millions of gallons of waste into a nearby aquifer and poisoning a water supply for years to come. Given the physics involved in injection wells, it is not a matter of if they will leak, but a matter of when.

In Los Angeles, California, there is an experimental project known as the Terminal Island Renewable Energy. The project has already injected over 200 million gallons of waste down several injection wells 5,300 feet deep. The project hopes the wastes will break down due to the heat in the Earth at that depth and cause the waste to produce large amounts of methane gas that will be pumped back to the surface and used for energy. Nobody can say with 100% certainty that all of the waste will stay sequestered where it is put until it breaks down to less harmful materials and that it won't at some future point contaminate underground water needed for the city. Because this isn't based on hard science, they are gambling with the health and drinking water of millions of people for quick and easy disposal.

Florida is trying to lead the nation in this disposal method with several very deep, newly-drilled injection wells. The hope is the waste they send down these holes will not return anytime soon and pollute the water. However, as discussed in the chapter on Injection Wells, Florida has a history of injection wells backing up on them. Back in the late 1990s, 20 injection wells were detected as having leaks, 17 of them in Miami-Dade County. The 3,000 foot deep injection wells showed evidence of treated sewage migrating upwards towards potential drinking water

aquifers and not staying impounded as were supposed to.

On the islands of Hawaii, Maui is sending their sewage waste into underground injection wells. Daily, three to five million gallons of treated wastewater is being injected down below the surface into supposed impoundment areas. Environmental groups sued to stop the injection of waste after dyes placed in the wastewater showed over half the waste from two of the wells was leaking into the ocean. The U.S. Supreme Court recently ruled that was a violation of the Clean Water Act. However, the ruling wasn't a complete win: the court only sent the case back to lower courts to reconsider some issues but stopped short of totally banning the practice.

The piles of sewage sludge we have all over are not going away anytime soon, nor are they going to get any smaller. For now, according to the experts, these disposal methods are the best options. However, many scientists and environmental groups dispute that, saying we are simply moving the problem from one place to another without really doing long-term planning. Like ocean dumping before it, our waste is really still there however we try to get rid of it. It still has the capacity to harm our health and environment.

Be cautious and thoroughly check out how your community handles the waste from its sewage treatment plant. Doing so is simply being smart and potentially preventing you winding up in an area or situation which is placing your health at risk. Research into better methods of disposal is ongoing. Maybe one day soon they will figure out a process to deal with the incredible volume we create. In the meantime, stay at least a half mile away from any property that sewage sludge is applied on. If you can put a mile between you and there, so much the better.

Chapter 26. Sewage Treatment Plants

Flush the toilet, and away the waste goes. People treat sewage much like they treat their other garbage. As long as it goes someplace else, they are happy to send it away. Sewage treatment plants are operations that take our untreated sewage and, through biological and chemical processes, treat the waste. Some plants are better equipped to do this than others. There are multiple stages in the treatment process. The sophistication of the treatment plant near you will determine how much waste is needed to be disposed of and what needs to be done with it.

While this subject is anything but glamorous or dinner table conversation, it's important to discuss. Sewage treatment plants deserve much of the credit for reducing diseases in many of our modern cities back in the early 1900s. These plants really are a marvel of modern engineering, and without sewage treatment plants, our society would most likely have continued to be plagued with the highly contagious and deadly diseases such as dysentery, cholera, and typhoid, which human waste caused throughout modern history. Incidentally, lack of adequate sewage treatment systems in many countries today is the single biggest cause for contaminated water supplies and diseases ravaging the people living in them.

The expanded definition of "sewage" from the Dover Township, Pennsylvania, website put it elegantly, "Sewage, also called wastewater, is the contaminated water from homes, schools, and businesses. It comes from toilets, showers, clothes washers, dishwashers, etc. The contaminants include fecal matter, urine, soaps, detergents, food particles, hair, rags, paper, toys, dead goldfish, and anything else that is disposed in a drain. A person creates an average of 60 to 100 gallons of wastewater every day."

The township's website also discusses treatment plants: "Sewers are a network of pipes that bring the sewage to the treatment plant for treatment. Treatment is the continual process of removing contaminants from wastewater and then processing the removed contaminants into a product that can be safely recycled."

As Dover Township's words make it clear, sewage treatment plants are a necessary part of any community. Without sewage treatment plants, disease would spread throughout the population. An effective sewage treatment operation also protects the public from contaminated water. In countries where there is no sewage system or treatment facilities for human waste, bacteria from sewage gets into the wells and other water sources.

Living in a neighborhood near the municipal sewage treatment can be trying at times. Almost every municipality that has a public water system connecting all the homes also has a sewage treatment plant. The system sends clean, safe water into the house and takes away the dirty water.

Usually, when we think about our wastewater, we think of feces. The reality is that much of the vast volume of waste that flows through the sewage treatment plant is contaminated water. The rest is water from showers, sinks, and drains--residential, commercial, and industrial. A single industrial user may discharge thousands of times more wastewater in one day than a residential home would in an entire year. Some of that water flowing through the sewage treatment plant is stormwater that ran off the streets. All added up together, only a few percent of the millions of gallons flowing through the plant are any kind of solids. Depending on how many industrial operations and commercial businesses are discharging waste into the system, once the solids are removed, much of the waste flow is made up of 95% or more of water. The solids though are made up of feces and industrial wastes, making it extremely hazardous.

However, this water may contain any number of hazardous materials ranging from radioactive isotopes to toxic waste. The wastewater flowing into the plant is, after all, coming from inside factories, stores, and restaurants in addition to all the dirt, oil, gas, and junk that washes off of the streets.

The vast majority of the liquid waste is treated in the primary and secondary process where it is separated from the solids through aeration and settling. It is then disinfected before it is sent from the plant into either a nearby body of water or used to recharge the water aquifer and go through the cycle again. The result of sewage treatment is clean water, but "clean" is a relative term. There are disputes as to what the discharged water still contains after treatment, but only a small number of chemicals are tested for. The solids that remain are separated out through the various stages.

Despite their valuable and necessary function, there are some hazards associated with living near a sewage treatment plant.

1. Odors

There is almost no way around it. Feces have an odor, and thousands of gallons of it in one place is necessarily going to smell. Many sewage treatment plants try to keep odors under control. Some have enclosed the giant aerator tanks, which is where much of the odor comes from. Many have not, though, instead having open air lagoons filled with sewage sludge, which stinks. There are chemicals that can be added to hold down some of the odor, but for those who live nearby, the odor can be overwhelming. On days when the humidity level is high, the odor may be especially bad. The moisture in the air keeps the evaporated water from the plant and the odor close to the ground.

Even treatment plants with covers do have an odor as well. When the plant is running well, the solids are separated quickly from the liquids and passed into another process where bacteria is used to further break down the sewage. As the breakdown continues, hydrogen sulfide gas is created in huge amounts. This smells like rotten eggs, only hundreds of times stronger because of the large volume of gas.

Though they try to minimize it, it is almost impossible not to have some odors from a sewage treatment plant. These odors can make the lives of people living nearby truly miserable. These odors can even come into homes through the cracks in doors and windows and give residents nowhere to escape from the smell.

While some plants do not smell all the time, all plants smell some of the time. Factors like the wind and the weather as well as the effectiveness of the treatment operation all affect how frequently and intensely a plant will smell. In almost every case the odors are caused by one of the gases given off from sewage sludge. While people may see odors as simply a nuisance, they are not without actual harm. Inhalation of the gas affects the respiratory system. Continuous exposure causes irritation of the airways and lungs, increases chances of asthma, and causes people who already have asthma to suffer increased attacks. Moreover, gases from the sewage contain pathogens which have the potential to make people ill in addition to actual particles of feces which is inhaled.

In Waterbury, Connecticut, neighbors near the city-owned sewage treatment plant complained that the odors coming from the plant were a horrible stench. Once the mayor himself checked it out, he agreed. The mayor made a public statement promising action on the odor and agreed with the residents that the odors from the plant were the "worst [he] ever smelled."

In Asheville, North Carolina, neighbors near the sewage treatment plant are subjected to strong odors almost every night. When the humidity is high, the odors are especially strong. Eventually, one of the neighbors asked the local news station to investigate. Not only do the odors come from the plant, but at the end of the day, all the sewage pumper trucks come to dump their loads, making the odors even worse. Neighbors of the treatment plant are continuously exposed to the pathogens that travel along with the odors as particles. This could lead to increases in gastrointestinal illnesses in addition to respiratory problems

In Mill Valley, California, a sewage plant is creating such odors that students in a school a short distance away are becoming sick. High levels of hydrogen sulfide gas are drifting towards the school from the sewage plant. Low concentrations of hydrogen sulfide affect the nose, throat, and lower respiratory tract. Continued low-level exposure can cause coughs and shortness of breath. Higher levels of gas may cause bronchitis in some people. Children are especially susceptible as their airways are smaller and breathe more frequently.

In Walker's Point near Milwaukee, Wisconsin, neighbors of the sewerage treatment plant have been subjected to odors from the plant for years. Currently, the plant releases the gases directly to the outside, bypassing an odor control filter system. The filter broke in 2002; the plant decided it was too costly to fix for over 15 years. The area is frequently beset with high levels of gases from decomposing sewage and composting sludge. Many neighbors suffer irritated throats, noses, and eyes from the high levels of sulfide gases emanating from the plant.

In Liberty Hill, Texas, neighbors of a sewage treatment plant say the odors are so bad that they can no longer enjoy their outside areas. One resident said she doesn't even want to let her

two-year-old daughter outside. The development was built within eyesight of the already existing treatment plant. For over a year, there has been a lingering odor of sewage throughout the community. The treatment plant has been cited by the Texas Environmental Agency for dumping chlorine and human waste into the San Gabriel River. One resident who called the Texas Commission of Environmental Quality while trying to take a sample of water from the sewage plant discharge line reported burning sensations in his eyes, nose and lungs after going near the water.

2. Vermin and Flies

Flies are always looking for their next meal and a place to lay their eggs for the next generation. A sewage treatment plant is the perfect hangout. Flies are a frequent visitor to the sewage treatment plant and the nearby neighborhoods. Flies thrive and feed on any waste available near the plant. Spills from trucks transferring loads into the plant attract flies. Plants with open air aerators are an invitation for flies to gather.

In Bryan, Texas, the sewage treatment plant had an infestation of flies that was breeding in one of the filters at the plant. The facility shut down the filter to try to get rid of the flies, but the flies infested homes in a two to three block radius of the treatment plant. One apartment owner counted 400 flies in her place. Flies are known carriers of pathogens and, especially as these flies were feeding on raw sewage, the area neighbors may have been exposed to a variety of bacteria. The scattered reports of upset stomachs and diarrhea most likely were caused by the contamination the flies carried and was transmitted each time they landed on the surfaces and food in people's homes. One person reported sneezing and having a fly come out of his nose.

3. Overflows, Discharges, and Releases

A big issue for these plants is what they do with the wastewater separated from the solids. Most sewage treatment plants are located next to a waterway when possible. Usually, the bigger the volume of the waterway for a drain, the better it is for the plant to use. The reason for this is twofold.

For one, large waterways provide a place to easily dispose of the millions of gallons of liquid that are removed. There are standards the government sets that the liquid is supposed to meet before it is discharged from the plant. These standards vary widely depending on factors including the age of the plant and the capacity of the nearby waterway to absorb and dilute the wastewater to safe levels.

For the other, sometimes even the best operated sewage treatment plants have incidents out of their control. Automatic monitors measure the flow of waste into the plant and watch for unusual events. But, when there is an incident caused by a pipeline that carries the waste into the plant, it creates a backup. Pumps that keep the flow moving through areas where gravity is insufficient sometimes malfunction. Other equipment inside the plant could have an issue. All these may cause the usually smoothly running system to back up.

Unfortunately, when it does back up, the first to see it are those closest to the plant. While these incidents are fairly rare, they do happen, and when they do, it makes the homes affected a rather unpleasant place to be.

In Craig, Colorado, the sewage treatment plant experienced an electrical shutdown that stopped the sewage pumps. The problem went undiscovered over the weekend, and by Monday, two million gallons of untreated sewage was backed up in the lines. Luckily, the backup that was headed for nearby houses and businesses was caught just before it reached them.

In Suffolk County, New York, a mom and her son both came down with aseptic meningitis due to a backup of sewage originating with the nearby sewage treatment plant. Susan Serrano, the owner of the home, told a local news organization, "You could see the feces, you could see things floating that came from the hospital," she said. "Our home was inundated with this raw sewage. Me and my son were walking in it." Again in 2014, her home was flooded once more with another backup from the sewage treatment plant.

In Silverton, Oregon, the local sewage treatment plant backed up raw sewage into one resident's house. The damage caused by the backup was so severe that the local health department told her she needed to move out until complete repairs and cleanup were completed. The city paid for one night in a local hotel. This unfortunate person wound up fighting with insurance companies. In the meantime, they were telling her temporary quarters were her responsibility. The remodeling company refused to do any work because the sewage treatment plan was still not repaired and another backup could happen.

4. Sanitary Sewer Overflow

Many towns and cities built their waste treatment plants before building booms hit America. The plants were sized to meet the capacity at the time they were built, with some extra for future growth. Very few towns and cities sized their plants for the explosive growth that came.

When a weather event increases the flow of wastewater coming into the plant beyond the capacity it can handle, the plant will allow the waste to bypass its system. It is not unusual for plants in larger cities to have several dozen of these events every year. For instance, in the Pittsburgh area, the Upper Allegheny Joint Sanitary Authority estimated that, for 70 days of the spring through fall recreation season, the rivers are unsafe from sewer overflow events.

The EPA says many of these overflows occur in the Great Lakes states and the northeast portion of the country where the sewer systems also are combined with storm water from the streets. Frequent heavy rain and snowmelt usually predict another bypass is coming shortly.

When the amount of waste exceeds the capacity of the plant to handle it, the excess may be diverted into a temporary storage area. That is, if the treatment plant even has a backup storage area. If they have one, it's usually nothing more than a lined lagoon. If they don't have one, they use the nearest river or lake. Unfortunately, the outflow is connected to the amount of untreated water coming in.

These events are so frequent that the EPA has a name for them: Sanitary Sewer Overflow (SSO). The EPA estimates there are between 25,000 and 75,000 of these SSOs each year. The government allows these plants to open up their overflow valves, which were designed to bypass the treatment system, and the sewage is allowed to be discharged directly into the waterway.

An example of an advisory issued by a sewage treatment plant in the Pittsburgh, Pennsylvania, area is as follows: "A combined sewer overflow (CSO) occurs when storm water and sewage carried in a single pipe overload the sewer system and flow untreated into the rivers and streams. CSO's can carry a variety of pollutants, including raw sewage, debris and bacteria." They go on to say that they have a complementary advisory program to warn the general public of possible river contamination. While they do not prohibit or discourage recreational river activities, they do encourage recreational users to limit contact with river water when boating, fishing, water skiing, or other river recreational activities.

No matter the fancy language being used, the reality is far uglier. The treatment plant is pouring millions of gallons of raw, unfiltered sewage into the river water. The complementary advisory should do more than discourage recreational activities. They should ask residents to use caution to limit any contact with river water.

In Pine Island, Minnesota, the sewage treatment plant was found to be dumping untreated waste directly into the Zumbro River. When a local resident saw the outflow from the pipes, he called local health authorities. The sewage treatment plant says there is nothing they can do about the dumping. The cost to upgrade the plant to handle the additional waste was too costly to fix. Any people who would normally use the Zumbro River for recreational purposes are exposing themselves to the raw sewage. If the water comes in contact with their skin, it could cause lesions and blistering. If any of the water was ingested could cause severe gastrointestinal illnesses.

Pine Island is not alone. In Nashville, Tennessee, the sewage treatment plant has dumped 4.7 million gallons of untreated waste into the Cumberland River over the past 2.5 years.In Omaha, Nebraska, a sewage treatment plant dumped 65 million gallons of raw sewage a day into the Missouri River. For over two weeks the Omaha plant continued to dump the sewage with no end in sight. They went on to state that communities downstream who relied on the Missouri for drinking water should be safe if they chemically treated their water before using any from their taps. Drinking water with bacteria from untreated sewage can cause severe gastrointestinal illnesses. The CDC had documented almost a half million cases of illnesses related to polluted drinking water in the U.S and tied 251 outbreaks of various diseases to polluted drinking water as well.

5. Radioactive Wastewater

A new issue has been hitting sewage treatment plants the past 10 years or so: wastewater from fracking operations. In most states it is illegal to discharge untreated fracking wastewater directly into the environment, so operators have been sending it to wastewater treatment plants. However, treatment plants do not have the ability to remove most of the contaminants the fracking waste contains. In many cases there is a concentration of radioactive particles in the waste. These are carried through the treatment plant and discharged out the other end into aquifers or rivers and lakes. Many states are banning the practice, now that they have become aware of it. Some sewage treatment plants continue, though, to take it anyway.

The same situation is also occurring from the liquid waste from landfills known as leachate. As liquids drain through the landfill and hopefully are caught in the liner at the bottom, they are removed through a leachate collection system. This toxic brew is then mostly disposed of at wastewater treatment plants, which are incapable of removing many of the substances it contains.

As a necessary result, the discharge of wastewater from treatment plants still contains high levels of radioactive substances.

At one wastewater treatment plant in Pennsylvania, the operator discovered that some materials discharged into their system were actually killing the bacteria the plant depends on to have the biological processes the plant depends on for treating the waste. As a result of decreased natural activity in the treatment plant, the treatment was unable to properly lower the level of the waste, and waste was being discharged with higher levels of contaminants than they should normally have had after processing, endangering the locals with a wide variety of health risks.

6. Airborne Hazards

Sewage treatment has several stages to take raw, liquid sewage and process it into a semi-solid state. Stages such as aeration and dewatering move the still-liquid sewage around. During those processes, the liquid has the opportunity to splash or form water droplets in the air around the treatment tanks. Especially on the days that are windy or very humid, the droplets get carried away, and with them, studies say, pathogens too. If these particles are inhaled by people nearby, it can cause respiratory and gastrointestinal issues.

Bioaerosols, as these airborne particles are known, may contain different types of harmful microorganisms. There is the potential for these airborne droplets to carry viruses, pathogenic bacteria, and fungi, just to name a few. Any one of these can cause respiratory, skin, nervous system, and digestive system disorders and diseases in whoever inhales them.

A study done by a group of researchers and published in the *Journal of Environmental and Public Health* showed concentrations of these microorganisms in neighborhoods near sewage treatment plants with outdoor aerators. These disease-carrying organisms are capable of traveling several miles from the plant, depending on winds and weather conditions. When the researchers compared occurrences of gastrointestinal illnesses with people living near the plant to those who were further away, they found increased reports of nausea, vomiting, and diarrhea occurring frequently in areas a third of a mile near the treatment plant. As the distance from the plant increased, the reports of illness decreased substantially although increased levels of pathogens remained in the air up to a mile and quarter from the plant.

7. Industrial Waste

To reiterate, industrial waste is also sent to the sewage treatment plant through the same pipes. All of the factories and commercial operations are also tied into the treatment system. There are supposed to be strict guidelines and regulations for pre-treatment of industrial waste before they are sent into the pipes to the plant. There are restrictions on certain chemicals and toxic substances being disposed of this way. However, because the pipes mix all the thousands of sources of wastewater together, there is no way to tell who violated these regulations. This can actually encourage shadier businesses to break the rules.

Even when these industries do follow the regulations, there are all kinds of acids, heavy metals, and other poisons they are allowed to send into the sewer. The sewer authority does not monitor for everything at the treatment plant, nor can many types of chemicals and substances be detected and removed. As a result, they go out into the stream or river as it flows to your house.

If you are downstream, take into consideration the type of waterway and the amount of flow that is usual. Obviously, a big river will carry more volume and be able to better dilute discharges. Seriously consider if you should use the water for recreational purposes. Do you want your children playing in it? Would you want to be downstream from a sewage treatment plant and swim in the water? If you are close to the discharge pipes, use an extra amount of caution.

We also suggest asking the town or state environmental agency about the water test results they have. Many rivers and waterways have a local environmental group keeping an eye on things and running their own independent tests because they trust their results more than the government's. Get to know these groups and see what they say. Personally, we would trust the test results from these groups before we would trust what the government or utility company tells us.

Sewage treatment facilities are absolutely needed to protect human health and the environment. They are needed to maintain the integrity of the public water supplies by keeping contamination out of our drinking water. However, the further away from residential areas and the more isolated they are, the better for the people living in the community.

Tony and I suggest that, if there is a sewage treatment plant nearby, you give it at least two miles of space to avoid the odor issues on all but some of the worst days and during upset conditions at the plant.

Chapter 27. Gas Stations

If popular expression is accurate, people in America have a love affair with their cars. According to one source, in 2018 there were about 276 million cars registered to be on the road in the United States. All of those cars need a tremendous amount of gasoline to get them to wherever it is they are all going. To help satisfy that appetite for fuel, there are about 168,000 gas stations across the country. There seems to be no shortage of places to fill up gas tanks most of the time. Of course, we all have had that stomach-churning experience of seeing our fuel gauge almost at empty and hoping we could make it to a station before getting stranded. For the most part, though, we see these gas stations on corners everywhere. Other than getting gas at them, what else can we get from a gas station? Those who happen to live near one also get toxic fumes and potentially cancer-causing pollution in the air and water they breathe and drink.

Is it a safe or good idea to live near a gas station? Based on the information scientists and researchers have reviewed from hundreds of studies, that answer is a definite no. It certainly is a convenience to always be able to get gas or a cup of coffee on the way out to work in the morning. Some people living close may even enjoy taking a walk to a nearby gas station to get one of those delicious snacks off the roller machines or a frozen fountain drink on a hot summer night. However, the constant exposure to the negative health impacts far outweigh any possible positives.

Regardless of all the rules and regulations in place for operating a gas station and minimizing the escape of toxic and hazardous fumes, there is a large risk. No matter how closely the gas station operators and their employees follow safety procedures, there are still too many uncontrolled issues. People living near gas stations constantly expose themselves to higher levels of hazardous chemicals than those living further away. Studies have measured the presence of airborne chemicals at gas stations and have found higher concentrations of harmful emissions up to 300 feet away. Toxic fumes have been found even further away from high volume stations.

We usually don't think of a gas station as having some of the same hazards as a factory, but there are similarities between the two. Gas stations have toxic emissions of many chemicals that are dangerous to our health. One of the most dangerous chemicals in gasoline and always present at gas stations is benzene. Exposure to this specific chemical is linked to many serious diseases and is the cause of several types of cancers. Locations where petroleum is burned (highways and gas stations) or manufactured are prime for high and dangerous concentrations of benzene in the air. Chronic exposure to benzene fumes affects blood cells frequently leading to anemia and potentially cell mutations.

Many studies show evidence that gas stations expose people to unsafe levels of hazardous chemicals, especially benzene. This is not an occasional problem but an ongoing one. Quite a few of the researchers saw results showing unsafe levels were exceeded numerous times for periods of over an hour or longer.

One of the researchers, first author Markus Hilpert, Ph.D., associate professor of Environmental Health Sciences at the Columbia Mailman School, made the following statement regarding his group's results: "We found evidence that much more benzene is released by gas stations than previously thought. In addition, even during a relatively short study period, we saw a number of instances in which people could be exposed to the chemical at locations beyond the setback distance of 300 feet."

When evaluating risk from gas stations there are several areas to look at.

1. Pump Vapors and Fumes

When we talk with people about potential health issues from a gas station, the first thoughts they have are the gas pumps. The most obvious concerns are about the gasoline fumes from pumping the gasoline into the cars filling up and from the tanker trucks delivering product. Most states have so-called low emission controlled pumps that work to limit the amount of fumes lost. However, even when they are working, they still leak some. Whether it is a few drops when a user places the nozzle in their car or when they put it back into the pump, some gas is usually dropped.

Gasoline has a very distinct smell. We notice this smell every time we pump our own gas. Some of us actually like the smell of gas and get a little pleasure rush when we breathe it. Even if we enjoy the rush, it comes at a cost. That rush is actually the effects of the brain being deprived from oxygen as the gasoline vapors restrict the flow to the brain. The odor of the fumes is caused by the hydrocarbons that make up much of the 150 chemicals needed to make gasoline being released into the air. When even small quantities of hydrocarbons enter the bloodstream through the lungs, this can result in a reduction of the functioning of the central nervous system (CNS) and cause organ damage. Health effects from exposure to petroleum products vary depending on the concentration of the substance and the length of time that one is exposed. Breathing petroleum vapors can cause nervous system effects (such as headache, nausea, and dizziness) and respiratory irritation. Chronic exposure to petroleum products may affect the nervous system, blood, and kidneys.

These fumes can and do travel away from the pumps and the ground near the station. A study examining the distances these chemicals can travel found fumes containing toxic chemical vapors in harmful quantities up to 300 feet from the station. The more pumps at the station, the further the vapors travel and the higher their concentration. Especially disturbing are the elevated amounts of benzene found. Benzene in particular is a known cancer-causing chemical, and long-term exposure causes leukemia and brain damage. Toluene is another compound found in high levels, and it can cause liver and kidney damage as well as threaten the health of babies and pregnant women.

Most gas stations in the past had six or eight pumps, but these days they frequently have 20 or more. The tiny amounts of vapors from spills and pumping add up fast. Tiny amounts of these chemicals have disproportionately large negative effects on the body.

Making matters worse, when the tanker trucks arrive to refill the underground tanks, fumes are released when they open the tanks to hook up the truck and also when they release the lines. This adds up to much more being released, and some stations pump and receive thousands of gallons each week. Repeated exposure to the chemical components in gasoline can lead to lung disease as well as brain and kidney damage. Children's small lungs, still developing body, and higher breathing rate make them especially vulnerable to long-term exposure and their ill effects.

2. Exhaust Emissions

One of the bigger issues with living near a gas station is the continuous exhaust emissions from the vehicles and trucks coming in and out of the gas station all day (and night if it's a 24-hour gas station). Some of the larger gas stations refill hundreds or even thousands of vehicles per day. Whether somebody just runs in to pick up a soda or a cup of coffee or to fill up the gas tank, they all emit exhaust while there. Sitting in line waiting our turn to pull up to the pump, our cars are running and putting out exhaust gas.

Not all gases and vapor in exhaust emissions are harmful. Water is one of the vapors that is emitted in exhaust, and water vapors themselves are not harmful. However, many of the vapors are harmful. Those include sulfur dioxide, carbon monoxide, and (the most dangerous hydrocarbon is this group) benzene.

But that is not all. In addition to these harmful gases is particulate matter called soot, the black smoke that comes from an exhaust. It is the residue left from incomplete burning of gasoline and emitted with the rest as exhaust. In diesel vehicles, which many pickup trucks and some cars use, much more soot is produced than regular gasoline engines. That is the primary reason some countries are trying to phase out new diesel engines.

There is no dispute that these vapors and fumes are dangerous to people exposed to them. Continuous exposure to these fumes over a long period of time has been shown to have harmful effects, especially in children.

3. Vehicle Repair and Service Work

Many gas stations still do repairs and service work on vehicles. In those cases, not only is there risk from gas, there is added consideration if there are any other chemicals used at the location for the repair shop. Solvents, antifreeze, and vapors from spray products may be present. Freon from air conditioners adds more toxins to the air. If the shop does painting, add even more vapors locals breathe in. In addition, there are leaking fluids from the vehicles themselves while they sit outside, needing repairs. Vehicle repair stations have a whole lot of serious health considerations. Tony and I never tell ourselves that if the business operates properly, there is no risk. We have been in this long enough to know there will always be fugitive emissions, leaks, and spills, no matter how vigilant and careful the place is.

If you are living near a service operation in addition to a gas station, this presents a double risk to you and your health.

4. Vent Pipes from Underground Storage Tanks

When I first worked with a group researching the subject of the dangers of gas stations, this hazard was a total surprise to me. After all the time I have spent looking at gas station hazards, it was news to me that they have a venting system attached to their underground tanks. Most people do not know about this, but it actually is sitting right in the open. Those who look closely enough will see several thin metal tubes on top of the building, or more likely, the gas pump canopy. The purpose of the venting system is to regulate the pressure in the underground tanks.

They need a mechanism to make certain the tanks do not have too much or too little pressure as the fumes expand and contract. The fumes regularly released by these vent pipes may exceed what is released at the pump. There is the hidden and almost always unknown danger of the fumes vented directly from the underground gas tanks to the air keeping the tank pressure stable.

Concerns over exactly what and how much these vent pipes released was raised by the scientific community because vapors from gasoline contain a large number of toxic chemicals. To reiterate, one of the most dangerous, and a known carcinogen, is benzene. A study was done by a group of scientists from Columbia University Mailman School of Public Health researching the vapors released from these vent pipes. Their results showed levels of emissions 10 times higher than deemed safe by the EPA. The levels of hazardous chemicals being evaporated at the vent pipes registered up to seven gallons of gas per day being emitted into the air from the pipes, an amount that would require an enormous number of little drops to evaporate from the pump. Other studies showed quantities of dangerous benzene fumes far exceeding those at the pump almost 1,000 feet away from the vents at the station.

Only some states have any regulations requiring a distance or setback between a gas station and another use like a school, apartment building, or park. In fact, some cities may have a gas station located right next to an apartment building or even a school. Living even 500 feet away from the toxic fumes released from these pipes can endanger people's health.

5. Leaks and Spills

Every hour, every day, many small gas spills occur at the gas stations as people fill up and use the gas hoses. These seemingly tiny and small spills over the course of the day add up. As the spilled gas evaporates from the ground, it adds to the levels of benzene and other chemical compounds in the air.

When we are at the gas station and have a few drips or a tiny spill, we usually shrug it off, figuring it doesn't matter. Researchers found, however, it really did make a difference. When there was precipitation, the runoff from the stations washed into storm drains, onto neighboring properties, and into the underground water supply for private and public water systems.

A study with a slightly different angle was conducted by Johns Hopkins Bloomfield School of Public Health. They looked at the cumulative effect of gasoline saturating the concrete. Results from their study determined that not only is runoff from the concrete a big problem, but some of the chemicals in the liquid gasoline will penetrate into the porous concrete material. More toxins and contamination will accumulate in the concrete pad under the pumps and fill areas. The researchers calculated that in a 10 year span, 1,500 liters of gasoline will penetrate through the concrete and get into the soil and groundwater. This creates contaminated water for neighbors and public water supplies with wells in the area.

Based on the amount of benzene in gasoline, a spill of 10 gallons of gas has the mathematical potential to contaminate 12 million gallons of water.

6. Leaking Underground Storage Tanks

All gas stations have underground tanks to store gasoline. Up until the mid-1980s, most tanks were constructed of a simple single layer of steel. The average lifespan of these tanks was anywhere from 15 to 30 years. The tanks simply corroded and rusted until they leaked. Back then, there were no monitors or devices to alert a station owner when their tank was leaking. Because most of the leaks started out small, they went unnoticed. The amount of gasoline being leaked quickly added up to thousands of gallons.

Once the gasoline was in the soil, it traveled downward until it hit the aquifer underneath and then would form a plume leading away from the direction of the leak. Leaks that soaked into the soil traveled as well, with the fumes from evaporation of the gasoline headed upward towards the surface. Gasoline-saturated soils also have plumes of contamination traveling through them and away from the source of the leak. Fumes work their way up to the surface to escape into the air. Depending on the distance the plume traveled and the size of it, fumes may also have been released underneath any buildings, similar to the way radon enters buildings.

Modern gas stations argue that they use the latest in technology to prevent any leaks, but leaks still do occur. They are much less common with the double wall plastic tanks and soil monitoring systems, but accidents occasionally happen. In particular, the nozzles and connections which attach hoses and pipes to these tanks are potential places for leaks to occur.

Moreover, even though many of the old tanks have been removed from the ground, there is still an existing threat from those that are still buried. Unfortunately, once the gas has leaked into the soil, removing the old tank does not solve the contamination problem that already exists.

Gas stations in areas that depend on their own wells for a water supply need to be extremely cautious about contamination from leaks, spills, and runoff nearby. In Towamensing Township, Pennsylvania, three houses next to a gas station had their well water contaminated by diesel fuel from a nearby gas station with a leaking tank. One resident ended up in the hospital after consuming some of the contaminated water and becoming sick.

In Longview, Washington, a 60 year old leaking gas tank at a gas station was removed, leaving behind contaminated soil and groundwater. Monitoring wells needed to be installed to track any further spread of pollution.

The past Attorney General of New York State said, "Leaking underground storage tanks are the largest single threat to groundwater quality in the United States." Already, in many areas, large aquifers supplying water to millions of people are polluted with chemicals from gasoline that leaked from these underground tanks.

A study by the U.S. Geological Survey sampled 225 water supply wells in Rockingham County, New Hampshire. In 40% of public wells MTBE a gasoline additive was found. Testing is being expanded, and the state has identified 228 high priority sites with past gasoline releases in the area in many of those. Over 100 wells tested positive for MBTE, which is linked to liver and kidney damage. The EPA considers the compound to also be a possible carcinogen.

According to one estimate, as many as 75,000 to 100,000 underground storage tanks leak millions of gallons of gasoline into groundwater each year. Some of the chemicals making up the gasoline mix with the water. People who live near these leaks would be exposed to these chemicals when they drink, bathe in, shower with, or otherwise use the water.

Living near a gas station is not a great idea because you are always being subjected to toxic chemicals, known as chronic exposure. Depending on the concentration of the chemicals and the length of time you are exposed, the cumulative effects may begin to affect your nervous system.

Some of the minor effects of living near a gas station include frequent headaches and nausea. Beyond that, there are links between long-term exposure to gasoline vapors and liver and kidney disease. Some of the more serious effects of long-term exposure include permanent organ damage, muscular degeneration, lung damage, brain disease, and several types of cancer. In fact, an analysis published in the *American Journal of Epidemiology* showed children living near gas stations had a higher risk of childhood leukemia.

Based on our review, Tony and I recommend 1,000 feet as the minimum distance you want to have from any gas station. Honestly, we are not even comfortable with that distance as safe. This is a dangerous place to live near. More distance is certainly better if possible. For people in higher-risk populations such as children and older people, the distance should be even higher. Hospitals, schools, and elder care facilities should never be closer than 2,500 feet at the minimum distance.

Chapter 28. Injection Wells

Wells are something we think of as a way to get water up from below the surface for us to use. Some industries, however, think of wells in the reverse fashion. To them, wells are an easy way to send liquids back down into the Earth from the surface. Instead of pumping them out, they inject them, hence their name. Injection wells are as creepy as they sound.

Taking liquid waste and disposing of it by sending it thousands of feet down into the Earth started back in the 1930s. The oil and gas industry started sending their waste liquids deep underground where supposedly they would be harmless. At the time, it seemed like an effective use of a well that ran dry; it put them to another good use. The logic seemed to be that they were filling the void left beneath the surface when liquid was removed, easily and cheaply disposing of poisonous waste, exchanging it for the profitable natural resources. In those days, the industry could pretty much pump and dump wastes wherever and whenever they wanted to.

In about the 1950s, the chemical industry, the people who discovered the miracles of chemistry, also discovered injection wells as well. The wells inspired the industry, most notably Dupont, to reverse engineer a well that would give them convenient places to dispose of and store toxic waste. They disposed of waste by creating a well several thousand feet deep and injecting from the surface incredible amounts of the most toxic by-products the industry makes. In some cases the wells used were those left behind by the oil and gas industry; others they created themselves.

Incredibly, the EPA thought this was a brilliant idea and approved it. Given some of their other policies, it's not entirely surprising, especially as chemical companies and their industry lobbying groups armed themselves with charts and studies showing just how safe this disposal method was.

The definition of an injection well from the EPA obscures its dangers: "An injection well is used to place fluid underground into porous geologic formations. These underground formations may range from deep sandstone or limestone, to a shallow soil layer. Injected fluids may include water, wastewater, brine (salt water), or water mixed with chemicals."

They are used to "place fluid underground in porous geologic formation." The ambiguity over what types of fluids are put into an injection well conceals the fact that they are used almost entirely to put toxic substances in the Earth. There are six different categories of these wells according to the EPA, and all of them are for the purpose of injecting some type of waste into the Earth or fracking, which still involves injecting poisonous chemicals in the first place.

My definition of an injection well differs greatly: "An injection well is a shaft that industry drills into the ground to find a place they can dispose of their industrial toxic waste, radioactive waste, contaminated fluids from oil and gas operations, and any other waste products they need to get rid of because it is far less expensive than other methods."

Before I move into the negatives I will point out two positive uses for injection wells but, use by industry for waste disposal outstrips the positive thousands of times over.

One positive use is injection wells are also sometimes used to pump excess surface water into underground aquifers to recharge them. If the water returned underground is uncontaminated, that is one positive purpose these wells are used for. Secondly, there is extensive investigation currently going on about using injection wells to sequester vast amounts of carbon dioxide in the ground similar to how they dispose of liquid waste. There is an entire class of wells the EPA sets aside for this. It was tried in the past but now may again become a way to reduce carbon dioxide in the atmosphere if it doesn't return to haunt us. With the exception of these two uses for injection wells, all of the rest are environmentally destructive.

One hazardous liquid being pumped back down almost as fast as it is being pumped out is fracking fluids. While there is supposed to be a difference between wells drilled for resource recovery and those for waste, the oil and gas industry is blurring the lines by reusing contaminated fracking fluid, which, instead of being disposed of in a regulated disposal well, are pumped back into the already depleted well. The gas and oil industry claims this allows them to get more yield. They also readily have permits for that purpose instead of a more difficult permit and regulation process for a well used strictly for disposal.

Another common fluid disposed of through injection wells is sewage waste from treatment plants. This has been ongoing for decades. Florida, for instance, had gotten rid of much of its sewage by injecting the waste deep underground, convinced it would never return. Surprise to them, it has returned in quite a few places.

Millions of gallons of hazardous wastes are routinely disposed of using injection wells as well. Some industries that are the largest producers of toxic wastes actually have one of these located on their property. It's a handy way to get rid of all the poisons they create right there.

For decades now, industries have used injection wells to pump over 30 trillion gallons of hazardous and contaminated substances underground. Across our country are tens of thousands of these wells being used as a toilet to flush industrial pollution down deep into the Earth. Scientists within the industry claimed the wastes would remain deep underground, never pose any risk to water supplies, never leak, and never travel through any fissures or cracks under the ground. It would not be possible for waste to travel back up to the surface either through the well they went down in or other wells that created pathways.

Those cracks and fissures that never were supposed to appear are now showing signs of strain. Up until fairly recently, industry and environmental officials continued to say that injection wells were a safe way to dispose of waste, incapable of failing, despite the growing information to the contrary that wells do fail frequently, creating large consequences for our water and people. All along, however, there were failures. There had been reports of cracked wells and wastes leaking into aquifers used for drinking water as far back as the 1990s.

The pressure needed to pump this waste down into the underground rock is tremendous. The walls of the wells may be rated to handle the initial pressure, but in many of these wells, they are not calibrated for the frequency and duration that the operators are using them for. Greed drives the industry to use the well for as long as they can until the well backs up and can't take anymore shoved down it. By then, it is too late, and waste already has created a path through the well into the surrounding ground long before the depth they were supposed to be sent to.

There are records filed with several state agencies such as Florida and California that people living near these injection wells are beginning to see more and more accidents happening from these wells. Some leaks date back to the 1990s. Quietly, over the past couple decades, people are being exposed to the contents of wells that have been coming back to the surface.

There are several areas of impact from these injection wells.

1. Water Contamination

Initially, the industry claimed that if waste is injected into a permeable area between two impermeable areas underground, the waste injected has no place to go. This perspective is rather absurd on the face of it. First of all, the Earth is not neatly layered like a brick wall with one layer of rock evenly stacked upon another. The various layers underground rise and fall like the surface, with high places, low places, and deep fissures and fractures opening up places where two areas are interconnected. Our water comes from aquifers trapped between layers but also from interconnected flows under the surface. Injecting fluids too close to the upper layers allows the waste to infiltrate the subsurface water and contaminate drinking water with this waste. That's what happened in Florida.

Near Bakersfield, California, some officials working with oil and gas companies made a massive error. Nobody will confirm whose mistake it is, but at least nine injection wells pumped over three billion gallons of contaminated waste into underground aquifers containing high quality water designated for drinking and irrigation. Upwards of 40 water supply wells are within a one mile radius of one of the injection wells pumping contaminated wastes.

In Northern Reeves County, Texas, an injection well used for fracking wastewater may be leaking thousands of gallons. Investigations show a sizable mound of land is rising near the site of the injection well. One of the lead researchers believes fluids from the injection well are not going down as they are supposed to but are spreading out and away. There may be cracks in the well casing that would allow this. It is also possible that pressure from the injection process may have blown out the well. In the meantime, there is growing concern about the contamination of groundwater in the area.

2. Return of Wastes

There is a growing number of instances of wastes coming back up, using the well shaft to travel to the surface. It is not just these injection wells that occasionally backup with waste; wells that have been previously used, perhaps ones used for fracking or old oil wells, are other routes the waste is using to resurface. If there are enough holes into the same underground area with continuous pumping of liquids and sludges at high pressure, these holes are almost like straws. The pressure will eventually cause the waste to find a release valve and rise. Because some of these places have hundreds of wells in a small area, it is a question of when it will happen, not if.

3. Leaking Wells

Injection wells are very susceptible to leak, especially if they are used for too long and with too much pressure. In Pennsylvania, EXCO Resources operates two injection wells in Clearfield County. In 2012, the company noticed the well was leaking brine, but continued injecting fluids into it for months without notifying the EPA. When they found out, the EPA temporarily shut down one of them and fined the company.

In 2010, contaminants from an injection well bubbled up in a west Los Angeles dog park.

Within the past three years, similar fountains of oil and gas drilling waste have appeared in Oklahoma and Louisiana.

In Romulus, Michigan, Environmental Disposal Systems began operating injection wells. After operating for 10 months, inspections revealed leaks in the piping of one well. The EPA shut it down. Additional investigations discovered multiple violations of the operating permit in the well's use, failure to provide records of monitoring, skipping required testing of the safety systems, and injecting wastes without an operator in case of an emergency.

Florida has long been the leader in disposing of their sewage waste using injection wells. They spend millions of dollars to drill thousands of feet down below the surface. Their scientists continue to state the waste will remain trapped where they are injected, that they will never escape. In the early 1990s their wells failed. A large release of partly-treated sewage contaminated fresh water aquifers that may be needed to supply their growing population.

In Miami-Dade, wastewater with levels of ammonia, chlorides, and fecal coliform from human waste have been found in the Upper Floridian aquifer, a source of drinking water for the area. A study of the wells by the Sierra Club concluded ten wells were not constructed properly and were missing extensive wall casing to enable the waste to travel into the confinement zone. The conclusion about the leaking wells and subsequent contamination of the drinking water source was that the supposedly impermeable geology was actually permeable. Unfortunately, once waste is injected and leaked, they are unable to be removed.

Florida, however, continues to insist it will use injection wells because this time, they say, these wells are going deeper, so the waste will stay where it is injected and never move. Failures, however, are inevitable. The waste will escape due to the nature of the Earth's structure and will travel upward and outward, contaminating everything in its path as well.

The safety record and serious violations these wells have is downplayed by the industry. If they lose access to this disposal method, it would create a financial disaster and force them to look for a real solution. There are many investigations going on by groups who are interested in exposing the truth about these injection wells. It takes a lot of time and digging through the data available, but the facts are indisputable.

One major review of information done by ProPublica showed an incredible number of problems with the wells. Routine inspections of the structural integrity of these injection wells showed a violation rate of one out of every six. During the three years of records reviewed, more than 7,000 wells had indications that the walls were leaking. Those walls are all there in between the waste and the aquifers we use for drinking water. Despite these problems, the operators never stopped injecting waste, even when they knew the wells were leaking.

One count puts the total number of injection wells at over 680,000. About 160,000 are used for toxic waste and sewage disposal. Every week more wells are drilled by the fracking industry. After using them to pump up oil and gas, many will be used to pump down waste. While the exact number of wells is not agreed on, there is agreement on one fact: the federal government and many scientists admit they have no idea how many are leaking, just that many of them are.

When it becomes so obvious a well is leaking and it no longer is safe to use, state agencies have no alternative but to close the specific well. Responding to these problems, Kansas shut down at least 47 injection wells in 2010 because their mechanical integrity could not be restored, while Louisiana shut down 82 and Wyoming shut down 144. Tens of thousands more may be leaking, but unless it is so severe, they remain leaking their poisons.

Out in an area known as Far West Texas, researchers from Southern Methodist University used radar imaging to examine a specific injection well that they thought exhibited signs of leaking. The researchers noticed the surrounding area near the injection well showed signs of the ground uplifting, and radar images showed as even more pronounced when the well was most active between 2007 and 2011. The conclusion was that the well was releasing the waste being pumped into it at a very shallow level, perhaps only several hundred feet deep instead of the much further depth the waste supposedly was being pumped to. Tests of the groundwater at the shallow level showed increases of chemicals found in the waste fluids and not occurring naturally in the quantity found.

The industry has run amok over the environment, and we the people will pay in the end. In 2017, the widely respected publication *Science Direct* published the largest study of fracking and its impact on health ever done. Researchers from prestigious universities such as Harvard participated in the documentation and analysis. The study concluded that without a doubt there are health effects on people exposed to fracking. The most significant finding was that infants within two miles of active wells had lower birth weights and more health issues than those further away; even a couple more miles made a significant difference.

How big does the disaster need to be before anything is done about the thousands of toxic waste dumps we created underground? We pushed the ocean to its limits and beyond before anything was done. It's rapidly becoming evident the Earth is showing signs that we've pushed it too far as well.

There is a whole new body of research on the connection between injection wells and earthquakes. In areas where heavy fracking takes place and much of the wastewater is injected into the underground, a higher number of tremors is being recorded. The fluids under high pressure are making the faults, which are already prone to slip, more likely to do so.

In North Texas, a series of quakes in a concentrated area of these injection wells is almost certainly related to fracking activity in the area. The pressure of waste disposal through injection wells could double the possibility of an earthquake. This gives even more urgency to getting real regulation of injection wells before huge catastrophic accidents happen.

Injection wells are in a class by themselves with regards to abusing the planet and creating a future catastrophe. The purpose of an injection well is simply to inject something an industry wants to get rid of below the surface of the Earth. Apparently, if we can no longer see it, it must have gone away. They may be wonderful to the industry as the handiest dumping grounds, but the costs to the environment eventually will catch up, and this time they may be uncontrollable and unfixable. Damage to the water supply will be measured in centuries and millennia, especially in cases of highly radioactive fracking waste disposal.

If you live in an area where injection wells are used, your water supply may already be contaminated with some levels of toxic waste. In time--maybe years, maybe decades--many of these wells will fail.

The best we can say is use extreme caution if your water supply is anywhere near the same place any type of injection wells are. The technology is unsafe and always has been. Failures will increase until the EPA will be forced to act against the desires of the industry. That will be too late in many places as trillions of gallons of waste are already underground.

Staying a mile away from these wells, regardless of what the substances being injected by them are, will afford you some protection from spills and leaks on the surface by equipment and transport vehicles. Fracking waste injection wells in which hundreds of thousands of gallons of radioactive and chemically contaminated wastes are injected are especially subject to spills. If the wells leak, either in the casing near the top of the wells or in the areas the wastes are injected to, aquifers supplying water to the area may be contaminated. Regardless of where your water supply comes from, test the water annually.

Chapter 29. Sand and Gravel Pits

Sand and gravel pits are in every state in the nation and in almost every county someplace. They are open pits that grind up rock and gravel and dig sand up from dry river bottoms and old glacial valley floors. Most of the sand and gravel is used for construction projects. Tons of the material is used as aggregate for making cement, asphalt concrete, roadway bases, fill material, and many other construction-related projects. Access to aggregate material for building and construction projects is critical to maintain economic growth and continued building and repair of critical infrastructure such as highways, bridges, airport expansions, and the like.

According to the US Geological Services, in 2018, over 970 million tons of sand and gravel were produced by approximately 3,800 companies digging in 9,350 sand and gravel pits.

When towns and cities were smaller, in some places just several decades ago, many of these pits were outside of town limits and not usually close to any neighborhoods or areas of dense population. As such, they could make as much dust and noise as they needed to. However, as the towns and cities grew and expanded, many of these pits became surrounded by residential neighborhoods and schools. All of that dust started to become a health hazard, and all of that noise started to drive people a bit crazy.

While it may seem that the locations of sand and gravel mines are haphazard, there are two driving factors why sand and gravel pits are located where they are.

Most obviously, deposits of sand and gravel in large quantities are usually only located where large river beds had been in the past. Many of these areas go back to the periods when giant glaciers came through the area, pushing and dragging incredible amounts of rocks and sand. When the glaciers receded, they left behind deposits, some of which are hundreds of feet deep. In addition to the old river beds, many valley floors are also rich with large deposits of this ground down material. So, there are limited areas where large enough deposits can be found to make it worthwhile to mine them. Many of these places run through areas that are quite populated and therefore can only be mined from still undeveloped land, making it harder to find places to mine.

Once the deposit is known to be large enough to use, the second factor in determining the location of the pits is whether it is economically feasible to mine the deposit. Feasibility, in turn, depends on the distance between the pit and where the material is going to be used. Transportation is the most significant variable in the decision. As a result, there is a financial incentive for sand and gravel pits to operate near developed areas.

The more developed an area becomes, the more the people living near sand and gravel pits try to close them down and stop any new ones from opening. Places like Boston, where there are no sand and gravel pits left, bring aggregate materials for building projects down from New Hampshire in trains. Without that aggregate, Boston can't do many of their construction projects.

Sand and gravel pits are difficult for a town trying to regulate them. Pits can't just open anyplace they like; they need to be where the deposits of sand and gravel are. At the same time, laws are very specific when it comes to natural resource development. Many towns find themselves out of options and legal recourse in controlling many of the operations. In many places, zoning regulations are frequently overruled when it comes to sand and gravel pits. Many times state laws allowing the extraction of natural resources take away the local right to restrict these operations.

Usually, all local governments are able to do is place conditions on the permit to protect the health and safety of the residents. Even these have been challenged in the courts and, when too restrictive, have been overturned. Local laws are able to regulate noise, hours of operations, travel routes, and some of these types of impacts. Apart from that, local towns most adhere to state laws.

Unfortunately, many state regulations controlling sand and gravel pits are quite lax. Some, such as Texas, do not even have requirements for onsite monitoring of air quality to measure particulate matter levels. They do not even ensure the pit operator restores the land after they are finished taking out what they want. In Texas, they can simply walk away leaving a deep hole in the ground even in the middle of a developed housing area. The industry does have to register with the state and obtain a permit for crushing rock and submit a dust control plan. However, no testing is done to see if the sand and gravel pit is in compliance with the plan they submitted. It does not seem fair that companies that are mining for natural resources are not subject to the same zoning rules as other businesses, but for now that is written into federal regulations, and the industry's lobby works tirelessly to keep them intact. There is a need for tighter regulations but a financial incentive to leave the laws as they are, creating a standstill.

However, in part because of real threats to locals' health, some pits are closed down. There's even a movement to not allow them to operate near populated areas. There are a number of specific risks from living near a sand and gravel pit.

1. Crystalline Silica (CS)

This danger has been around for decades, but with the explosion in fracking and the need for specialized sand to blend with fracking fluids, this has become a larger problem than ever before. Many new sand mining operations are opening up to supply the needs for this sand, and with that are exposing thousands of people to the dangers. The biggest danger to people working at these pits as well as living nearby is from CS. This material is one of the basic parts that makes up sand, soil, granite, and many other types of natural materials processed for our uses for sand. According to the federal government's

Occupational Safety and Health Administration, OSHA, "When these materials are crushed and ground, the CS becomes particle matter (PM). Much of the Crystalline Silica in particulate matter is small enough to be inhaled. This material is listed by the EPA as a human lung carcinogen. It is known to cause cancer. Also, exposure to and inhaling CS may cause silicosis. It gradually causes lung tissue to scar. The scarring reduces lung function and the ability to take in oxygen for the body. It can also lead to other respiratory problems."

The serious health effects of exposure to CS dust have been known by the medical field since the 1930s, but regulations and some protections were only put into place by OSHA around 2017. For decades, it was known that working in mining operations and other types of industry that generate large quantities of CS particulate matter has caused cancer, lung disease, and other health issues in many workers. Now, however, there are CS-generating mines in much greater numbers and much closer to large populations than ever before. Instead of the industry being alarmed about the potential for widespread harm to people, they are, in fact, doing the opposite by denying there is an issue and moving full speed ahead to get the profit they can before there are any new restrictions and regulations placed on their industry.

While research into the effects of these sand pits is limited, enough already points to dangerously high levels of silica particles in populated areas nearby. New research and reviews of past studies are being done by many universities and groups such as the Civil Society Institute and Environmental Working Group. The results are expanding and reinforcing the knowledge that CS and other particulate matter are extremely harmful to people living nearby. Sand mines for fracking are one of the worst environmental hazards of recent times. Unfortunately, it is already too late for the thousands of people who are being exposed to the silica and have it deep inside their lungs.

In Chippewa, Wisconsin, researchers tested air samples near sand pits for silica particles less than 2.5 microns. Remember that particulate matter this size or less does not get trapped on its way to the lungs but penetrates the lungs going deep into the air sacs.

What the researchers found was levels of tiny silica particles two to three times the safety level recommended by other states.

Westerly, Rhode Island, is famous for their beautiful beaches. They also are famous for the extremely high quality granite the quarries in town have produced for almost 200 years. The granite mined from the quarries is composed of almost 90% quartz, which is the most common form of crystalline silica, a known carcinogen, and a recognized hazard since the 1950s. The dust and particulate matter from these quarries infiltrates the neighborhoods around their borders. One man who, at age 47, was diagnosed with silicosis, the lung disease associated with high exposure to CS, never worked at the mine or any industry where he came in contact with it. He does however, live at the boundary of the quarry. Across town near one of the elementary schools, children play outside. However, in Westerly, the number of children who have suffered asthma attacks needing emergency room visits is the highest in the entire state. In 2016, Westerly Hospital noted almost 50% of their patients have chronic asthma or chronic obstructive pulmonary disease.

The industry likes to refer to this material they coat the nearby areas with as "rock dust" rather than the carcinogen it is.

2. Particulate Matter

CS is not the only thing locals are breathing in. Dust and particles are continuously churned up as a normal part of sand and gravel pit operations. Overall, particulate matter is the largest component of air pollution from the pit. While silica dust is one part of the dust, the rest is made up of other compounds released when the rock is crushed and ground. The particulate matter runs the spectrum between grains as fine as talcum powder, which easily gets absorbed into our lungs, up to the larger particles, which seem to coat everything outside with a dusty cover no one can keep up with.

When the mine is in operation and the grinders are running, depending on how the winds are that day, people living nearby may feel like they are at the beach trying to get the tiny grains of sand out of their mouths. It seems unstoppable. Small particles will find their way through every opening, every tiny crack, and into their houses, their cars, and everything else.

The EPA does try to enforce the Clean Air Act when a sand and gravel operation consistently violates it. However, it takes many months and a sizable amount of work by the people being affected to even get the ball rolling. With thousands of these operations and very limited enforcement staff, many sand and gravel operators are left to self-compliance.

In Barre Town, Vermont, a quarry was denied a permit to operate their rock crusher as the state Environmental Commission monitored it and said it made too much noise and emitted carcinogenic dust into the surrounding areas. For six years the quarry operated the rock crusher in spite of not having a permit to do so. For six years the community and state agency battled with the quarry operators. Finally, the Vermont State Supreme Court told the quarry to shut it down. The quarry closed down the rock crusher but almost immediately started running a new one they installed 3,000 feet away, also with no permit to operate.

3. Noise

Conveyors, crushers, and grinders run all day at a pit. If their size warrants it, some have multiple units. Noise levels can easily exceed 100 decibels at times. Rock grinding equipment and dump trucks generate noise levels comparable with a motorcycle or leaf blower running 50 feet away. While there may be some limits on hours that the pits can operate and how loud they can be, constant loud noise can be very unnerving on a body assaulted by it all day. Many times there are noise barriers and buffers set up at the quarry to mitigate some level of the noise; however, they are only so effective at reducing it. Some regulations that regulate noise levels use average daily noise levels, so the noises could be 120 decibels for a few hours but 100 for the rest and still be in compliance.

Almost similar to some other hazards such as airports, the continuous noise will eventually take a toll on the nervous system and entire body.

In San Diego, California, neighbors brace themselves for the weekly explosions at the quarry next to their neighborhood. One time, rocks from the explosion flew through the air and fell on four homes in the nearby neighborhood.The people living in the area say dust coats the insides of their homes and everything else, but despite all that, they find the noise to be the worst part. The noise from the equipment starts up at 7 every morning and runs through the day.

In Antioch, Tennessee, neighbors near the quarry are dealing with more than just the negative effects on them from the noise. They also talk about the cracks in their foundations and walls that blasting at the quarry has caused. Despite ten complaints filed against the quarry in one year's time, the owner says he is not the cause of the damage. His quarry follows all state guidelines, and the house damage must be environmentally caused. In the meantime, the neighbors wait for the next round of explosives to go off.

4. Truck Traffic

Obviously, the sand and gravel needs to be shipped off site. The trucks carrying the materials try to get as much into them as the road weights they travel on allow. Volume is the key to profitability. A constant flow of trucks entering and leaving with their loads can be expected, as can the traffic they cause on local roads, the dirt and dust from the load inside, and any damage and wear and tear on the roads from the sheer weight of the loaded trucks. Some towns will require that all loaded trucks be covered with a tarp to prevent sand and gravel from blowing out, but they're the exception. Even in towns like that, nothing, however, will prevent rocks from falling off the trucks as they bump along the road. The diesel fumes the trucks will put out contain the ultra particulates the EPA and WHO label as known carcinogens.

The roads the sand and gravel trucks use are the same roads families and residents travel. They are the ones children walk along to school, and these same roads run through quiet residential areas on the way to and from many of the quarries.

While flying rocks breaking your windshield may not seem like an environmental hazard, the potential for having an accident certainly is worth pointing out. The quarries point out that it may only be 90 trucks a day. That amounts to 20,000 truck trips per year, a not so small number after all.

5. Water Impact

There are two types of pits: wet and dry. While all operations need water in the processing, the mine or quarry is referred to as a wet operation or a dry quarry. Some sand and gravel pits are naturally full of water, and materials need to be dredged out. Others are totally above ground and the water table. Either type can lead to water contamination. In many sand and gravel pits, chemicals known as flocculants are added to the sand to remove unwanted minerals and other particles during the washing stage. Then the wastewater is disposed of usually being routed into a nearby one or ditch running away from the pit. Once in the water supply, they are ingested by people. Long term exposure to many of the flocculants used can lead to blood problems, nervous system issues and increase risk of several types of cancer. There are currently no regulations on these chemicals in drinking water to protect you. Gravel as well is usually washed before getting loaded onto trucks for transport to its destination. Any small particles of other minerals or compounds are washed away and drained.

Many gravel and sand deposits are sometimes found in the beds of old rivers and valley floors where the water table is naturally high. When the mining operations start removing sand and gravel, the pit fills in with water. The sand and gravel are dredged out and run in conveyors to the grinding and crushing equipment. When water is pumped out of the pit, it needs to go someplace, and that usually ends up being any body of water in the nearby area. A small stream could be turned into a small river of water tearing through a farm field down the road from where the pit is emptying out thousands of gallons of water. The addition of all this new water expands present ponds and streams to where they may flood lands they had been on and leave people's property muddy and under water.

In St. Croix, Wisconsin, a quarry saw its water containment area destroyed after a rainstorm and the resulting torrent of water that escaped into local creeks eroded a berm on the railroad tracks and covered a creek in sediment. Both needed immediate repairs to avoid further damage and flooding.

In some locations the gravel or sand mines may come from hills or the sides of mountains that are blasted with explosives and dug away. Large quantities of water are needed to wash and add to the materials as they run through the conveyors and sorting and crushing equipment. When water is pumped out of the ground to use at the operation to clean sand and gravel and to help movement in the operation, the local water table is also affected. People with wells higher than the level the pit is pumping their water from could be lowered and potentially run dry.

On Yacolt Mountain in Washington State, people who have lived in the area for years experienced issues with their wells once a new gravel quarry opened up operations. A retired hearing examiner for the county with knowledge about other similar operations elsewhere fueled concerns about well issues even before the mine was permitted. Now, apparently, blasting into the fractured rock and water withdrawals for the gravel processing has led to wells running low and in some cases, dry. The company denies there are any activities at their quarry which would affect their wells.

In West Rockville, Pennsylvania, 40 homes within a mile of a large gravel mine have had their wells run dry over a four year period. In a six month period, ten of those wells dried up. The quarry denies they are the problem and the state Department of Environmental Protection is backing them up. While the quarry was responsible for a couple of them and reimbursed the property owners to install new wells, the rest, they say, are not their fault, and they will not pay for any other new wells. The quarry does withdraw extensive amounts of water from underground aquifers in the area of their operation. In the meantime, homeowners are dry.

6. Future Restoration

In better regulated states, there are legal requirements when the original permit is issued to include plans for closure of the pit. Sometimes, a bond must be posted to ensure this will happen. However, in many places, there is no requirement to do any restoration and the pit is simply abandoned once they are finished with it. When the mine pit fills in with water, they cover up the hazards beneath such as open shafts, old broken machinery, or how deep the area the water may actually be. According to information on the website Geology.com, each year dozens of people die in abandoned open pits and quarries. Drowning is the number one cause of death, followed by ATV accidents. These open pits look like inviting places for recreation, but they have potential to kill people playing in them.

In 2019 in Farmington, Connecticut, a father and son were enjoying a day riding their ATVs around an old quarry. The quarry was abandoned and unused since the 1980]s. Both the father and son were killed when they fell off the edge of the quarry and down 75 feet to its bottom.

Some people blame the company for walking away and leaving the site a dangerous place for people who see it as a place for recreation. Others blame the individuals who get hurt for being in a place they really should not have been. Either way, sand and gravel pits that are not restored create a hazard for the community.

7. Carbon Monoxide and Diesel Exhaust

Sand and gravel pits use many pieces of equipment. Conveyors, crushers, grinders, and all the trucks emit constant flows of carbon monoxide. The levels of this can at times be as high as a well-trafficked roadway. Living near one of these pits, one is bombarded by the constant smell of engines, forced to breathe in high levels of carbon monoxide. Exposure to carbon monoxide can cause headaches, nausea, and vomiting. Long-term low-level exposure may lead to neurological issues including difficulty concentrating, emotional issues, and difficulty breathing.

In South Amherst, Massachusetts, a sand and gravel mine was fined by the EPA for violations to the Clean Air Act. The operation had been cited for emitting excessive particulates into the air surrounding the pit and for creating high levels of carbon monoxide from the equipment exhaust used at the operation. Neighbors in the area were grateful to finally get relief from the health effects of the emissions.

Sand and gravel pits are a hazard that have the potential to cause permanent damage to your health. There is no way a sand and gravel operation can be compatible with residential property nearby. Due to the nature of the business, it is impossible to entirely stop all of the particulate matter, especially the CS and the fine particulates, from blowing away. Attempts have been made by operations to continuously spray water onto the piles and over the roads in the pit to hold down the dust. However, that is only effective to a point.

Living near these pits is a hazard and should be avoided as much as possible. If you happen to be in an area where these gravel pits and fracking sand mines are proposed or operating, hopefully you can get the town to propose very strict regulations to at least minimize the extent of damage you are going to inevitably have to endure. Stay at least a half mile away but we prefer to see one mile between you and the operation.

Chapter 30. Contaminated Fill

The soil we see around our house has a past. Sometimes if we look into that history, it would tell us things we may not want to hear, but it could save us from major health problems later.

One of the first steps in any construction project is to make sure there is a nice base to build upon. Sometimes that means needing to bring in fill and other material from someplace else. This is an expense the developers have to add to their costs unless they are able to get the fill for free or perhaps even have somebody pay them for allowing fill to be dumped at their construction site. This sounds too good to be true. How could it be possible for the developer to get fill for free or get paid for fill?

The answer to that question really depends on why the fill is needed at the site. If all they are going to do with the material is dump it into a large area to fill in a hole, then some old demolition debris would work fine. If some roadway was torn up recently, the old concrete and asphalt would be fine to use as fill. They could most likely get that for free. Maybe a very large office complex is being built nearby and there was a lot of extra dirt from preparing the site that is not needed. That could also be used on the site.

In the construction industry, it is not unusual for waste from somewhere else to be used as fill for some needed purpose on another project. Part of it is always about the expense of disposal, but part is the fact that there is nothing wrong with reusing old concrete or debris from a demolished building, for example, as fill. Some construction debris landfills have equipment to break up materials so they can be more easily used as fill by contractors. When clean fill is used, all is good.

There may be small differences between towns and even different states as to what exactly clean fill is. The generally accepted definition is uncontaminated material including rock, soil, used asphalt, and construction or demolition debris such as concrete or bricks. Depending on the construction manager, testing may be required to confirm the fill is indeed clean and not contaminated with substances or materials that create a liability later on.

Then there is the other type of fill: contaminated fill. This may contain almost anything. Old insulation, household trash, chunks of metal and wood, cardboard, plastic, dry wall, dirt removed from other sites, and hazardous substances of every type imaginable. Many neighborhoods and homes were constructed using contaminated fill. Some of the people did not know the fill contained harmful substances, but others did.

There is an entire industry that has grown up around providing fill. Some are simply brokers who find it at one place and sell it for a profit at another. However, there are the shady opportunists who operate a business making big dollars disposing of contaminated fill by passing it off as clean fill. Toxic dirt brokers look for opportunities to pose as legitimate contaminated soil removal businesses. They get paid large sums to take away the dirt under the guise they will dispose of it according to environmental regulations. Many times, the dirt may not even make it off the back of a dump truck before it gets sold and dumped at some unsuspecting construction site and passed off as clean fill. Where there are millions of dollars to be made and low risk at getting caught, many disreputable businesses will be born.

Unfortunately, there are way too many people who put profit above people. Tony and I thought we had seen most of the ways unscrupulous people dispose of hazardous waste, but we were mistaken. We do believe that honest mistakes are made sometimes. We also believe there are instances somebody truly has no knowledge about something and, therefore, is not purposely doing something that they know will cause harm. Sometimes, though, people are responsible for doing something so outrageous it defies explanation of how they could be so cruel.

To combat these opportunists, many states are passing bills meant to stem the disposal of contaminated fill. New Jersey recently passed theirs after state hearings found the problem found the practice to be a bigger problem than ever expected. In 2019, one of these dirty soil brokers was sentenced to a mere 90 days in court for dumping contaminated soil for over eight years at sites across the state. The contamination included the pesticide chlordane, chemicals from petroleum products, and PCBs. Only be a fraction of the sites the convicted soil broker dumped the waste at may be known; the others are still sitting poisoning the unsuspecting property owners and occupants.

Because the practice of selling and using contaminated fill is rampant, the examples we have here are a small cross sample of the different ways contaminated fill gets used.

In New York City and in New Jersey, hearings and an investigation into soil brokers found an entire industry making millions in the toxic dirt trade. In one example contaminated soil was removed from projects in New York City and sent over to New Jersey, supposedly for cleaning. The same toxic dirt was sent back to New York City and used at other construction projects.

Contaminated dirt disguised as legitimate fill is nothing new. Disposing of highly toxic and contaminated soil has been going on for decades. No place is exempt either. Sometimes the toxic secret stays buried until a lucky accident causes it to be exposed, such was the case at Greenwich High School in Greenwich, Connecticut, one of the fanciest and most expensive towns in the entire country. Back in the 1970s, more than 28,000 tons of fill dirt containing PCBs, arsenic, lead, chromium, and other toxins were dumped to build athletic fields. For years, school children played and rolled on top of this mound of toxic chemicals unaware of what lay below the surface.

The worst part of this is, the town was told at a public meeting at the same time the fill was dumped that it was old incinerator ash and power plant coal ash. Eyewitnesses reported recently they saw all the black sludge ash dumped and reported it but that nothing was done at the time. The fill was just covered over and used. It is currently unknown if the exposure and contact with the PCBs and other toxic chemicals can directly be tied to any illnesses and diseases the people who were exposed experienced. One fact is certain though, the investigation into how this happened and went undetected for decades will continue as will the confirmation of medical issues due to this horrible example of neglect and disregard for people over profit.

In Florida, the location currently being considered as a site to build a one billion dollar soccer stadium and shopping center involving David Beckham, was suddenly shut down in mid-2019. The site was an existing golf course that was operating for years. While many old golf courses are known to have issues with arsenic and other toxins from years of pesticide applications, this golf course was also built on top of a dump containing tons of old incinerator ash. The local government says they are working to clean up the locations that are heavily contaminated. Their plan is to cover them up with topsoil only where needed, and not remove the toxic contaminants. Since the officials say it was typical to use incinerator fill back in the 1970s for construction projects, there are parks and other public areas that have this arsenic and lead contaminated soil too. Regulators with the county and environmental health departments acknowledge they have no idea how many other polluted locations may be out there waiting to accidentally be discovered.

On Long Island, a toxic fill dirt ring was broken up in 2018, and over 30 people were arrested. The ringleader was sentenced to two to four years in prison. The ring placed ads on Craigslist advertising free screen fill, but the trucks would dump solid waste, usually contaminated with arsenic, lead, and mercury. The homeowners were stuck with the clean-up bill while the toxic dirt dumpers made money from finding places to illegally dispose of the toxic fill.

In Pennsylvania, the Department of Environmental Protection found haulers from out of state were trucking in contaminated soil removed from toxic sites elsewhere. Over sixteen sites were identified as having been victims of the dumping scheme. Hazardous materials ranging from petroleum-soaked soils to cancer-causing cobalt, the banned pesticides aldrin/Dieldrin, lead, and vanadium (another cancer-causing material). The DEP is warning people to be vigilant as these soil brokers are getting more brazen.

In the town of Portsmouth, Virginia, homes were built on top of an area where foundry sand had been disposed of for years. Most foundry sand contains very high levels of heavy metals, including lead. By the time the magnitude of the problem was realized, many of the children in the area had high levels of lead in their blood. Before the health problems arose, the town had never done any testing on this site, and the people whose health was affected will pay for it for a long time.

In 2014, a Wisconsin environmental group traced drinking water contamination to the use of coal ash as fill material. There are hundreds of school children who are not allowed to use the water in the school due to contamination. In a rural, four-county area in southeast Wisconsin, many of the homeowners have had to install costly water filtration systems or depend on bottled water for a safe supply. The wells throughout the area are pumping from aquifers where the groundwater has unsafe levels of molybdenum in it. This toxic metal is found in coal ash, now known to have been extensively dumped throughout the area as fill. Almost 1,000 wells were tested, and, of those, forty five percent had levels of the metal at high or very high levels.

In several neighborhoods in North Birmingham, Alabama, people had a flooding problem. When there would be heavy rains, the streets and all around their homes would flood. The solution the local government came up with was to bring in fill dirt and use it to reduce the levels of flooding. There were some nearby industrial facilities where the fill was obtained for use around some homes as well as some future developments. Unfortunately for the people living there, the fill was contaminated with industrial chemicals, including lead and arsenic. Every time it rained, toxic chemicals spread all over the area. It took years of fighting, but finally the government declared these neighborhoods, which people had been living in for years, a Superfund site.

In Davidson, North Carolina, asbestos was found in a local park. Previously, a factory that manufactured asbestos products was located nearby. Town officials have said this now-closed factory created health and environmental problems in the nearby neighborhood for years. People in the area of the factory used its soil as fill dirt for all sorts of projects. The EPA tested the soil in 77 of those properties surrounding the factory, and 32 of them had asbestos levels so high it required removing more than 6,000 tons of contaminated fill.

In Philadelphia, Pennsylvania, back in the 1970s there was a big plan to build a housing community in an available marshy area of the city. The original plan was to develop 12,500 homes. The old city landfill was the closest neighbor. Because the land for the housing was a marsh, fill needed to be brought in to build up the area. Fifteen million cubic yards of fill dirt was brought in for the project along with contaminated material from the landfill. All of this was spread out and used as soil for the neighborhoods. Of the original plan, only 4,200 of the homes were ever built.

Over the years many of the people living there became sick. Some of the original residents filed a lawsuit against the city, but it never really went anywhere. Neighbors sold their homes and moved on, and new people came to take their place. All of them had one thing in common: the fill used on the land under their houses, in their gardens, and all over their backyards was heavily contaminated. For decades, people lived there and raised their families. Many decades later, the EPA finally investigated. Tests showed poisons such as PCBs, lead, mercury, arsenic, pesticides, and petroleum residues. Almost 200 homes needed to have the soil dug up and removed. Over 4,000 tons of PCBs and contaminated soil have been taken out, and there is still work to be done. The people who were harmed by the unknown contamination are trying to do the best they can.

In Huntersville and Mooresville, North Carolina, residents in both towns share something nobody anywhere should ever have to: thyroid cancer rates far above the national average. If that was not enough, they also share clusters of ocular melanoma cancer also above national average. They also share coal ash. A troublingly long list of hazardous chemicals is in the ash. The developers of these homes in North Carolina used coal ash as free fil. And they used as much of it as possible to help the coal plant dispose of this assumedly free fill. Residents in the area tell stories of when they would plant or dig outside and a gray substance would come up from the dirt. They had no idea it was toxic and filled with hazardous chemicals. For years they played in it, ate food from their gardens that grew in it, and breathed the dust it created. Now, the people in these two towns wonder who will be the next one in their town to get sick.

In Atlanta, Georgia, on the west side of town is a group of houses in a low-income neighborhood. They just discovered the soil around their houses is contaminated with high levels of lead. Many of these homes have young children, for whom lead is especially hazardous to brain development. The lead was part of the byproduct from the metal smelters that used to operate in the area. A material called slag was used as fill when the area was built. Now, the people are unsure what will happen to them and their children. The EPA is looking to see if the problem can be removed. In the meantime, the neighborhood waits.

Contaminated fill is a big problem and is only going to get worse. Many sites where this fill was used in the past are being discovered all the time. One of the main hazardous materials allowed to pass as fill is coal ash. It is known to contain many hazardous chemicals and harmful compounds, but the EPA continues to allow its use in places where people will be in close contact with it.

Add in all the construction that has already taken place on top of contaminated sites where nobody ever tested, and there are plenty of potential possibilities for problems. Legislation, testing requirements and vigilance are what it will take to get control of this problem.

The only known way to make sure you are not living on top of contaminated land is to have your own basic soil test done. Moreover, if you are buying fill for a project on your property, you should test that soil too. A "clean fill" sign in front of a business is not a guarantee the material is safe. Even if the business itself is above board, because of the toxic dirt market, reputable establishments may unknowingly wind up with contaminated fill.

Chapter 31. Junkyards

Known as junkyards to some people, salvage yards to others, and recycling centers to others, the environmental risks are the same no matter what it's called. Depending on your age, you may remember a television show called *Sanford and Son*. It featured a father and son team who ran a junkyard together. In that junkyard, they had piles of every conceivable item that modern society has used and thrown away over the past 80 years.

While *Sanford and Son* was a television show, the depiction of a junkyard was fairly close to reality. There are tens of thousands of junkyards scattered across the United States. Many are literally just a pile of junk dumped someplace out of the back of a truck as a cheap place to dispose of unwanted stuff. One large example is along a creek in Trenton, New Jersey. The pile started small with somebody dumping a mattress and a few pieces of old furniture. Over time, tires, appliances, more mattresses, and demolition debris followed, making the pile into a large illegal junkyard filled with throwaways.

Some junkyards, though, are real businesses that serve a purpose as intermediaries between people getting rid of something and people repurposing it on the other end.

For an easier discussion here, we will talk about three different kinds of junkyards, which all together make up almost all of the ones operating across the country today. There are junkyards that specialize in selling automobile parts from cars and trucks no longer running. Another common category is junkyards that tear apart old automobiles and trucks into pieces and separate them into metal for recycling to the metal industry. Lastly, there are junkyards like *Sanford and Son* that have piles of almost anything that may have any value at all but found a place among old piles of washers, refrigerators, toilets, baby carriages, etc..

1. Automobile Parts Specialist

The first group of junkyards are automotive parts specialists. The history of automobile junkyards goes back almost 100 years. Over the past century, cars and trucks have been manufactured by the millions in factories all over the world. When their useful life is over, many of these have simply been dumped or left in a field, backyard, or building to rust away. The idea of operating a business by accumulating these old vehicles for other uses can be traced back to at least the 1930s and '40s. Some historians say it was the Shipman family in Minnesota; others say it was the Lewis family in Georgia who had the first auto junkyard business. Many more got into the business during World War II for the value of the scrap metal and rubber these old vehicles had. Others realized there was more money to be made in selling the old individual parts that were still useful even though the car no longer ran. Whether it was the Shipmans or the Lewis family, a whole new industry was born around the disposal of these millions of no-longer-running vehicles.

These junkyards are a great place to replace a car part at a cost much lower than a dealer. Many body shops buy salvaged parts from junkyards too. If original, new equipment is not specified on the contract when they work on a car, there is a good chance they will order a used part. Insurance companies also encourage the practice because they love being able to use a salvaged replacement part for a car instead of the cost of a new one. There's nothing wrong with installing used parts other than that they usually don't tell the customer. There is not much difference between a used car door and a new one.

Many of us, myself included, have walked up and down the rows of these cars and trucks looking for a vehicle like the one we own. From the old junkyard car, we too can salvage a needed part at a price much lower than we would have to pay to buy it from a car dealer or regular auto parts store. I have replaced door handles, trunk lids, car seats, and other items over the years, saving myself hundreds of dollars and recycling these parts back into use.

This is the main purpose of this type of junkyard. They are for recycling parts from wrecked and no-longer-running vehicles. Whether for a two year old Honda or a 1932 Plymouth, salvage yards are sometimes the only place we can find an original replacement part. Some of these salvage yards specializing in certain classic cars do hundreds of thousands in business each year. Many of them are known across the country as the go-to place for a Corvette part or other classic vehicle. Many of these salvage yards are run as professional businesses and take every precaution to be a good neighbor and protect the environment.

These automobile junkyards can be a small group of a dozen rusting cars to a huge metal recycling operation spread out over 20-30 acres. There are some vast junkyards, such as Desert Valley Auto Parts, claiming 10,000 autos to choose parts from, or Turner's Auto Wrecking in Fresno, California, with 100 acres of old cars to pick parts or bodies from. Others specialize in motorcycles such as All Bikes in Rye, Arizona, owned by Ron Adler, estimated to have over 9,000 motorcycles in his yard before they went up in smoke in 2013.

2. Automobile Junkyard

While some of the millions of cars disposed of each year may go to automotive parts yards, far more go a different way. According to statistics from the Bureau of Transportation Statistics,11.047 million cars were scrapped in 2014 in the United States. They can no longer serve their purpose to provide us with transportation, so they are broken down to what they are made of: plastic, metal, rubber, foam, and other materials.

This is the second type of junkyard. Old cars go here to die and have no value but scrap value. Automobiles are made up of about 50% metal. The rest is plastic, glass, upholstery, and other material. These junkyards are dedicated to getting every dollar in value they can with the least effort from the remains of the old vehicle. The metal is the most valuable material to recover and make money from. The rest is usually garbage.

What happens in this scrapyard usually depends on what type of equipment the junkyard has. Some of the largest junkyards can take an entire car, place it onto a conveyor belt, and crush it into a small, dense package to reduce the volume. Then, the entire crushed car is sent into a shredder that rips it apart into small pieces. The pieces are sorted out in another conveyor line to metals, plastics, and other materials for recycling or disposing of.

One machine like this operates in Jersey City, New Jersey, and can reduce a car to tiny metal scrap in 10 seconds. Some days the giant machine shreds 4,000 tons of metal into scrap. In Elkhart, Indiana, another monster machine dubbed the Mega Shredder chews up about six cars per minute. It was even featured on *Modern Marvels* on the History Channel. The Mega Shredder only operated for a couple of years before it was shut down and the company went bankrupt. There was tremendous opposition to the shredder from the community because the vibrations were felt all over the area. When the company closed down, the property where the junkyard sat was declared a hazardous waste site with the groundwater polluted by materials from the vehicles it recycled.

However, most junkyards can't afford these large machines, so they use a more labor-intensive method. Some have small shredders to tear the vehicles into smaller pieces and sort out the metals using magnets. Others use blow torches to cut away the biggest chunks of metal they can and remove the engine blocks. Hammers, torches, grinders--they can use anything that will separate out the metal from the rest of the car. The large portions of metal that are cut, crushed, shredded, or chopped apart are sent out to be recycled. The rest of the material, 50% of the car by volume, is shredded remnants or piles of other materials, known as fluff in the industry. This lovely mixture of foam, plastic, rubber, glass, fabric, and more is also contaminated with rust, fluids, dirt, and anything else that was in the vehicle and not removed with the metal.

The metal scrap is sent off to giant smelters to be melted down and reused. Some junkyards actually have smelters operating on their property that present significant environmental risk for the area. Luckily, most don't, so the metal gets shipped off, and the fluff either gets dumped into a big pile or sent off someplace out of the way.

3. Junk Junkyard

These types of junkyards usually will also accept any type of metal or recyclable material of value from whoever can drag, pull, or push it in. Cars are usually the favorite as they are heavy with metal and easily able to be sold to dealers for scrap. Other items of interest to them include washing machines, scrap wire, furniture, old shelving, pallet racks, metal ducting, and so on. The scrapyard's main interest is to buy anything at as low a price as possible and sell for as high as possible to the recycling broker or someone else who may want the item.

Scrapyards such as these are usually located in urban areas located near industrial sections or within poorer neighborhoods. Bags of aluminum cans and plastic bottles all come in with washers, dryers, and broken air conditioners. Many of these junkyards also handle industrial scrap, like old machinery, empty metal drums, and broken parts. Along with the industrial scrap may also come old transformers containing PCBs and residues in metal drums from solvents and toxic chemicals. Sometimes, even, there are junkyards that will accept drums of unknown liquids and material and disappear it among the piles.

In the end, a junkyard is a junkyard. It doesn't really matter what type of junkyard is nearby. The environmental risks for all of them are quite similar. It does not matter whether the cars are used for parts or scrap metal, or if the yard contains washers, refrigerators, or whatever; they still sit outside exposed to the elements. Over the course of time, they lay there rusting, rotting, and leaking. These junkyards are not simply eyesores to the community and neighbors. They are also potential hazardous operations. In many cases, they are handling materials that are toxic and hazardous to people and the environment. In almost every community, there are laws carefully licensing and regulating these operations because, in the past, serious issues created health problems and very expensive clean-up bills. As is almost always the case, it was the people living nearby whose health and property were affected, and it was the taxpayers of the town who had to pay to clean up the mess. Most of the sloppy junkyard operators, when harassed too much, simply close down, file for bankruptcy, and move on, leaving their destruction and toxic messes behind.

While there are strict regulations governing the procedures of draining and disposing of the liquids, operators may look for shortcuts. Economics dictates compliance. It costs money to properly dispose of these wastes, and the chances of getting caught is very low. One of the huge problems with scrapyards is that even though there are regulations in place with respect to air emissions, hazardous waste storage and disposal, and water and soil contamination, there is usually no enforcement or inspections. Inspections usually only take place after a complaint is made related to a bad odor, sloppy appearance, or the smell of Freon being released into the air.

Unethical operators have become excellent at covering their tracks. Investigations have shown a variety of cost-cutting and potentially dangerous ways salvage yards do this. Their records may show a regular pick up of hazardous waste materials by a licensed disposal company, but this may only represent a portion of the actual volume of liquids drained and collected from the vehicles. The rest may be drained inside of a building directly down a drain leading to the municipal sewer system. There may be small drums of these poisons thrown into the salvage yard dumpster to be collected as regular solid waste. At worst, it could just be drained onto the ground to seep into the groundwater and run off to the adjacent properties. There are simply too many places that need to be inspected but very few inspectors. When inspections do happen, many times the operators do not correct the violations, and it could be a few years until they are inspected again.

An article written by *American Recycler*, a large recycling industry news organization stated, “Scrap metal and automotive recyclers handle a wide and complex variety of hazardous substances that could potentially trickle down from storm water runoff and contaminate drinking water or pollute rivers, lakes and oceans."

Much of the negative environmental hazards occur when the cars are being dismantled for the reusable parts the scrapyards want to resell. Toxic materials travel to you in several ways.

4. Water Pollution

This is without a doubt the greatest danger from these junkyards. This is not just the potential for a little bit of pollution in the water; this is the possible contamination of an entire town's aquifer.

A major problem with these salvage operations is when old vehicles are first brought into the junkyard, the fluids in these vehicles are not properly drained. Over time, an assortment of toxic liquids seep from these rusting, rotting hulks into the ground. Antifreeze, oil, transmission fluid, and especially highly toxic battery acid create a puddle of toxic soup that works its way into the ground beneath the junkyard. If the yard is simply a processor and quickly crushed the cars for scrap, the fluids simply mix together into a toxic soup.

Groundwater and surface water pollution are of primary concern. Many of the toxic materials in automobiles find their way through the soil into aquifers below, contaminating wells. Freon from air conditioning systems, asbestos from old brake linings, lead from batteries, and many other metals and toxins are in old vehicles, and those chemicals can make their way into the soil. Oil leaking from old cars is a common pollutant found in groundwater near junkyards. Surface pollution under the vehicles runs off during wet conditions and winds up in streams and creeks nearby. Health effects of ingesting water contaminated with oil, gasoline, and these other pollutants can lead to respiratory system damage, liver damage, and increased risk of cancer. Hydrocarbons and heavy metals in gasoline also may cause reproductive problems and birth defects.

Acid from old and cracked batteries leaks. Since lead acid batteries are not usually accepted at landfills, they wind up at scrap yards. If ingested, lead causes cognitive and developmental issues in children. In adults, it causes neurological issues.

Antifreeze and coolants that are not drained simply leak out of the vehicle as it sits or when it is crushed. Antifreeze, when it gets into groundwater, can quickly contaminate many gallons of a water supply.

Many city and town water supplies are also contaminated with the gas additive MTBE. This additive was phased out in the late 2000s and replaced with another product for oxygenated gas. However, many of the existing cars in junkyards had it in their gas tanks when they were disposed of. Health effects of MTBE were studied and still remain somewhat unknown in humans. The EPA did conclude though, there is evidence showing it is carcinogenic in animal testing. Children with asthma, the elderly, and people with autoimmune diseases may be more susceptible to toxicity at lower levels. Because of a strong odor and taste in water contaminated with MTBE, it makes much of the water supply undrinkable when present.

Mercury is one of the most dangerous pollutants found regularly in junkyards. Not many people are aware that many switches in vehicles have mercury in them, which by law are supposed to be removed, collected, and inventoried. Even if the fluids are removed properly, the switches usually are not. These switches rot and will eventually leak mercury onto the ground. There currently are quite a few states that encourage salvage yards to remove switches by paying cash in exchange for collecting them and turning them in. In the meantime, the rest of that mercury is ending up in the water supply. Mercury is highly toxic and causes damage to the brain, liver, and kidneys.

Asbestos from brake linings in older cars commonly gets into the water supply as weather corrodes the car. Ingesting levels of asbestos in drinking water has been known to cause intestinal polyps.

Heavy metals used in the car itself are subject to corrosion of the body and parts. Metals such as arsenic, zinc, copper, chromium, aluminum, lead, and cadmium are commonly found in measurable amounts in water near these facilities. These metals, if ingested by people through contaminated water, accumulate in their body tissues. A person does not need to ingest a large amount at one time. Very small amounts over time will make a person sick too. Copper can cause kidney and liver damage as well as anemia. Cadmium can lead to lung disease, bone defects, and renal failure.

5. Soil and more Groundwater Pollution

As these liquids seep into the soil, they eventually find their way into the groundwater. This contamination can spread out like a plume from the junkyard and wind up in the nearby properties' wells. There is very little that can be done at the operation site to correct the problem once it has occurred. Remediation would necessitate the removal of all the soil from the area, as well as pumping out and purifying the groundwater beneath the site.

Licensed and unlicensed facilities like these exist everywhere across the country. These junkyards number in the tens of thousands. New regulations in quite a few states recently addressed this problem by requiring that a car, prior to being placed in a junkyard, must have had all its fluids drained and the battery removed. This is a great way to keep any new facility that is just beginning operation in check, but for all the existing facilities out there, it may already be too late. The surrounding area's water may already be contaminated.

In Kingsbury, New York, a junkyard operating a car crushing business was fined over $350,000 for pollution and violating state regulations. The junkyard was located in an area with residential properties on three sides and a school on the fourth. They were repeatedly cited by the state environmental agency for illegally discharging harmful chemicals into the surrounding air and water. Nearby neighbors complained about headaches, difficulty breathing, nausea, and noise from the junkyard. The state also contends the junkyard discharged oil at the site itself and the surrounding area

In downtown Phoenix, Arizona, an auto salvage yard operated for years recycling old cars and shredding the remaining 25% of material left into fluff. Unfortunately, the fluff contained many toxic materials, and after years of operation, extensive contamination was discovered in the fluff piles. Dangerously high levels of PCBs and various heavy metals such as lead, cadmium, arsenic, and chromium were present in the soil. The site was so toxic that the EPA declared it a Superfund site. Cleaning the site required the removal of almost 6,000 tons of hazardous waste. A cover of concrete was eventually poured over the top to contain what was unable to be removed.

In Lynn, Massachusetts, a junkyard was shut down for illegally dumping hazardous waste on its property. Police went to the site and said gasoline, antifreeze, and oil were being dumped onto the ground everywhere. A member of the town council expressed concern the contaminants were going to run into the Saugus River, which is along the property border. The town ordered the junkyard to cease operations, but there are reports people continued to do work there and have been seen in the middle of the night. One town council member witnessed people taking a load of oil drums, tires, and other materials and dumping them out back. After police responded, many buried items were discovered. Crushed up fiberglass, gas, and oil appears to have been buried as well. In 2018, the city finally was able to shut down the junkyard. Environmental evaluation of the damage is ongoing. Several communities depend on the Saugus for part of their drinking water supply, so the stakes are high.

In Potsdam, New York, a junkyard where metals were prepared for recycling is being cleaned up after years of toxic chemicals contaminated the ground and wetlands nearby. The state estimates the cleanup will take a year and cost upwards of $10 million. The company who operated the junkyard contaminated the soil with cancer-causing PCBs and many other toxic chemicals. The groundwater is not used for drinking or other household purposes, so for now the potential for poisoning the water supply is minimal. How much of the contamination traveled off the site during the years and landed on nearby properties is an open question nobody is looking into. The plan for cleanup is to remove the most contaminated soil and leave the rest on the site. The lesser contaminated soil the state wants to place in a lake on one area of the site. Most likely, this property will be classified as a brownfield down the road and allowed to be used for a project another developer proposes.

6. Air Pollution

There are quite a few ways the air near junkyards carries hazardous materials to the surrounding area. Freon from old air conditioners, refrigerators, and other appliances is emitted into the air as the old materials are drained and released. There are regulations strictly governing how to deal with old refrigerants because they are so harmful to the environment, especially the ozone layer which Freon gas destroys. Freon can also be dangerous to people. Inhaling Freon and other refrigerant gases cuts off the oxygen supply to the brain. Other symptoms of exposure to the gas include irritation of eyes and throat, dizziness, and nausea. Usually, the exposure to gas vented at a junkyard is not severe enough to cause poisoning, but the gas certainly is a hazard to inhale.

Old types of Freon such as those found in vehicles at junkyards are usually the ones no longer in use. Some junkyards follow the law by capturing the old Freon and disposing of it properly, but many licensed and illegal yards just bleed it out into the air because it's free disposal.

In addition, most junkyards operate shredders and crushers that are constantly creating airborne particles of all the metals and contaminants as the machines buzz along, cutting through chunks of metal. The metal that appears to go missing when someone uses a blowtorch or other cutting or chopping equipment to make a cut goes someplace. That place is into the air.

Another major junkyard concern with later model cars is the airbags. Airbags contain sodium azide, which is an explosive. It will burn the skin if it comes into contact with it and causes major irritation to the lungs if inhaled. It is considered a hazardous waste when disposed of. When a car is crushed in a compactor, if the bags are not properly removed first, they explode, releasing their chemicals into the air. Because airbags are worth money in the parts market, junkyards try to recover them when possible, but many end up exploding, and many more were already blown apart. The sodium azide just blows off or gets washed away.

Some junkyards also operate onsite smelters in order to melt down recovered metals to recycle them. This may sound eco-friendly, but, factoring in human health, it is one of the most dangerous operations a community can be exposed to. Smelters send heavy metals, especially high levels of lead, into the air. Quite a few Superfund sites are in locations where smelters operated for years, blowing toxic levels of various poisons into the community for everyone to breathe. Many people have documented cancer, kidney disease, lung ailments, and other illnesses directly related to the operation of a smelter nearby.

In Fairfield, California, a recycling center that scraps appliances was caught burning hazardous waste at their facility. In addition to illegally disposing of hazardous waste, they were also cited for improper storage of the waste. Further investigations determined the scrapyard also had been draining refrigerant from appliances for years by cutting the lines on the appliances to let it drain out. They were cited for removing hazardous materials they were neither trained to handle nor permitted to remove. Refrigerants need to be removed in a special way with special tools to ensure they do not escape into the atmosphere, releasing the harmful compounds. This scrapyard was using a bolt cutter and a container of water. A fine of over $145,000 was given to the company.

In a residential neighborhood of West Oakland, California, a scrapyard/recycling center was melting down aluminum and other metals for the repurposing market. Elevated levels of lead, mercury, manganese, nickel, and arsenic were all detected by local environmental health officials in the air at emission levels over the safe limit. Health effects of exposure to these include asthma, kidney disease, cancer, and neurological disorders.

In Claymont, Delaware, a plant that recycles scrap metal was sending high levels of particulate matter into the neighborhood. Heavy metals were detected in the air, including lead and mercury. These metals cause developmental disorders in unborn babies, learning disabilities in children, and blood diseases and nervous system disorders in adults. The residents knew something was wrong with air quality in the area, but they needed to call in an outside environmental consultant to do air monitoring. Armed with new information, the residents got the state to finally do something. The state environmental agency found the plant had been melting down switches taken from scrapped cars and emitting mercury in levels much higher than those reported to the state. This plant ranked in the top 20 emitters of mercury in the nation for several years.

In Houston, Texas, a single metal recycling facility had a total of 41 fires and explosions over a four month period. An overall investigation by the Texas environmental agency determined many facilities that process metal for scrap emit high levels of cancer-causing chemicals and heavy metals.

Some scrap yards go so far as to burn the rubber coating off of metal wire so they can recover the wire for recycling. Boilers onsite use old oil from the car to heat their buildings. Open air burning may be used to dispose of items that have no scrap or resale value to avoid disposal costs. Sometimes they use an incinerator to dispose of the useless scrap. Most times, practices like these are unregulated. However, when solid waste, especially plastic, is burned, it creates dioxins.

These are only most of the potential reasons a junkyard is a risky place to live near. But these are just the most likely reasons; there are a few more Tony and I could probably have added, but it would be overkill. Junkyards do not make good neighbors.

Give any type of a junkyard at least a half mile distance to protect yourself from any airborne contaminants. If you use well water, have it tested twice a year, especially for volatile organic compounds frequently found in petroleum products and heavy metals. While a half mile is a good buffer, Tony and I actually prefer a full mile distance as these places present so many possible problems. Further is always better when possible.

Chapter 32. Trash Transfer Stations

The journey of our trash is a long one. From our homes where we throw out our household waste to the place it finally goes, the entire trip can sometimes be hundreds of miles. One town located outside Albany, New York, has a landfill where thousands of tons of New York City trash head every week. Shipments of trash from New York to Alabama and other places over a thousand miles away are fairly routine cities all over the country. The trash transfer station is one stop along the way.

A trash transfer station is exactly what it sounds like. It is an operation where materials to be disposed of, specifically trash, are brought for future shipment to another destination. It helps to consolidate smaller amounts of trash, making the transportation from that point more efficient. Trash transfer stations are found in both densely-populated, urban areas and sparsely-populated, rural areas alike.

In the urban areas, these stations are usually used for household and commercial waste picked up by the garbage trucks we are used to seeing on the roads. Since the final disposal facility for the trash is usually a long distance away, multiple garbage truckloads of trash are dumped and transferred onto much larger trucks, or even trains and barges, for the trip to its final resting place. This also allows the garbage trucks to fill up several times a day closer to where their customers are instead of wasting time driving further away to dump.

In many rural areas, trash transfer stations usually are also used for garbage trucks to drop loads for transfer, but many communities may not have regular trash pick services available. In that case, the residents need to have a place to bring their household trash on a regular basis. Depending on the amount of trash generated, giant dumpsters may be set up at the transfer station and hauled away when full.

In both rural and urban locations, trash transfer stations may also be collection and drop off points for local hazardous waste collection drives, spring cleanup drop off, yard waste, unwanted junk too big for pick up, and others. Sometimes, these transfer stations are located inside a building, and others are completely outdoors. Urban transfer stations tend to be mostly enclosed, but the surrounding yard may be extremely messy and loaded with tell-tale signs of what is going on inside. They usually have an endless parade of garbage trucks dumping load after load of trash into a huge pit where it is compressed and shoved into huge trailers or rail cars for transport. Some of these facilities have very heavy truck traffic during operating hours and an extremely bad odor much of the time, especially during warm and or muggy weather. Transfer stations in places like New York City fill barges with trash from their transfer station for heading down the river and who knows where.

While these facilities only accept non-hazardous waste to transfer, over the years there have been cases of trucks of liquid hazardous waste being poured into the solid waste and disposed of quietly. It was a frequent but illegal practice that was exposed by a major investigation into the solid waste industry back in the 1980s. Federal and state investigations in New York, New Jersey, and Florida determined that organized crime had taken over numerous solid waste operations and was using them to dispose of hazardous wastes. The hazardous waste was being mixed with household waste at transfer stations in many cases and was then dumped at landfills. In Florida alone, state law enforcement officials documented at least 18 solid waste companies that organized crime had taken control of and may have been using transfer stations and landfills for illegal dumping. In New York and New Jersey, dozens of waste companies and several large landfills were controlled by organized crime. The investigations covered the transfer stations as well as the landfills under the control of organized crime.

In Bergen County, New Jersey, there was a big investigation into the trash industry. At the time there was a transfer station near where I had lived where hazardous wastes were being brought in and soaked into the trash before being moved to the landfill.

In spite of tighter regulations over hazardous waste disposal over the years, there remains many opportunities for continued illegal activity. With all of this activity going on, there are some unavoidable issues that will happen. Due to the nature of the transfer station and the amounts of material they handle, there is no way to totally avoid or fix some of the issues.

1. Truck Traffic

One of the biggest complaints from people, especially in urban areas, is the amount of extra traffic the transfer station generates. After school and in the morning, there usually are children on their way to or from school. Trucks and kids do not always mix well as both are trying to get someplace, usually in a hurry. The streets are unsafe with the hundreds of trucks all rushing to dump their loads to go out and pick up more trash. Children frequently dodge the trucks as they walk along the streets to and from school. A motorcyclist was killed by a garbage truck earlier in the year in Jamaica, Queens. Sometimes transfer stations are located close to a highway exit or in an industrial neighborhood, but that is not always the case for many.

At many of the larger transfer stations, there will be a daily flow of trash trucks using local residential streets to dump their loads at the transfer station. Depending on how the collected waste is being shipped to the disposal site, there may be additional large trucks entering the facility to be filled before heading to the disposal site. While the truck may only take ten minutes to dump its load into the pit inside the trash transfer facility, there may be several trucks lined up idling, waiting for their turn to dump. The average trash truck holds 12-14 tons of trash, which averages out to roughly 800 homes of residential pick up. How many homes are on the truck's route determines how many times the truck needs to enter and exit the facility daily.

The exhaust fumes from the trucks creates a health hazard because their diesel engines emit particles known to contain chemicals that can cause cancer. Very few trash companies use anything other than diesel trucks although several are slowly moving towards methane-powered ones. The diesel fumes hang low in the air, especially in narrow city streets where much of it mixes with the air residents breathe. Effects from the constant addition of these fumes aggravates asthma, causes respiratory issues, and irritates nose and airway passages, causing sore throats. People with chronic obstructive pulmonary disease are especially susceptible to diesel particulates and may experience more attacks as a result.

In September, 2019, a Jamaica, Queens, resident recalled first moving to the neighborhood and enjoying walking down the street each morning to grab a coffee and sit in the park to watch the people and happenings. Then the trash transfer station opened. Daily, an estimated 350 garbage trucks rumble down the street to the station to dump over a thousand tons of trash. One family had to move after their children developed breathing issues from the particulate matter in the diesel fumes. Once relocated to a new neighborhood, the children no longer had respiratory problems.

2. Odors

The problem of odors from transfer stations comes from wet and rotting food waste. Contaminated packaging, spoiled vegetables, old meat, and food scraps all contribute to an unavoidable stench, no matter how hard the station tries to suppress it. Some are more successful than others with enclosed areas with ventilations systems, odor suppression systems, and special air curtains for the building. There will be bad days and not-so-bad days, but the good, odorless days will be far and few between. If the neighborhood is lucky, though, the transfer station operators will try to be a good neighbor as much as possible.

Odors, while they may be offensive to the nose, are actually composed of gases and particulates from chemicals and decomposing wastes carried into the air. That smell of rotten eggs is actually sulfur hydrogen gas, which, when inhaled, is an irritant to our lungs and respiratory system. Small particles of food waste, dust, and dirt from the garbage all mix in the air and are inhaled. Bacteria growing on food waste and germs from disposed tissues or personal care items are contained in the air near these transfer stations. Increases in asthma, coughing, eye irritations, and colds are all possible health issues from exposure to these germs.

In rural areas, the transfer station is usually an open air yard. These transfer stations tend to be located on the edge of town. Depending on the local population growth, they may already be surrounded by residential neighborhoods. Many rural transfer stations have several large roll-off containers residents can use to throw away their trash. Sometimes they are enclosed with doors on the side that slide open and closed to minimize open trash. If there is also a garbage company in the area using the transfer station, there may be a ramp with large open containers the trash trucks can dump into. Usually, stations who receive garbage truck traffic have personnel to assist with dumping the trash. The truck traffic is much lower due to fewer pick-ups, but the trash may sit at the transfer station longer, waiting until the larger transport vehicle is full enough to take the trash to the disposal facility. In the warmer months, the decomposing trash may get awfully disgusting between pickups to take it away.

Even in urban areas where the trash is removed from the transfer station daily, it may have been sitting out behind the restaurant or your house for several days before the regular weekly or bi-weekly pickup is scheduled.

3. Trash in Neighborhood

There are usually regulations requiring all trucks carrying trash either in or out of the facility to be tarped and covered. No matter how well the drivers comply, there will always be some trash along the roads from the trucks. It is a common sight to see a garbage truck driving down the street with plastic bags blowing off or a few foam cups flying away.

There is also the trash that blows away from the site off the incoming trucks or during the transfer and transport. Most times there are perimeter fences to contain loose materials, but some trash, like plastic bags and foam cups, are especially easy for the wind to catch. Even if the trucks are tarped and the building fully enclosed, there still will be noticeable trash escaping its boundaries.

Many well-run transfer stations will send a crew out to pick up trash along the main routes into the facility, but not many places have the budget to be able to do this on any consistent basis. Many will also say that is neighborhood trash and all their trucks come in tarped, so it's not their problem. Whether it falls off the garbage trucks on the way in or blows off the trucks on the way out, expect garbage in the neighborhood along the streets. While the thought of plastic bags and old styrofoam food containers may not sound very threatening, there is more. Neighbors have reported finding large chunks of rotting meat, leachate leaked from the truck on the streets that smells so bad it makes them nauseous and in one case, used hypodermic needles apparently from illegally disposed medical waste.

4. Birds and Vermin

Anytime there is a place where garbage collects, there are going to be animals because all they see is a free and easy meal. Open transfer stations have to deal with birds flying in and grabbing exposed trash. Rats, raccoons, and other vermin slip through fences and inside buildings in both urban and rural stations to feast on the large amounts of edible food waste available. Traps may regularly be placed around the perimeter and inside the facility, but they are only somewhat effective at reducing the numbers of animals looking for the easy handout. Sometimes, these facilities drape netting over the large doorways where trucks enter and exit the building to keep birds out, but, again, it is impossible to completely control this problem.

Birds are often seen circling overhead at the landfills; they know where the food is and are going to try their best to hang around to eat.

At a transfer station in Washington, DC, the problem got so bad and the issue went on for so long that the district finally used eminent domain to shut down the private transfer station. Since it opened back in 1988, it has been a constant target of complaints and lawsuits. In spite of new ownership taking over the operation, neighbors are fed up with the flies, rats, and other vermin the place has attracted. Rats are known to be carriers of diseases, and the trash attracted an increasingly large number of them into the neighborhood for years.

In Evanston, Illinois, people have been living with a transfer station for years, and very little has improved since the odors and rodents started infesting the neighborhoods. The neighbors have been trying to have the transfer station closed but have not been successful. Worried about the germs carried on the rats and other vermin in the neighborhood, the residents keep many of the children off the streets after school, children who normally would be out playing. The neighbors, like other places with transfer stations, suffer from respiratory issues from the dust and dirt and inhaling the sulfur fumes in the air.

5. Neighborhood Decline

Let's face reality. No neighborhood is going to welcome a trash transfer facility. Most of the facilities are in areas that are already run down and predominantly industrial or in poor neighborhoods where operators feel the poor people have no political clout to stop them. As property values in the area decreases, even more commercial businesses may move in, bringing other hazards with them. With those industries come a whole host of other risks posed to the neighborhood.

While the vast majority of these transfer stations are located in poorer neighborhoods or industrial areas, they are sometimes near neighborhoods many would not think they belong by. Availability of real estate and location to transportation of the collected waste may drive the decision where the transfer station goes. On the upper east side of Manhattan, a 10 story high waste facility is being built. The location is along the river so barges will be able to be loaded with waste from the transfer station. The city of New York will operate the transfer station 24 hours a day, six days a week. Upwards of 500 trucks per day are expected to dump trash there. The neighbors tried to stop the facility from being built based on the traffic, noise, pollution, diesel fumes, and odors, but lost their battle. Many residents sold their apartments in buildings nearby for $2-3 million, glad to leave, but unhappy property values decreased several hundred thousand dollars as a result of the transfer station. Those who stayed behind or rented these once-lavish apartments are now stuck with the consequences of living near a trash transfer station.

6. Noise

Depending on the size of the operation and how much trash it handles, the equipment needed could be quite significant. Larger transfer stations frequently use large bulldozers and front end loaders to move trash around. Mechanical floors and giant crushers churn away inside, preparing trash to be reloaded. During operating hours there will be noise from equipment within the facility as well as the ever-present trucks with diesel engines delivering the trash. Conveyors moving trash from the tipping floor into containers runs almost non-stop throughout operating hours.

The noise of the giant diesel engines on the garbage trucks, which weigh about 33,000 pounds empty and upwards of 50,000 full, rumbling past all day affects the people in the neighborhood. The nervous system in the body responds to these noises before anyone can realize and stimulates the system into thinking it needs to react. The continuous triggering of a response affects the entire nervous system and causes stress on the body. Communities such as Richardson, Texas, East Williamsburg, New York and Evanston, Illinois know first hand about the rumbling of the trucks on the road, the constant whirring of the machinery and the banging from the bulldozers that assaults their bodies every day taking a toll on their nerves, increasing blood pressure and interfering with rest.

Transfer stations serve a valid purpose in the chain of trash disposal needs. Without them, not only would it be more difficult to get the trash to the disposal facility, it would also be much more expensive to do so. At the same time, much more could be done to control at least some of the problems and issues they create for their neighbors.

Living near a transfer station can make your daily life less than pleasant. Even when the operators do their best, there is still unavoidable truck traffic and noise. Odor control at best will still not be enough for you not to notice this is your neighbor.

If you have a choice, being at least a half mile away from these transfer stations would help limit your exposure to the effects of the odors, hazards from gases, and vermin in the area. Truck traffic may still go past your home, so checking on the routes these transport vehicles use should still be checked. If you are already living near a transfer station, installing an air filter in your homes and keeping your windows closed may minimize intrusion of odor carrying hazardous gases and dust particles. If the permit for the transfer station has requirements for hours, noise levels, and odor controls, make sure they are following them. If they are not, report the facility to the health department and document what is wrong.

Chapter 33. Railroads

Sometimes when Tony and I are doing research into a subject, we like to ask people who are totally unaffected by what we are looking into their opinions. It helps us get two views of how people see the same thing. One of the questionnaires we put together was focused on gathering information from random people about trains. We asked people what they think when driving along and seeing a train moving down the tracks with a string of cars behind it. We heard a bunch of different answers from them. They brought up thoughts of "I always wonder where that train is going," "How many cars are on that train," and, "I wonder what is in those cars." Most responses were fairly neutral. There were even a few people who thought about what it would be like to just get on the train and go wherever the train was headed.

Then we asked people who specifically lived near train tracks the same set of questions. Most of the people living near the railroad track think about the noise associated with the trains when they pass. They think about how long the train is going to last until it stops vibrating. They talked about how long they sit at a railroad crossing near their house, frequently waiting 10 minutes or longer while the train passes through. Sometimes, they said it could be almost a half hour as the freight train goes back and forth, hooking up to new cars and dropping off others.

Living near railroad tracks can have many more negative impacts on people living nearby than simply being noisy. There have been quite a few studies on the risks and potential health effects caused by proximity to a railroad track.

However, just as there are types of roads, there are types of railroad tracks. Some railroad tracks are more dangerous than others. Factors such as the volume of traffic makes a big difference in pollution levels of air emissions and noise. Freight train lines, as opposed to passenger lines, present risks from the hazardous cargo they move and the potential for accidents.

For those who near railroad tracks, the following issues need to be taken into consideration when evaluating the potential harmful effects.

1. Ultra Fine Particulate Matter

The majority of train engines burn diesel oil for fuel. Diesel oil is not a clean burning fuel. When combusted, the portion not burned is given off in the form of exhaust containing ultra fine particles. On top of that, diesel exhaust contains many toxic chemicals. The size of the particles is especially significant for the toxic ones. Most larger particles get caught by our nose or outer breathing areas and are expelled from the body. Once the particle size decreases down to fine and ultra fine, these particles are inhaled directly into our lungs. Once inside the lung, they attach themselves and do damage. The particles are also transported into the body from the lungs where they then travel through the bloodstream. The health risks associated with these are lung disease, cancer, blood diseases, and others.

The World Health Organization has linked diesel exhaust from train engines directly to lung cancer. Their science panel also upgraded diesel exhaust from a "probable carcinogen" to the highest level: "carcinogen."

In San Francisco, an area known as Mission Bay is subjected to particulate matter from the trains' diesel engines up to 20 hours a day at times. The fine particles cling to everything and get into the houses through the smallest cracks. The residents in the area have much higher rates of asthma and respiratory issues.

The Mayo Clinic conducted a study on children living close to railroad lines. They concluded that there was a 40-70% higher rate of asthma in children near the tracks than further away. The high levels of fine particulate matter usually found in diesel engines triggered onset of asthma symptoms and more frequent attacks in children involved in the study.

A study done by researchers at the University of Washington examined if any additional pollution was added to the air by the type of cargo the train was carrying. Since millions of tons of coal move by train every year in Washington state, that was a great place to look. They conducted additional research into the specific types of pollution that would only be found if coal was present. The study indeed found when coal carrying trains passed through the targeted study area, levels of particulate matter from coal were also found in the air.

2. Right of Way Areas

The areas on either side of the tracks and of course the track itself are considered the railway company's right of way. They need to keep the area clean and trimmed of any vegetation that may interfere with a train passing.

The usual method of clearing the track area is using herbicides in a broad application across the tracks and out several yards on either side. Every living plant needs to be killed, and the railroad usually only does this once a year, therefore needing a highly effective, long-lasting killer. A blanket spray applicator is usually mounted on a work car running on the train lines.

Herbicides have been known to drift into people's yards and schools along the tracks. It frequently also drifts and kills private gardens and plants in yards.

This haphazard application of herbicides is one of the more dangerous situations about living near the tracks. The railroad company claims they use control methods for drift. However, because of the large area they need to spray and using work trucks on the tracks that they are shooting the herbicide from, drift from the herbicides in the air to surrounding properties is almost unavoidable. The usual method is to have large tanks of chemicals in the bed of pick up trucks specially designed to run on the railroad tracks. The truck moves along the targeted area using their jet sprayers to hose down all the vegetation and areas on either side of the tracks.

The railroad and their contractors like to use the most powerful herbicides they can because the longer vegetation stays dead, the less maintenance they need to do and the less they need to spend. Past herbicides have included similar defoliants used in war zones to kill every living piece of vegetation.

Because many of the best chemicals are now banned, one of the favorite chemicals used now is known by its market name: Roundup. This chemical has already been labeled by the EPA as a known/likely carcinogen because it has caused bladder cancer, kidney cancer, and breast cancer in studies with laboratory animals. However, if the government allows the use of the chemical to keep tracks clear, the industry will use it until concerned citizens force them to stop.

Other herbicides used in right of way applications include Triclopyr, which is highly corrosive to the eyes and skin. Short-term exposure may cause difficulty breathing and tremors; long-term exposure may cause reproductive and developmental issues. There are several other chemicals used to control vegetation, and most of the studies on health effects are done by the manufacturers themselves, seeking approval to use these chemicals in widespread applications.

Such was the case with Roundup and the active ingredient glyphosate. It is now known after many years of use to be linked to liver and kidney damage, reproductive and developmental issues, health risks to children and pregnant people, and cancer. In 2018, a jury unanimously agreed the product was responsible for causing the health issues and the company was guilty for failing to warn people of the risks associated with the product and awarded over $250 million in damages to the plaintants.

In Virginia, children playing along a fence in the back of their day care were directly sprayed by railroad workers with a blend of herbicides and pesticides, including Roundup. Three of the chemicals used for the spray have been linked to reproductive problems, neurotoxicity, and cancer. The pesticides in the air were inhaled by the children and blanketed their exposed skin. Some of the children experienced mild pesticide poisoning with irritated nose, throats, and eyes in addition to nausea and skin irritation. Three of the children experienced acute pesticide poisoning with symptoms including bloody noses, diarrhea, and skin blistering. The company that sprayed the chemicals on the children had previously been fined by the Kentucky Department of Agriculture for 809 violations.

In Rhode Island, a railroad company was caught spraying herbicides along their tracks in violation of state regulations. The poison they sprayed affected wetlands and also killed off vegetation in people's yards along the tracks. They were ordered by the state to stop, but much of the damage was already done. The pesticides in the wetlands killed off some of the protected vegetation areas. People who had gardens sprayed with the pesticides were unable to use those areas due to chemical contamination caused by the spraying.

3. Noise

Intense, loud noise can have a significant effect on our mental and physical health. This may lead to higher levels of stress, which in turn create other issues. Disruptions to sleep, cardiovascular issues, and hypertension are a few of the effects. It is impossible to overstate how big an impact noise can have on the well-being of people subjected to it continuously. For children it can be even worse. Over 20 studies have shown constant exposure to environmental noise affects cognitive performance and learning. The CDC says noise levels of more than 80-85db can cause hearing loss. Most of these compressors exceed that.

In Oswego, Illinois, the railroad would idle trains on the railway behind a residential neighborhood. The noise and diesel fumes inside the houses became unbearable. The local police came out in the middle of the night once and measured the noise at 90 decibels. Imagine the sound of a gas powered lawn mower or a leaf blower outside your house. Or even the sound of a motorcycle at the end of your driveway. That is comparable in level to these compressors. This went on for months until finally a city resolution was made.

In Garden City on Long Island, New York, idling trains, construction, and trains backing up were a nightly problem encountered by residents. The neighbors knew they were located near train tracks; however, the tracks were part of a train line and not an overnight workyard or transfer station. The railroad was working on a construction project further down the tracks during the daytime hours and decided the quiet track behind these neighborhoods was the perfect place to do all the overnight preparation of the construction for the next day. Cutting rails and moving materials all through the night became the norm. Neighbors were becoming ill from the disruptions in sleep from all the noise and vibrations caused by the crews. Changing the use of the tracks in the residential neighborhood from simply a lightly used track to a construction zone was a surprise to the neighbors.

In Bradford, Massachusetts, a residential neighborhood sits right next to a train station where the railroad parks them overnight. Many of the locomotives are older designs, and the railroad needs to keep them running all night in the colder weather. Nearby houses say the noise is unbearable with multiple locomotives running all through the night for days on end. The noise is affecting the locals causing sleep disorders, nervous system disorders, and hearing loss. An idling locomotive runs at about 90db, the same as a lawn mower. In addition to the hazards caused by the noise, diesel fumes and smoke from the trains permeates the homes, bringing cancer-causing particulates from the diesel exhaust inside, from which there is little escape.

Because there are no regulations on how long a train is allowed to be, there may be 200 or more cars attached to five or six engines on a freight train. Depending on the speed of the tracks and the area the train is traveling through, it could take ten or fifteen minutes for the train to rumble through your area while the noise and vibrations are affecting your body. While not continuous, the ten or fifteen minutes is more than enough to disrupt your nervous system, sending your body into a defensive state which may last an hour or longer.

4. Hazardous Cargos and Accidents

The communities that freight trains pass through while carrying hazardous and dangerous cargo have no idea what is rolling through their towns. Tanker cars filled with chlorine gas, ammonia, propane, oil, and gas are but a few of the potentially hazardous materials rolling only several hundred feet from people's houses and backyards. Some of the cargo regularly transported would be deadly if there was an accident or if a leak or rupture occurred. If there was an incident, emergency responders would also have to quickly determine what the hazardous materials are and what risk they present the public. In most cases, they would be unable to respond as quickly as they would like to a specific threat based on this delay.

Tankers and freight cars carrying chemicals and hazardous materials do carry placards indicating what material is inside. But, in a situation when a railroad car catches on fire, leaks, or derails, emergency personnel first need to get close enough to the area to be able to read the placard. In too many accident scenarios, this is not possible due to the contents being on fire or the car being on its side with the contents spilling and leaking out. Further complicating matters, there is currently an ongoing debate about removing some of the placards for some hazardous materials because of fears the signs could be a potential target for terrorist attacks.

To avoid a potential accident or leak in an undesirable area, one idea was to reroute trains carrying hazardous cargo around large metropolitan areas to keep them away from large population centers. Train companies have pushed back because of the high costs of rerouting and detouring: billions of extra dollars.

Trucks that carry hazardous materials and potentially deadly chemicals are required to use specific routes when passing through metropolitan areas. Many tunnels and bridges also have strict requirements for passage or outright bans on moving hazardous material. Trains, however, have no restrictions at all. Radioactive waste can move through a city on the tracks almost as easily as soybeans can.

The Federal Department of Transportation acknowledges accidents involving trains carrying hazardous cargo are becoming much more frequent. In addition, the department also added that they are increasingly concerned about the sharp increase in oil spills along railway tracks.

Near Dupo, Illinois, a train derailed, and a tanker car containing chemicals went up in flames. The car was carrying a flammable liquid solvent. The solvent flowed into a nearby creek, and flames were seen on the water's surface.

In Graniteville, South Carolina, residents still talk about a train crash involving tanker cars filled with deadly chlorine gas back in January of 2005. One of three tankers containing chlorine gas leaked, killing eight people. Another 234 people needed to be treated at local hospitals, mostly with respiratory issues from chlorine inhalation. A mile wide area of the town was evacuated due to the toxic plume from the leaking tanker car.

In New Jersey, a train carrying highly toxic chemicals was traveling over a bridge when the structure failed, sending several of the chemical tank cars plunging off the tracks and derailing. The chemical vinyl chloride leaked from at least one of the cars into the nearby creek. Nearby towns were placed on lockdown, and 71 people reportedly went to the hospital with respiratory issues. It was estimated about 180,000 pounds of vinyl chloride had leaked into the surrounding areas. Two other train cars containing the toxic chemical were hanging from the bridge, and rescue workers were able to remove them without the chemicals leaking.

5. Railroad Crossings

The majority of accidents occur at railroad crossings. Many of them occur when people decide to drive across the tracks to beat the oncoming train. Many others occur when people living in the area of the tracks use them as paths and shortcuts. Children especially use them to walk back and forth in the area in addition to playing near them. Every year hundreds of people are killed either at these railroad crossings or are hit by a train while walking on the tracks or bridges.

While the idea of living near a train track and hearing the whistle blow may sound romantic to some people, it quickly becomes old and very annoying when every train passing through town does the same thing. The constant noise and rumbling from the cars rolling by as they vibrate the foundation of your house and everything in it also gets to be extremely irritating after a while. If all that wasn't enough, breathing cancer-causing train exhaust, getting sprayed with toxic herbicides, and living with the ever-constant threat of accidents are constant factors. Living next to train tracks most likely is not a great location.

There are several considerations to make to determine how far from a train track to be. It is important to be far enough away so the noise from the train itself is at a level where it does not cause negative reactions in the body each time a train comes through. That includes the train's warning whistle. Hazardous cargos and the potential for an accident, even though low, would make leaving a distance of at least a mile a safe choice. I live two miles from a fairly active freight train line and hear the whistle, especially at night. It is not at all disturbing and actually at times I enjoy it wondering where the train is headed to.

SECTION THREE

Most of the hazards in this group have the potential to present a risk to you from low to extreme danger. That will depend on the specifics of the hazard near you and what you are exposed to.

- Radon
- Past uses
- Well water
- Public water supplies
- Waterways
- Dry cleaners
- Golf courses
- Gas pipelines
- Crematoriums
- DPW yards
- Tire dumps
- Cell towers
- Electric power lines
- Cemeteries
- Zoning
- Proposed Uses

Chapter 34. Radon

Radon is widespread and dangerous. The sheer number of people in the U.S. potentially affected by radon is enormous, and radon testing should be more commonplace.

Radon is not produced in great quantities as a commercial product or as a by-product of any industrial operation or process except in nuclear reactors and labs that work with radioactive materials. Some radon may be created during certain other industrial processes during the decay of radioactive materials used in the operation.

The vast majority of radon, though, is naturally occurring and found almost everywhere in varying amounts. When uranium and radium break down or decay, new elements are created, one of which is radon. All over our planet, radon is found in certain types of rocks and the surrounding soil. While the elements uranium and radium are decaying, those two elements stay put, trapped inside the rocks they are part of. Since radon takes the form of a gas, it is free to move through the soil and cracks and fissures in the rocks. It moves upward through whatever it encounters until it reaches the surface. Radon enters buildings and houses in the air.

Many areas of the United States have naturally high concentrations of uranium and radium in them. Some areas have so much that it is mandatory to do a radon level test before a real estate sale of a house or building can take place. Radon is invisible and odorless, so testing for its presence is the only way to check for it.

Once radon gas is in a building, it continues to decay into several radioactive elements itself. The most deadly elements created are polonium and bismuth. Even though radon is a gas, the new elements created when it breaks down and decays are very small, solid particles. Because these particles are solid, they are easily inhaled along with the air and drawn into our lungs when we breathe. These radioactive particles, like all radioactive materials, have a half life during which further decay happens. These specific elements break down so quickly, our lungs do not have the time to exhale them. They break down in our lungs where they are trapped in the air sacs and sent out into the body through the bloodstream. The same process happens when we breathe in radon gas.

Cancer.org says radon is the second leading cause of lung cancer in the United States. Only cigarette smoking kills more people with lung cancer. The EPA has similarly made an enormous effort to let people know about the dangers of radon.

These dangers are widespread because radon can be found everywhere in the U.S.. However, everywhere it is found, it is not always in hazardous concentrations. The higher the concentration of uranium in the ground, the higher the level of radon. According to the EPA, states such as Montana, North Dakota, and Pennsylvania are in high risk zones. Pennsylvania, in fact, is trying to pass a bill mandating radon testing in all school classrooms. An estimated 40% of homes in Pennsylvania exceed the EPA safety level. In a study done of schools that tested for radon levels, 28% of those had levels exceeding the EPA threshold.

Many other states have high risk areas running through parts of them. There have been extremely high and dangerous levels of radon found in some of the houses in these high risk zones throughout the United States. In the state of New Jersey, for instance, the governor has issued a memorandum telling people of the importance of testing for radon. State officials said one in six homes in the state have elevated levels of radon but only 30% ever get tested.

Radon was not thought to be a widespread health issue affecting everyday people until relatively recently. Back in the mid-1980s, it was only workers in uranium mines who were thought to be exposed to radon gas. The first person with a known radon issue in his house was Stanley Watras. Before Watras, nobody knew there was a problem, and hundreds of thousands of people most likely were exposed and affected. Watras worked at a nuclear reactor in Boyertown, Pennsylvania. One day, when he went through radiation monitors at the end of his shift, alarms went off. Unable to find the source of the radiation anywhere at the plant, the safety specialists went to check Watras's house. The radon levels at his house were 2,700 times normal background levels. His house was on top of an area venting radon up from the ground in an area of Pennsylvania now known to be a hotbed of radon.

Interestingly, there are big variations in detected radon levels. In different areas of the country and in different areas in high risk zones, there are natural variations, but even building to building in the very high risk areas the levels of radon vary. Where the largest amounts of radon are rising up out of the soil near the surface matters greatly.

Areas of the country that have granite, sedimentary rock, and shale, many times contain higher levels of uranium in them. The higher the level of uranium in an area, the more likely the buildings nearby will have increased levels of indoor radon. That is not the only factor, though, as some buildings in high uranium areas have low indoor levels of radon, and homes in low uranium areas are found to have high levels. There are other geological factors that greatly affect radon levels.

Since radon is a gas, it is able to move easily in soils that are more permeable such as sandy or gravelly areas. Soil that is more tightly structured, such as clay-rich areas, makes it more difficult for the radon to rise to the surface even though the underlying rocks may be rich in uranium. Any fractures in the soil and rock structure create a pathway for the gas. For these reasons, buildings and homes in areas with drier, more permeable soil and underlying rock structures, such as along the sides of hills and at the bottom of canyons where old glacial deposits of gravel may underlay, may have quite high indoor levels of radon due to the gas being able to quickly rise through and get to the surface before decaying.

The EPA estimates at least 1 in 15 homes has radon levels high enough to be a potential health problem. Buildings act as a natural trap for the gases to accumulate in, which elevates their level until they are dangerous to those who spend extended periods of time exposed to the air inside.

Radon enters the body in two ways.

1. Air Contamination

Radon has become a huge issue because of how easily it is able to enter buildings and combine with the air inside. Once the radon is inside, it can be inhaled by people in buildings with high levels of radon. The radon inside buildings comes up from the soil underneath and enters through any small cracks and holes in the foundation. There are places the radon will find to permeate through the foundation of a building. The gas builds up pressure under the building around some small opening until the gas is able to work its way through. Older houses, buildings with no poured foundation, and those with open drains in them are very unprotected. Radon will continuously make its way into a building unless stopped.

Homes are the most dangerous place to have high levels of radon because of the large amount of time we spend there. Still, any place we spend more than several hours a day puts us at risk for exposure.

It was quite uncommon to have schools actually test the classrooms for radon levels before many schools began to discover they had problems in their buildings. Dangerously high radon levels have been found in school classrooms as well as in office buildings.

Not every area of a building is likely to be equally affected by radon. Because radon gas almost always enters the building through the lowest area, the lowest floor of the building, such as a basement, is always the place where the readings will be the highest. That is also the area where testing for radon in the air is done. The radon gas builds up inside on the lower floors and continues to rise to the higher parts of the house. Radon levels on the upper floors could be a third of those down below, but how much time is spent in the lower areas and how high the radon level is determine the extent of any remediation needed.

The time of year can even influence the amount of radon in a building. During the times of year when the heat or the air conditioning are running, radon levels tend to be much higher. The air conditioning or the heat is running inside the building with the windows closed actually helps to pull the radon up through the cracks and holes into the building. It creates a negative pressure inside the bulging, pulling outside air in through every crack and hole possible.

The amount of insulation can even affect the radon levels of a building. Many newer buildings are better insulated than older ones, which actually increases the levels of radon in the house. The tighter the building is of any leaks and drafts, the less air exchanges it has. Therefore, it retains more of the radon gas. The same thing happens when a window or door is opened inside. It pulls the radon-laden air up from the lower level to another floor in the building, but the open window or door does allow for some of the gas to escape. Older houses with their leaky doors and windows actually may have lower levels of radon due to more air leaking out.

2. Water Contamination

Radon can also get into our bodies through the water we consume. When radon gas comes into contact with underground water, it dissolves and accumulates in it. The radon now trapped in the underground water is unable to evaporate, so it stays dissolved. When the water is pumped up to the surface either by the public water utility or private wells, most of the radon will evaporate into the building it is released into and some remains in the water. Whenever the radon-contaminated water is exposed to open air when used for cooking, showering, or other purposes, the radon is released. The people inside then are breathing the radon gas and the polonium particles it creates. The drinking water still contains radon and enters the body through its consumption.

Large bodies of water do not have the same issue with high levels of radon contamination. When radon rising from the ground comes in contact with surface water such as a lake or river being used as a reservoir, the radon usually has time to evaporate through the water's surface. The movement of the water helps to continuously agitate the radon out of it. Some, however, will remain suspended in the water itself. Most public water supplies will test for radon levels as part of their regular testing. So, those on public water usually can be assured the water delivered to their homes is safe from radon. However, it can't hurt to double check.

In Wake County, North Carolina, the county Water Quality Division issued an alert to people in the eastern half of the county. It is suspected that as many as one in five homes, or as many as 6,000 people, could be drinking and bathing in water dangerously high in radon. Water Quality is urging people in the high risk areas to get their well water tested. One well test came back showing levels of uranium (the element radon comes from) were five times the level considered safe. The high risk zone is now being extended even further. They are urging everyone who uses well water to get theirs tested.

Exposure to radon and the horrible effects of it are almost entirely avoidable. Usually, a simple test can check the levels inside a residence. If the test results come back indicating high levels, there are ways to remediate and eliminate the radon fairly inexpensively.

Testing for radon is as simple as ordering a radon test online and setting it up in the basement or lowest area for a few days. Many test kits are a simple, small metal container that collects the radioactive particles in the air. The container is sent back to the lab and analyzed to see if the level of radon detected is dangerous. Test kits usually cost under $20 and are widely available at hardware stores and other big box retailers. While Tony and I do not endorse any product, the First Alert radon test kit is inexpensive, effective, and very easy to use.

No matter where you live in the country, it can't hurt to test for radon. If you live in a building and are on the third floor or higher, there should be no issue at all. However, if you live in a basement apartment or on the ground level, invest the $20 and test. Absolutely, positively, if you are buying a house or even moving into another rental property, test before you move if the new residence is in a high risk area.

If you are the one doing the testing yourself, you have total control over placing the kit where you need to and letting it sit for the proper amount of time before packing and sending back to the lab. If, however, you are buying a property, the people there are the ones in control of the kit. The test results are easily manipulated simply by moving the kit from the area where the radon level would be the highest to another location in the house where it would show a lower result.

Tony and I do know of times that the kit was moved from the basement of the house being tested and placed upstairs for a couple of days before being sent to the lab. People who suspect or know they may have radon issues and are selling their house could possibly do this to make it seem no problem exists. If high radon levels are shown, the owners could have to pay to install a radon mitigation system before the buyers close on the home. If I was buying a house and it had a radon problem, I would expect and insist the sellers take off money from the price to fix it. The only real way around it if you are a suspicious person like me is to pay an inspection company to do the testing using their equipment. In that case, they would be liable for any false readings if that happens.

Health studies done by the World Health Organization put the percent of lung cancer attributed to radon in the United States at up to 14%. That is about 21,000 annual deaths. Many of these are avoidable if only people would take this risk more seriously. It only needs to be a low or medium level of radon to cause damage. The continued exposure is the largest issue as many exposures are not extreme levels. Testing for radon levels and remediation if needed is all it takes. Unsafe levels in any place are dangerous.

Chapter 35. Past Uses

What do golf courses, old landfills, toxic waste sites, and orchards have in common? Developers seeking to make a quick dollar have built residential homes directly on top of these. In more than a few cases, the past uses of the land beneath our feet has a toxic legacy which came back to bite the people above. As way too many homeowners are finding out the hard way, what is below our houses is as important as what is in them.

Many parcels of property once considered unbuildable just a few years earlier are being snapped up by developers eager to place new projects and, sometimes, residential housing on them. In several cases, the land has switched ownership multiple times, and the actual developer is claiming not to have known about the previous uses that contaminated the property. Disclosure laws requiring the selling party to share all information known about the property with prospective buyers are being skirted. Admittedly, in some cases the past uses have honestly been buried during multiple transactions leading up to the development, but in others, non-disclosure was avoided due to loopholes in enforcing the requirements.

As land for building new projects is becoming scarcer in suburbs and inner cities, open spaces such as golf courses and agricultural properties are giving way to residential housing. Contaminated sites such as brownfields or old industrial parks are being developed as well. As a result, many more homeowners are finding surprises from the past. Some of the most common contaminated past uses are discussed in other chapters. We do want to go a bit more in depth on a few of them here.

1. Agricultural Lands

As many urban areas spread out further for new development, pressure is mounting on any farmland left near them. In many of these areas, farmland has already shifted from growing crops to growing houses. Old fruit orchards have been razed, and housing developments have appeared with cute names like Citrus Acres and Groveland Estates. Acres of cornfields and other crops have sprouted developments of hundreds and sometimes thousands of houses.

Lands previously used for growing crops are not always contaminated. However, depending on what was grown and what chemicals and fertilizers were applied to the land, there may be an issue. It is always better to be aware of a potential for problems and seek to limit exposure to it both from a health risk standpoint and financially.

Over the past 50 years, many of the pesticides, herbicides, and other chemicals used on crops have been phased out. Sometimes, new research developed better, easier to use chemicals. Many, however, were phased out because their toxicity to humans was found to be a health risk. DDT was widely used and promoted as one of the best chemicals on the market until *Silent Spring*. Other heavily used chemicals included many arsenic-based ones. Chemicals like these do not break down in the soil; instead, they accumulate. The same is true of many of the replacement chemicals. Each time they are applied to the soil or the plants, the concentration left behind increases. In some cases the contamination extends to the groundwater below as well as remaining in the soil near the surface. Private wells in some developments are showing high levels of toxic agricultural chemicals, which makes the water unsafe for the residents to drink. Soil samples being tested also contain high levels of chemicals applied to the land long ago.

In Montville, New Jersey, one family in an upscale development received a letter from their town telling them their house was built on an old apple orchard that had been contaminated with pesticides. Other people in the developed area received the same notice. Now, the homeowners are involved in a lawsuit trying to find out why the homes were allowed to be built on contaminated property and who allowed it. In the meantime, property values have dropped tremendously, and residents are unsure what negative health effects can be attributed to the toxins.

Developers use laws to their advantage, not necessarily disclosing information they should. Several states, including New Jersey, require disclosure if the property is located near a toxic waste site, but no requirement to disclose if the property was agricultural land even if it was contaminated with toxic pesticides. In addition, states only suggest soil testing on old farmland, but do not require it. This creates a loophole for developers to omit telling anyone about potential issues most likely to be found down the road. After all, a developer can ignore potential contamination as long as they do not look for it and find it, turning a blind eye to the problem.

2. Landfills

There are thousands of old closed landfills across the country. Many of them were old town and county dumps located on the outskirts of towns. Explosive growth has long ago pushed development beyond where the old landfills sit, making the land desirable to develop. With available land difficult to find and lots of money to be made by developing it, old closed landfills are frequently eyed for their potential.

Ignoring the risks, developers build on these closed landfills. All over the country, residents are suddenly finding their homes with cracking foundations, sinking backyards, and splitting driveways. The closed landfills that their homes were built on shift and move as the waste continues to decompose and gases continue to rise up. Methane gas, smelling like rotting eggs, permeates homes, and explosive levels of this gas have been measured in some places.

Homeowners blame developers, who in turn blame town officials, ultimately bringing many legal cases to the courts. In the meantime, though, it's the homeowners who are paying the mortgages and watching their investments and health rocked at the foundations. It is certain there will be many more developments both knowingly and unknowingly built near and on top of these rotting piles of toxic trash long ago covered and supposedly forgotten.

In the town of Mesquite, Nevada, residents in one development found themselves wondering why they were suddenly having issues with foundations fracturing and driveways breaking into multiple pieces. Turns out, the developer of their homes built them right on the edge of the old town landfill. At the time construction started, the town knew that the developer was building on property closer to the landfill than the town thought was proper. Unfortunately, the town approved the lots for residential houses. Now, the people living there are paying the price with their health potentially compromised and their property values in decline.

In Havelock, North Carolina, one resident has been finding much more than weeds coming up from his ground. Just working in the yard, he has pulled spark plugs, shards of glass, and the hood of an old truck out of the soil. Throughout the neighborhood, people's previously flat yards are sinking down. Even trees planted in the neighborhood are no longer pointed straight up but in all directions off to the sides to where the land has sunk.

The town admits the land that the neighborhood was built on used to be a landfill through the 1940s and 1950s. When the town started to grow back then, houses were built over the closed landfill. Some residents who have been in the town since those days even remember some of the items that were buried in the landfill, such as an old school bus. At that time, it was legal to bury toxic poisons in landfills, so the potential risk to this town is high.

In New Orleans, Louisiana, residents are hoping after years of living on top of a closed landfill filled with toxic chemicals they may finally have a chance to move. In a lawsuit against the city of New Orleans, the residents said the city knowingly built housing on top of the old landfill and never disclosed to prospective owners the toxic chemicals they knew were underneath. The EPA previously tested the site years before the homes were built and determined there were high levels of toxic substances. Known as the Gordon Plaza site, an outside environmental expert and chemist tested and found 140 different toxic chemicals. There were 49 associated with increased risks of cancer. Lead was present at the site in levels that may cause developmental issues in children and harm almost every organ in the body in all the residents. The chemicals in the landfill under the homes people have lived in for years cause lung, liver, and bladder cancer. Unfortunately, many of the residents are unable to move elsewhere and hope the lawsuit may help them to fund a relocation desperately needed.

3. Toxic Waste Sites

Many toxic waste sites are located in areas where the land is quite valuable and developers would like nothing more than to be able to build a project on them. We have seen these sites become shopping malls, offices, factories, and housing developments. Toxic waste sites that qualify to be included in the brownfields program usually get fast tracked by communities eager to get the property back onto the tax rolls and by developers eager to make money off a project. Many of these sites, once investigated for contaminants, are allowed to do a limited cleanup based on the intended future use. In most cases, that means not all of the contamination will be cleaned up and some may even be left onsite. Often, there may be a simple cap of asphalt or dirt allowed to cover the hazardous materials underneath as a barrier for what is built on top. Other times the level of cleanup was only required to bring the toxins down to acceptable levels to have the land able to be reused but the actual levels of contamination are still not safe ones for people to be exposed to.

Surprises await the new homeowners who find out there was more toxic waste buried under their homes or nearby that was not remediated for some reason. In the haste for local governments to get these old toxic waste sites clean enough to be returned to the tax base, they allowed the toxic legacy of the property to be carried forward to the new owners or occupants unfortunate enough to purchase or set up business there.

In Richmond, Virginia, the city is looking at building 4,000 apartments on the site of a heavily contaminated brownfield. The clean-up plan is to do nothing but pour a cap over the property and build the apartments on top of all the chemical contamination. It is already known the toxic mix will leak into the wetlands next to the development and the nearby bay. There also may be vapors coming up from the sealed area under the apartments, so the developer is proposing to vent the air in the lower levels to the outside so the residents aren't poisoned. The developer of the project offered the city of Richmond a contribution of $52 million for parks and city programs. If the toxic site was cleaned up the way it should be, the cost is estimated by some people at over $100 million.

In Bradenton, Florida, the town approved an affordable housing development. Normally, this action would be applauded as helping solve a housing problem. In this case, however, the town is approving the homes be built on top of a toxic waste site. An old dry cleaning operation nearby managed to poison the groundwater for over 16 acres. The state environmental protection agency is telling people it is safe to put houses there and have people live on the waste site.

It is safe unless, they add, groundwater comes up to the surface during construction or after people are living there. Apparently, they are expecting a problem because the developers are even restricted from putting in stormwater ditches to divert flooding even though the frequent storms in the area necessitate some type of drainage on the property. Florida already has a very high water table in most of the state. All it needs is a slight change in water level, and then toxic waters will flood the homes.

In Great Barrington, Vermont, a hazardous waste site that was contaminated with toxic chemicals including dioxins from a wood treating company is being turned into affordable and low-income housing. The clean-up plan to make this toxic waste site ready to have families living on top of it does not include removal of all the toxic chemicals or excavating to ensure there are not more underneath. The plan is to push much of the toxic dirt into piles and cover them with a material to hopefully seal the poisons from being exposed or leaking. The rest of the toxic waste site will have clean dirt dumped on top of it to cover the toxic contamination below.

The situation and risk from the hazardous waste is so serious the site will need to be monitored forever but, they are allowing people to live on top of the site and not clean it up.

4. Golf Courses

There is a large number of golf courses being developed into residential communities. Sometimes a larger golf course that has two complete 18-hole courses may close one of them, using it for a housing development and keeping the remaining golf course open. Others have undergone complete changeovers, and the land previously maintained for golf is now all housing developments and green space.

In Aventura Isles, an exclusive housing development built on top of an old golf course in Miami, residents are suing the developers. At issue with the more than $60 million of houses sold is the fact they were built on land contaminated with Dieldrin, arsenic, and other toxic materials from the prior golf course and possible agricultural use even before the course was built. Because of the contamination, the developers were required to notify prospective buyers of the ongoing cleanup, and major land use restrictions were supposed to be in the buyer's deeds. The developers neglected to do any of the notifications prior to the sale and did not place the restrictions into the recorded deeds as required. The homeowners, who had no idea of the contamination, were not supposed to use the groundwater for wells or dig in the soil any deeper than 18 inches. They also were not allowed to remove any soil due to much of it containing toxic chemicals.

Golf courses being made into housing developments are a large concern. Living next to the golf course exposes people to airborne toxins, but living on an old golf course exposes people through soil and water as well. Arsenic and other residues from pesticides remain in the soils for decades. Disturbing the soil while building or even making a garden in the yard could be hazardous. Many towns and developers dismiss the problem as minimal or even nonexistent. In every case, soil and water samples should be taken and tested for a range of chemicals used on golf courses through the past decade. You should especially test for the chemicals such as arsenic and Dieldrin, formally banned before the year 2000 as they are deadly.

5. Mines

Surprisingly, housing built on former mining sites is not as uncommon as people think. Unsurprisingly, it poses a risk to those living on those former mines. Property where mining had taken place previously, sometimes over 100 years ago, still has remnants of materials with levels high enough to be hazardous.

In Highland, Wisconsin, a farmer planted a field of soybeans on his plot at the edge of town. While most of the field did well, he noticed one section where the plants died off. Curious as to what caused this, he had a soil scientist at a local university perform a soil test. It turned out the soil had zinc in a concentration of 15,000 parts per million, far more than the normal 300 it should contain. That means 1.5% of the soil was made up of zinc. Investigations into past uses of the land determined that the section of Wisconsin had been a big mining center with lead and zinc mines all over. In fact, many of the existing towns there today were built right where the old mines still sit. The soybeans weren't growing because the field was covered with mine tailings from a nearby zinc mine.

At one point the area was the largest producer of lead in the entire country. Researchers have located old records for over 2,000 mining sites and are now searching to see if other people are living on top of contaminated soil, some of which would have high levels of lead in them. Many more piles of old mine tailings are out there as well, most covered up with brush hiding a deadly secret underneath.

On Navajo Nation land near Arizona and New Mexico, many towns are known to be contaminated with old mine residue from uranium mining. In some cases, ore from the mines was used for construction in the houses. The federal government continues to ignore this issue for the most part while Native Americans are subjected to levels of contamination that would be unacceptable in the suburbs of Virginia and Maryland.

Tony and I could go on listing potential past uses you need to watch out for, but it is possible to have housing built on top of almost any of the hazards in this book. You need to check for these risks at each step. At some developments, it will be quite obvious there was no past use on the land although we always encourage some checking. Others may have had hazardous activity on them 50 or a hundred years and require some deeper investigation to turn up what that use may have been back then. From closed military bases to brownfield developments on contaminated factory sites, there are many dangers to watch for. Nobody may disclose all the truth to you when you ask, so always check and be cautious when someone tells you something you are unable to verify for yourself.

There are simply so many possible past uses and multiple past uses that may come up. The most important item to remember is to use your own judgment in evaluating the location and to not rely on information that may come from a source with a personal interest in you purchasing property or moving there. If you find you are living on or next to contaminated fill, there is no safe distance.

Chapter 36. Well Water

We all rely on having a clean and reliable water supply for many of our daily needs, from cooking and drinking water, to washing our clothes and bathing. Practically all of us take for granted that the water we get when our faucets and showers are turned on is clean. People who depend on their own wells like to think the water they get from it is pure and pristine, fresh water from beneath the ground, completely untouched by human hands.

Most of my adult life was spent living in places where we had well water. One was in a developing area with mostly horse farms and wooded areas. When we drilled our well, we checked for radon because we were located in a high risk area. We also checked for bacteria because of the concentration of horse farms. We included a check for a range of volatile organic compounds just in case there was a past operation someplace nearby that may have had spills onsite or leaks from storage tanks holding petroleum products or oil for purposes such as heating. There was a heating oil dealer nearby, so that made sense.

Other places we lived near large agricultural operations or concentrated animal feeding operations. At those places, we focused our testing on herbicides, pesticides, nitrates, and coliform. When we lived in Lancaster County, Pennsylvania, we had a hog operation operating literally in our backyard. The farm disposed of all the manure from the hundreds of pigs by spraying it over the farmland that was used to grow animal feed. Yes, using manure for fertilizer is good practice in place of chemical fertilizers. However, in many of these large animal operations, the amount of manure the animals produce is far more than the land could possibly absorb and use. The farmer would set up his giant sprinkler sprayers, and there would be manure shooting out 75 feet in the air. The sprayers would continue until the ground was so saturated that the manure was running off the field and down towards the creek nearby. Our levels of nitrates and fecal bacteria were sky high when we moved in and had our well first tested. This is the level of thoroughness it takes to make sure well water is actually safe.

It is a harsh awakening to many people when they find out their water is not as clean as they thought. Instead, it may contain any assortment of bacteria, chemicals, radioactive isotopes, and other harmful substances. Private wells, unlike public water supplies, have very little oversight and regulation. According to the EPA, there are over 13 million households that rely on private wells for drinking water. The U.S. Geologic Survey says that it comprises 43 million people on well water in the United States. The EPA does not set standards or criteria for private wells, nor does the agency regulate them.

In most cases, believe it or not, very few state or local governments regulate private wells either. There are some basic regulations that may or may not be followed or enforced. The fact that well water may taste and smell fine in no way indicates there is nothing wrong with it. Vigilance is important because of issues with well contamination.

1. Chemical Contamination

Chemicals that occur naturally can sometimes be found in higher levels in water than is safe for human consumption. Iron and uranium are two examples, depending on where the well is and the level of those chemicals in the aquifer that the well draws water from. It is much more likely, though, that the chemicals are coming from other known or unknown hazardous waste sites or industrial sources that have infiltrated the underground water source the well uses.

A study of potential arsenic contamination in well water was done by the U.S. Geological Survey and Centers for Disease Control and Prevention. Since arsenic is a naturally occurring heavy metal, the assumption was areas with high arsenic in well water were areas where arsenic is in abundance under the ground. The agencies determined where these possible areas were located. The results estimated there were about 2.1 million people in high risk areas potentially using private wells where concentrations of arsenic would be considered high enough to be a concern. A spokesperson with the USGS hoped people in these high risk areas would have their water tested and monitored for arsenic levels as a result.

Hazardous wastes from illegal dumps have polluted water supplies across the country. Underground, rusty old storage tanks number in the thousands, and many of these have been leaking gasoline for months and years. Radon and heavy metals leach from giant piles of mine tailings and seep into the water supply we use. Anything that can contaminate the local water or soil can contaminate a private well.

Outside Baltimore, Maryland, in Salisbury, trichloroethylene, a toxic industrial solvent, was found to have contaminated almost 270 wells in residents' homes. Environmental officials investigating are unsure what the source of the contamination is. There are no known old landfills or dumps in the area or any types of businesses that would have been using the chemical in their operations. Possible explanations included intentionally disposal by dumping the liquid chemical onto the ground where it found its way into the underground water. Another explanation, and one which seems even more likely, was sewage sludge that had been applied to nearby farmland was already contaminated with high levels of trichloroethylene. Back in the '70s, septic cleaners frequently emptied tanks from businesses using this chemical and may have been simply disposing of it down the drain where it then wound up on the farmland.

The chemical contamination was only discovered when one of the residents thought the well water had an odd odor and had it tested for contaminants. Based on the fact the area wells are 60-80 feet deep, the chemical had to have been spreading through the groundwater for years. As more wells in the area were found to also be contaminated, federal health officials were brought in. Drinking or breathing trichloroethylene can cause damage to the kidneys, liver, and nervous system. The residents were exposed for an unknown time before the chemical was discovered, both through drinking the contaminated water and by breathing in the vapors that evaporated when they were showering or using the water to wash. The federal level for exposure is fve parts per billion. Many homes tested higher than that, with one resident's water having 550 ppb. He shared that he had never noticed anything wrong with the water before during the 13 years he had been using it. He added, though, that once he switched to the clean water being trucked in, there was a very noticeable difference. The EPA is installing carbon filtration systems in some of the houses, but hopefully a source of safe water will be available to the area.

In Hanover, New Hampshire, one family found out their water was contaminated from the chemical 1,4-dioxane and radionuclides. Where they lived in town there were not many possible sources for the contamination. Instead of industrial pollution, they received a phone call from Dartmouth College informing them the landfill the college had operated about 800 feet from their home was leaking and chemicals moving past their home were contaminating the groundwater. The landfill had been used to bury hazardous wastes, dead animals, fetuses, and human tissues. Radioactive materials the college used in research were also buried at the site. By the time the college had notified the residents to stop drinking their water, they were already having health effects they had no clue were caused by consuming the contaminated water. Loss of hair, peeling sores in their mouths, and sores between their fingers were symptoms both had been experiencing. Their dog, who also was drinking the water, had been urinating blood. The college equipped the home with a supply of bottled water, and the symptoms stopped. They worry, however, there will be long-term effects from drinking the water before anyone knew there was an issue. The groundwater plume continues spreading towards other residences in the area. The college is spending $5 million to clean up the area, but the groundwater contamination is another issue that will remain. The house has become unsafe and unsalable, and the people who live there feel helpless.

In the town of Marengo, Illinois, private residential wells were to be found contaminated with several industrial solvents known to be cancer-causing. The chemicals dichloroethane, trichloroethene, and vinyl chloride were all used by a nearby company, Marengo Magnet. None of the residents knew anything about the contamination until state officials tested wells in an area near the factory the contaminated groundwater was traveling. Only then were neighbors informed they had been consuming and using water containing these cancer-causing chemicals and needed to cease using it immediately. As of 2018, attempts to connect the residents to the town municipal water supply continue to be delayed by issues, many of them legal ones over ensuring the responsible party pays the expenses. The Illinois Department of Public Health admitted exposure to the chemicals in the private wells may have had adverse impacts on residents' health during the length of time they consumed the water.

In New Hampshire in the town of Leverett, the residents all have well water as the town has no public water supply system. Many places, that would not be a problem, but in Leverett, the old unlined municipal landfill, which closed back in the '90s, used to also take in anything thrown into it by the community. The chairman of the town's Select Board said, "For years people just threw stuff on the ground and covered it up." Now, a plume of toxic chemicals from the leaking landfill is migrating through the groundwater and contaminating wells in the area. Chemicals include 1,4-dioxane and manganese. Some manganese is usually found in groundwater, but the levels found in the contaminated wells is ten times higher than normal.

For ten years a combination of water filters and bottled water had been used by the residents with contaminated wells, but a better long-term solution is really what is needed. Two of the residents said that when they purchased their homes back around 2006, the realtor downplayed the issue saying not to worry about it. Realtors, home inspectors, banks, and lawyers were all involved in these transactions, but no one, including the home inspector, raised any concerns about the contamination but, instead, let these homeowners move ahead with the purchase and deal with the problems afterwards.

In Hereford Township, Pennsylvania, wells have been contaminated with chemicals including trichloroethylene disposed onto a farm almost 50 years ago. Back in the 1960s and ‘70s, the farm used a pit to dump household waste and at least 1,200 drums of an industrial solvent known to cause cancer from a local manufacturer. At the time, it was still not illegal to simply dump toxic waste anywhere, so the drums were buried in the pit. The EPA believes when the farm owners still had found the drums around 1980, it was then illegal to have them on the farm, so they dumped the contents of the 1,200 drums, each containing 55 gallons, onto the ground. Over time, the toxic waste traveled through the soil and contaminated the underground drinking water source. In 1983, residents were finding the water had a strange smell and taste to it, so they notified the state environmental agency. Trichloroethylene affects the liver, kidneys, and nervous system. It is also a known carcinogen. The EPA contaminant level is five ppb, and up to 700,000 ppb were found in the area the chemical was originally disposed of.

The town of Wedron, Illinois, has had issues with contaminated water on several different occasions. Starting in 1982, residents suspected their water had something wrong with it when it started to smell like gasoline. Environmental officials believe the source of the pollution was from two leaking underground storage tanks used by a nearby gas station for almost 50 years and since removed. The soil and groundwater at the site had been contaminated and supposedly cleaned up. In 2011, residents again noticed the smell of gasoline in their well water, and after the county health department found high levels of benzene and other volatile organic compounds, they told the residents to stop drinking their water. The wells in Wedron are contaminated with a petroleum based product known as BTEX, which contains benzene, toluene, ethylbenzene, and xylene at levels as high as 70 times the maximum allowed level. Benzene exposure is linked to cancer; ethylbenzene is also a possible carcinogen. Long-term exposure to BTEX, according to the Ohio Bureau of Environmental Health, can cause damage to the kidneys and liver. Nobody knows for certain how many years the residents had been consuming the water prior to the second discovery, but they're sadly confident it had still been polluted that whole time.

2. Nitrate Contamination

The most common source of nitrate contamination is from overuse of fertilizer or large amounts of manure. It almost always can be traced to a farm nearby. Fertilizers applied to the soil have excess nitrates that leach down into the water below. In the cases where hundreds or thousands of animals are together, such as in a factory farm or even a small livestock operation, runoff from manure will also make its way through the soil into the aquifer below. Infants are especially at risk to high levels of nitrates as they cause the oxygen levels in blood to drop. The high nitrates can cause blue baby syndrome in infants, a toxic reaction in their bodies to the nitrates in water they are drinking.

In Millsboro, Delaware, Mountaire Farms, a huge chicken processing plant, was thought to have been responsible when numerous wells near the plant tested high for nitrates and bacteria in the water. The company denied these claims, and politicians were quick to jump on the side of Mountaire, a large area employer and tax payer. The company regularly disposes of the waste excreted by the hundreds of thousands of chickens by spraying it onto 900 acres of nearby fields after some basic treatment. Neighbors maintained their case and had photographs to document that thousands of gallons of improperly treated chicken waste were being sprayed. State environmental officials found the processing plant had been discharging wastewater for months containing levels of nitrates 5,500 times legal levels for disposal and fecal coliform bacteria. This resulted in extensive nitrate contamination of area wells, and possible bacteria contamination as well. The company denied all allegations for a while but eventually admitted their wastewater treatment plant had an upset event in which waste was not being properly processed. After a two year court battle, the company agreed to pay a settlement, however, still maintained they were not responsible despite the facts presented by the state and the residents to the contrary.

In La Cienega, New Mexico, a subdivision was built next to where a horse racing track was located. For almost thirty years prior to the development being built, manure from up to 1,500 horses at the track was piled in an area adjacent to the subdivision. Two years after moving into their new home in the development, Melissa Wright's one-year-old began to have medical issues: specifically, he would stop breathing and need to be rushed to the hospital. Finally, after multiple episodes of her son almost dying, it was discovered the well water in the house was contaminated with such high levels of nitrates from the manure it was causing the child to be unable to get oxygen, a condition known as blue baby.

In Nekoosa, Wisconsin, levels of nitrates from fertilizer were found in the well water of 40% of the town's residents. The nitrate levels in those wells was significantly higher than legal limits. The terribly unfortunate and unnecessary event that led to the discovery of the toxic levels of nitrates was when a mother, Celina Stewart, had her baby daughter pass away due to a fatal brain malformation directly associated with high levels of nitrates

3. Bacteria and Microorganisms

Fungi, bacteria, parasites, and viruses frequently contaminate well water. The main source of these comes from human and animal waste. Usually, the source of the bacteria is from farm animals. The animals leave a tremendous amount of manure on the ground, and farmers add more through the additional application of the animal waste and sewage sludge on the land as a fertilizer. All the bacteria from this waste saturate the soil and seeps into the aquifer. Outbreaks of giardiasis, shigellosis, and salmonella have been seen in private wells in areas with very high animal populations, such as on factory farms.

In Batavia, New York, residents' wells were discovered to be contaminated with a potentially harmful intestinal bacteria E. coli. A group of the affected homeowners reported having stomach issues possibly related to the contamination. A nearby factory farm was initially thought to be the source of the bacteria. The farm had over 5,000 cows and regularly applied manure to the local fields. Lab tests done on the bacteria confirmed that the farm was the source. However, it was later discovered in addition to the farm, some of the wells were being contaminated from the residents' own septic systems.

The residential wells in the area were old and did not meet the standards that newer wells have regarding protections for infiltration. Because of that, bacteria from their septic was able to penetrate into them. In addition, setbacks from septic systems and the depth of the wells also played a factor in the dual contamination. The farm agreed to pay for a water system in the homes with contaminated wells, and, hopefully, the homeowners will be either installing new wells or more frequently testing their water. Older wells do present problems; they need to be tested more frequently to ensure the water is not contaminated.

The county of Morrison, Wisconsin, is home to more than 41,000 dairy cows. Every year, these cows create 260 million gallons of manure. Much of that manure is spread onto the fields. Some of the farms grow grain crops; others get paid by the cow farms to let them dispose of the manure on their lands. There is a balance between the amount of manure that is fertilizer on the land and an excess that simply can't be absorbed into the soil. This excess runs off or seeps down to the water below. In Morrison, over 100 private wells were polluted in a period of a few months. Parasites and bacteria infiltrated into the wells, causing people drinking the water to get stomach illnesses, ear infections, and diarrhea. One resident was quoted in the *New York Times* as saying, "Sometimes, it smells like a barn coming out of the faucet." Water in her water and nearby areas tested positive for E. coli bacteria, coliform, and other contaminants found in manure. Her son wound up with an ear infection so serious it required an operation, and the doctor believed a likely cause was using the polluted water to bathe.

The problems and examples of contaminated and poisoned wells can go on and on. The key here, though, is to take precautions to ensure that the well that you and your family depend on is not compromised and creating health issues. The possible sources of pollution are wide-ranging. However, there are several things you can do to minimize the possibility you may be drinking contaminated water.

The most important thing is to make an assessment of what is near where you are located. Since most homes with private wells are located in more rural areas, it is important to know the potential sources that could contaminate your well and the water. Especially in these rural areas, check for any factory farms in addition to agricultural operations. Take special care if you are near a landfill, toxic waste site, military base, or any specific hazard.

In areas where well water is common, the health department usually has a required distance between a septic tank and its field and the well. A home inspection company can make certain this was followed. They would or should also ensure the septic field or drainage area from the septic tank flow away from the well area to avoid groundwater contamination. Also, check any neighboring septic tanks as well using the map that the property filed. Maps for these are usually filed with the county health department and are easy to find.

The depth of the well is also important. Shallow wells tend to be polluted from surface runoff and shallow contamination sources much more than deeper ones. The chances of contaminants reaching the upper aquifers is higher than deep down contamination. It is quite expensive to drill deep wells and install the well casing, but a properly drilled well should last for about 30 years. The best material for the well casing is steel as it does not corrode or crack easily, both of which would allow contaminants to enter the water. Plastic, especially PVC, is less expensive; however, it may crack much more easily, and you may not even get ten good years from it.

Once the casing allows intrusion from groundwater, repairing it or even needing to drill another well is possible. The well pump usually will last for about 10 years but is easily replaced. Shallow wells tend to be polluted from surface runoff and shallow contamination sources much more than deeper ones. The chances of contaminants reaching the upper aquifers is higher than deep down contamination.

One word of caution, though. In areas where there are industrial waste injection wells and oil and gas wells, a deeper well isn't always a safer well.

Once you know what the potential contamination risks are, the other important step in protecting you and your family is testing. If you are drilling a new well for a new house, you usually are required to run a basic water test package on a sample for the health department. In most places, it's part of getting the certificate of occupancy. A usual basic test package will vary depending on the location but usually consists of a group known as primary pollutants, such as nitrates and VOCs, short for volatile organic compounds. VOCs are frequently used in manufacturing, cleaning, and gasoline products. Tests also commonly look for harmful microbial bacteria that may indicate contamination from nearby farms or even your own septic tank.

Testing a well once isn't enough, though. Test your well for contamination and pollutants on an annual schedule at minimum. This practice is critical to catch any issues that may have arisen since the last time it was tested. Too many people don't retest their wells to save some money. Another frequently cited reason to skip retests is that there didn't seem like there was any need to check again. The first test came back fine, and nothing seemed to have changed.

Even when we can't see them, many things change. Pollutants in underground water supplies travel. Plumes of gasoline, dry cleaning fluids, and chemicals from a toxic waste site have all been known to travel over a mile from the source of contamination. Movement of the plume and any new ones may contaminate your private well water even though the past tests indicated there was no problem.

Once you determine what contaminants you are dealing with or watching for, you can decide if the levels need to be remediated. There are filters on the market that can do almost anything. The question for you is what is in your water and how clean do you want your water to be?

Chapter 37. Public Water Supplies

No matter where you live in the country, one of the most important things we must have is access to water. In order to live, we all depend on having a reliable water supply. Many of us take it for granted that when we turn on the tap, the water we get from our utility company or our well is safe and clean. However, as we have seen by stories out of Flint, Michigan, and Newark, New Jersey, to name only two, we know that is not the case. As documented in the records for many water districts, this appears to be an ongoing issue that never was really addressed. Our water in many cases is not safe for us to drink.

The federal and state governments handle the monitoring and regulations for the water utilities to ensure they are supplying their customers with clean and safe water. It would be reasonable to think part of those regulations includes testing for the wide range of chemicals and contaminants most likely to be in the water.

Disappointingly, this is not even close to being the case. The regulations that govern what the utilities need to test for, how often they need to perform these tests, and how much contamination is allowed are insufficient to protect people. There are many chemicals and contaminants that regularly show up in our water that are not tested for at all. The government says many contaminants are safe at a certain level. The research from the independent scientific community differs with that and has the toxicology reports to show the risk to humans.

The EPA only regulates approximately 94 chemicals in drinking water that water companies supply to its customers. Those are the only ones there are enforceable regulations if there are violations. No enforceable standards of any sort are set for many other chemicals the EPA knows are present and also may cause harm. According to a review of data from the CDC, there are more than 286 million people who use drinking water containing contaminants that go unregulated. Many of these contaminants come from industrial waste and from agricultural operations

We have a situation here. We have hundreds and thousands of chemicals not tested for at all but frequently found in our water. We have a small number being tested for, which the government considers safe to consume if below a level the government approved. The remaining hundreds, for which there are no real regulations, the government has a category it calls "guidelines." Exceeding these guidelines has no consequences and is not enforceable.

In practice, these aren't guidelines; they're suggestions. The government suggests water companies do not supply tens of thousands of people with water containing more than these levels of chemicals in it, but there are no consequences if they ignore the suggestions. Many of these unregulated chemicals are toxic.

The Environmental Working Group has collected evidence that tens of thousands of water utilities are delivering water to our taps that exceed the guidelines.

There are no federal limits on half of the contaminants detected in drinking water, according to the report, and establishing such regulations takes years. One of the best quotes I have heard in a long time came from Sydney Evans, who works for the Environmental Working Group as a science analyst: "Legal does not necessarily equal safe."

The majority of our country gets its household water delivered through a public water supply company. Some of these utility companies are owned by cities and municipalities, and others are owned by private businesses. Regardless of ownership, there are rules and regulations in place supposedly to protect you. That is about as far as it goes, however.

The rules and regulations are not written equally for all the water supply companies. The government applies higher standards to larger companies and lower standards to smaller ones. The logic is the larger ones can afford to do a better job of protecting its consumers than the smaller companies with less revenue. The larger ones should be able to remove a wider range of toxic contaminants and deliver water with lower levels of these. Even though all consumers are equal, they are not entitled to the same quality of water.

Both large and small utility companies sometimes try to game the system. A different investigation done by *The Guardian* discovered that both sizes of utilities, when testing for lead, routinely ask their customers to thoroughly flush their pipes by running the water for a while and removing the aerator before taking a sample. This is directly the opposite of how the EPA tells people to do it. The flushing lowers the level of lead that is in the pipes and faucet and, therefore, does not show a true report. In spite of the EPA asking utilities to stop telling their customers to use this method to get a sample, some continue to do so. According to *The Guardian* report from 2016, utilities in some of the largest cities used tests that minimize the amount of lead and have been doing this for over a decade.

In Michigan, a lawsuit filed to require local and state authorities to follow the EPA guidelines for testing tap water has succeeded in making the changes sought. The state has amended its water-testing rules to eliminate the instructions to run the tap for several minutes before taking a water sample for testing. That allowed the lead and other continent levels to decrease and gave false readings.

In Bloomfield, New Jersey, levels of lead higher than EPA standards have been detected in samples. The town says after running the water for a short time and flushing the pipes, the level goes down. But, that is exactly what we are not supposed to do. Whenever we get a drink or fill a pot of water, we do not flush the pipes. What the town is requesting people to do is causing false test results. If there is lead in the water when the tap is turned on, then that is a problem people are exposed to.

There are issues with the system that need to be fixed. Public water companies are supposed to deliver water free of chemical contamination to their customers. There are several reasons this is not always being done.

1. Chemical Contamination

As already stated, there are only so many chemicals that a water company is required to test for. Out of the total 84,000 potential chemicals identified by the EPA as possible risks, only a tiny fraction are regularly tested for, and many chemicals have no test designed to detect them. Spokespeople for the chemical industry argue that these numbers are a lie and exaggerated. In spite of my feelings about the propaganda the chemical industry promotes, I will agree with them on one point they claim on their website. In their own words from chemicalsafetyfact.org the say, "EPA's Chemical Data Reporting (CDR)...requires manufacturers to provide the EPA with information every four years on the chemicals they produce domestically or import into the United States. The 2016 CDR showed 8,707 chemicals used in commerce."

I agree with the probable accuracy of the fact that 8,707 chemicals were produced domestically or imported into this country for use in commerce in 2016, all of which needed to be reported to the EPA. That statement, however, has nothing to do with how few chemicals are tested for in our water supply, whether the number the chemical industry likes to use or the number of potential chemicals the EPA lists. The fact remains there still are the same 84,000 chemicals in the EPA database that have been used and have potential to continue to show up in the water.

Additionally, many of the EPA's 84,000 chemicals are no longer being manufactured or in use because they turned out to be deadly to humans. However, they were used for decades before being banned. As a result, many of these chemicals are still in existence in toxic waste sites, leaking landfills, in soil on golf courses and farm fields, and in old drums stored in warehouses. In other words, they all have the potential to still poison people and should be tested for.

Further complicating the chemical industry's complaints about over-regulation is the simple truth, already stated in this chapter, that many chemicals are not regulated by the EPA with regards to being in public water. Even if there are tests to detect them, the water companies do not look. That is a horrible way to protect the safety of people. If these chemicals are present in water on a regular basis, they need to be tested for.

Unfortunately, it takes years and millions of dollars to get any change like this to happen. We are up against the entire water supply industry and the entire chemical industry as well.

Furthermore, even when tests show a certain chemical is present in the water, if the amount detected is less than the threshold level set by the EPA, nothing is done even when the chemical is a known carcinogen. What many of these public water companies find when they check the levels of chemicals the EPA requires them to test for is equally shocking

In California, hundreds of public wells that are the water source for a million people are contaminated with the chemical TCP. TCP got into public water wells serving central California because Dow Chemical and Shell Oil took TCP and mixed it into agricultural pesticides. Despite the fact this hasn't been widely used or applied to cropland since the 1970s, it is still present in the water today. The EPA concluded this synthetic chemical is a likely carcinogen to humans. In the city of Arvin, California, TCP levels were over six times the level the state says is acceptable.

Many more sources of chemical contamination are industrial facilities that send their effluent to municipal wastewater treatment plants that don't have the technology to remove the compounds. Industrial facilities are supposed to pretreat their wastewater to remove many of these toxic compounds, but a lack of real monitoring and enforcement means many companies take a pass on the regulation.

North of Pittsboro, North Carolina, a group of industrial operations send their wastewater to the local sewage treatment plant. The chemicals in the industrial wastewater are not able to be removed at the treatment plant due to inadequate technology in their equipment, so the sewage plant discharges the wastewater containing the chemicals into the Haw River. Downstream from the discharge, Pittsboro water company has an intake where they get the town water supply from.

Tests of Pittsboro water have found perfluorinated compounds, or PFAS. This group of chemicals is suspected of being carcinogenic. The Pittsboro Town Board of Commissioners has agreed to mail notices to local water customers to alert them of contaminants in the town's drinking water.

In addition to PFAS, the tests are showing the chemical 1,4-dioxane and bromide. A researcher at North Carolina State University said, "I wouldn't want to drink that water or just give it to an infant without having some additional treatment at home."

In Miami-Dade County, Florida, the public water supplier discovered that several of their main wells have been contaminated with the cancer-causing chemical benzene. The levels were measured at five times the federal level for safety. The utility shut down those wells and shifted over to others to fill in the gap. Then, those wells were also discovered to contain high levels of benzene as well.

In Assumption, Illinois, the city sent out notices to its residents warning that allowing infants under six months old to consume drinking water from the public water supply may be fatal to them. The city advised not to use the water to make baby formula or allow any consumption by these infants. High nitrates in the public water supply may cause blue baby syndrome and possibly death.

Near Charleston, West Virginia, an industrial chemical storage company operating on the banks of the Elk River spilled a toxic chemical into it, contaminating the water supply for 300,000 people. The spill occurred two miles up the river from where the city water plant had its intake for the municipal water supply. A federal disaster was declared by the White House, and people were warned not to use their water for anything, not even bathing. Several dozen people were in area hospitals with symptoms such as vomiting and skin rashes. Up to 900,000 people were under a no-use ban for a while.

In Hoosick Falls, New York, industrial operations contaminated the groundwater the public wells depended on with PFOAs. While the poisons were not found until fairly recently, the companies involved had disposed of these toxic chemicals for decades. Once the local government discovered the municipal water supply had been contaminated, they kept that information to themselves for a while before making the public aware.

A survey by a state university nearby found dozens of residents with various types of cancers and diseases high above the levels which should have been found. PFOAs are a suspected cancer-causing chemical that were phased out of production due to health and environmental concerns. Because of foot dragging and inaction by the federal EPA to set regulations on the amount of PFOA allowed in drinking water, many public water supplies are largely ignoring its presence.

Meanwhile, downstate in Newburgh, New York, the water supply was found to contain the toxic chemical PFOS. The contamination is reportedly to have come from firefighting foam used at Stewart Air National Guard Base. The city declared an emergency in 2016 and switched to a new source for their water. Unfortunately, residents had been drinking and bathing in the contaminated water for years. The issue continued while the local government and the Department of Defense bickered over who was responsible for the cleanup.

In North Windham, Maine, two public wells supplying 2,000 residents had to be closed and a new water supply hooked into. A new gas station using the latest state of the art equipment contaminated the town's well source protection area. Once the water underground is contaminated, it takes a very long time and is very costly to clean up.

2. Contaminated Water Supply Pipes

Even though the water may leave the water treatment plant clean, it still needs to travel through a network of pipes to reach our taps. Many towns and cities have a water infrastructure that could be upwards of a hundred years old. Newer parts of town may have had new pipes installed when the water supply was extended to new areas and developments, but the old main lines along the way were not. Leaks in the pipes and growth of bacteria and corrosion in the pipes contaminate water as it passes along.

A government-funded study on contamination in water supply pipes, the largest ever conducted on public water systems, concluded a larger risk than thought was caused by leaky and corroded old supply pipes. The pipes that bring drinking water from public wells to home taps may let in viruses that cause more than a million cases of stomach illness every year.

In Ranger, Texas, lead had contaminated the public water supply pipes serving the town's population. The town hall knew about it as did state and federal agencies whose job it was to protect the people. Everyone apparently knew, except for the people being poisoned by it. The Walton family was part of the group of people who were never told their water contained 28 times the amount of lead that was considered safe by the EPA's figures. Their son, who was two years old by the time the family found out something was wrong, had been terribly affected by the poisoned water. According to the doctor who tested him, the blood in the two-year-old had high levels of lead, which directly affects the development of the brain. According to the parents, the toddler was only able to say three words even though he was already two years old.

3. Inadequate Enforcement

Data from the EPA shows there are about six million people getting their water from 2,000 public water systems qualifying as serious violators. The water companies have continuous, serious violations and are not complying with regulations in place to protect people using their water.

Estimates are that about 99% of the water companies on the list are small utilities serving fewer than 50,000 people. Millions of people are at risk of a variety of dangers, ranging from water contaminated with toxic chemicals to bacteria-laden water causing intestinal issues and potentially severe health issues.

In addition, CNN reported that 5,300 water systems are violating lead and copper testing rules. Violations run from not testing properly to not reporting test results. In other words, some of these water companies are not telling the people drinking the water when a problem is found with lead and copper in it.

A review of water company testing by *USA Today* found 9,000 small water companies serving almost four million people failed to test properly for lead in the past six years.

Moreover, many school districts have water safety people who are supposed to test the water for lead levels annually. Many of the districts like Klondike Independent School District in West Texas and Orange Center School in California skip testings. Some skip even though high levels of lead were previously found. School officials still let children drink the water in spite of having high lead levels.

In more than 2,000 communities, lead tests were skipped at least once. Over 850 small water utilities who already had a documented history of lead contamination failed to properly test for that substance at least once since 2010.

Lead is far from the only contaminate that may be in public water. Pesticides and herbicides that go untested by the utility are frequently found when private testing is done. Industrial chemicals present in the water prior to the utility processing are found in large numbers in tap water due to the fact that so many contaminants are not tested for.

Sometimes, the contaminants in the water are not addressed by these companies even after they have been found. As a result, unsuspecting customers continue to use it. In a few cases, this even happens with the approval of the EPA and agencies supposedly in charge of protecting us. The public water supply company in Sanders, Arizona, was one of those small, local companies that fell off the radar of the EPA for years. Even though the water company did not test for bacteria or lead, nothing was done to enforce the law. A simple notice of violation was sent along with a non-compliance notice. All the while, trusting people were drinking the water. High levels of radioactive uranium were found in the public water supply. The contamination was most likely coming from the many mines in the area.

A notification to the residents about the radiation contamination did not come for ten years after it was known by the government that the water was contaminated. In those ten years the residents continued to consume and use the poison water. Outrageously, once the public finally found out they had been drinking contaminated water for at least 10 years, the Arizona Department of Environmental quality told the residents it was all right to continue to use the water. That was despite also telling residents the radioactivity in the water was over federal safety levels for drinking water and could lead to liver and kidney damage if consumed for an extended period of time. Even with that reality, the state DEQ maintained it was fine to consume until another safe supply could be found.

The residents were also not told the water supply company was given over 200 notices of violation. Infuriatingly, many of those violations were never taken care of. Some of them were for high bacteria levels in the water. In essence, the residents were drinking tap water under the assumption it was safe, but it was contaminated by both radiation and bacteria.

It is a frightening thought to think we are drinking water that is potentially unsafe. Testing for contaminants that are potentially harmful to us but are allowed to be in tap water just seems wrong. Hundreds of contaminants are present but never detected because tests are not required even though these contaminants are just as bad, if not worse.

There are several steps you can take to limit the chances you are exposing yourself and family to potentially harmful contaminants.

The first is to use a fantastic search tool set up by the Environmental Working Group. They have compiled a database of thousands of water utilities across the country. This database houses the records of the water utility companies' test results that were submitted to the EPA. You can look up the company serving your water district and see if there are any contaminants to be concerned about. This website will also show the levels of the dozens of other contaminants that have no enforceable safety levels or regulations that are not regulated for levels but are present and are known to be carcinogens or other risks to your health. This would give you an idea what you need to watch for in the second step.

With an idea of what contaminants are in your water, a test would be the least expensive way to tell what is coming out of your tap. Some municipalities will offer tests for free or a very reasonable fee. There are also many testing companies where you can physically take a sample of your water and others where you can mail it off for analysis.

We absolutely do not suggest letting your water company do the testing for you. Remember, it is in their interest to make certain your water meets standards and regulations. You, on the other hand, have a vested interest in determining what exactly is in the water. An omission of a chemical on the tests or a simple change in how the sample is taken can make a huge difference in the results. Do not clear your lines by running the water for several minutes first. Instead, follow the same process you'd use when you get a drink.

Once you receive the test results back, you can see if there is any need to take action on your water. The red flag items are usually lead levels, any VOC contaminants, and arsenic levels. Beyond that, you can make an educated decision if you need anything additional.

Depending on what you find out from the lab tests, you may or may not want or need additional filtering. My collaborator and I do advise testing your water on an annual basis just in case something has moved into the groundwater and the utility either has not yet found it or does not care enough to remediate it because they don't have to.

If the tests do find contamination, you're going to want to invest in a filter. It is always a good idea to use a basic carbon filter that can remove many impurities and will usually make the water taste better. I have nothing to do with any water filter company or product. I use these because I know they help, so I am passing on the information to you.

Filtering your water is the single most effective way to prevent using and consuming contaminated water. There are many types of filters on the market, ranging from single tap filters for drinking water to whole house filters for bathing and washing in addition to drinking.

At the same time, use your critical reasoning. The water filter industry has spent millions to convince us our tap water would be better when using their pitcher filter or water system. Some of their claims are not always entirely true, but our tap water is in more cases contaminated than it is not. Given that reality, filters have a critical place in ensuring that the water you and your family drink is safe.

This chapter is not a wholesale condemnation of the public water supply in America. Instead, the goal is to highlight the shortcomings of it. There is a failure in the oversight and enforcement of the water companies charged with providing clean and safe water to millions of Americans. We can't depend on our water being safe; we need to ensure that for ourselves.

Chapter 38. Waterways

A lot of people really enjoy living near water. There is just something peaceful and happy about it. Whether it's a lake or the ocean, a big river or a small stream, living along the shore is nice. It is wonderful to be able to sit and watch the water go by. Maybe you remember going to a summer home on a lake when you were young.

In addition to people enjoying living near the water, businesses also like to situate themselves there as well. Their reasons for being along the water are quite different in most cases. Aside from a restaurant or a hotel located there for the view, many businesses usually want to use the available water as a resource. Some businesses want to take the water out from the source for use in an industrial operation such as cooling water for a power plant or for mixing materials in a paper mill. Other businesses want to use the water for discharging waste from the industrial operation into the waterway.

There are very large industrial operations along the nation's rivers and lakes. Taking a look at what hazards they could possibly be causing is a bit alarming. Even more alarming is the sad reality that in most cases these harmful emissions are allowed.

For those who live along a body of water, there is quite a high possibility of being in the pathway of outflow from an industrial operation. It may be from a sewage treatment plant, an industrial operation, or any other number of places that discharges any number of harmful substances into the water.

Looking closer at each of these groups gives a better idea of what we may be exposing ourselves to. We want to make it very clear that living next to any type of water, while very pleasurable, comes with some inherent risk. It does not matter whether it is a lake, river, ocean or even a small stream--you need to be aware of potential sources of pollutants.

1. Industrial Wastes

As unbelievable as it may sound, industries treat our lakes and rivers as their own personal sewers. They always have. Every year, millions of gallons of waste are discharged into the nation's waterways from industrial operations. Everything from arsenic to acids pour out of pipes and into rivers, lakes, and any body of water large enough to be used as a disposal site. There are actually permits to pollute. Companies tell the EPA how much and what wastes they wish to discharge into our nation's waters. The permit is reviewed and either approved as is or modified.

The amount of hazardous chemicals the EPA allows industries to dump into our water is staggering. Over a four year period, permits were issued to allow more than 14 billion pounds of chemicals to be released into waterways. For larger rivers like the Mississippi or Missouri, for example, the length of the river system allows all of the industrial waste to multiply as it moves along further downstream. There are those who think dilution is the solution, but this is outmoded thinking used as a cover by polluters.

Rivers pass through big, industrialized sections in many large cities. As the river passes by, all those factories pour in their chemicals and wastes. By the time they reach the ocean, most rivers are so heavily polluted with industrial waste an analysis of the water would read like a list from a Superfund site.

This type of pollution on a massive scale has always gone on, and it likely always will. The only way it can possibly ever change is to force the government to enact and enforce stricter guidelines and regulations limiting, and eventually eliminating, industrial discharges. Unfortunately, at the present time and for the foreseeable future, that does not appear to be happening, so we need to be vigilant and aware.

One of the worst disasters and examples of such industrial waste came from what General Electric knowingly did in upstate New York. They polluted well over 150 miles of the Hudson River in New York with PCBs. For 30 years, the company dumped an estimated 1.3 million pounds of this highly toxic chemical into the river from two manufacturing plants they operated. Traces of these chemicals are found along the Hudson River all the way down to New York City over 200 miles downstream. The chemicals bioaccumulate in the environment and the body, meaning they simply stay put continuously increasing in quantity. People are at the highest risk because of being at the top of the food chain. PCBs in humans increase rates of cancers. Cancer of the brain, liver, gallbladder, melanoma and others are all linked to PCBs exposure. PCB exposure in pregnant people may affect unborn children with developmental issues, physical issues, and deformities. For years, people in the Hudson area ate fish from the river, and many communities along the 200 mile stretch depended on fish as a food source. Nobody told people they were being contaminated for many years.

The public tried for years to pressure the EPA and GE to clean up one of the worst stretches of river where there was some of the highest levels of contamination. Despite the fact that they dumped four to five million pounds of toxic chemicals in the water, GE mounted a public campaign of misinformation for years, telling people that cleaning up the wastes would cause even more pollution and instead, everyone should just leave it sit where it was. The EPA and public pressure, however, forced GE to clean up about 40 miles of the river. At least 40% of the PCBs are still in the river, and GE and the EPA are not willing to clean any more. Much of the river will remain polluted and a hazard to people and the environment for decades to come.

A report reviewing the issue of industrial waste discharge in Michigan revealed that over 40% of the major industrial plants poured amounts of industrial wastes into Michigan's waterway in excess of what their permits were for. Environmental regulators, whose job it is to enforce the Clean Water Act, were accused of lax enforcement, which risks the health of the public. In one 21 month period, groups reviewing the data reported to state and federal regulators that they had counted 196 discharges of pollution beyond what the permits allowed. Environment Michigan and the Frontier Group noted the discharges included highly toxic mercury and fecal bacteria. Both of these groups counted 70 instances where the industrial facilities discharged twice the legal limits. Three beaches along Lake Michigan were closed because of cyanide and ammonia wastes dumped by a steel plant along the shores. The worst part was the plant never notified the authorities about the spill. It was discovered by a fisherman who noticed dead fish floating in the ater and notified the local environmental agency.

In another instance in November, 2017, hexavalent chromium, a highly toxic chemical, was dumped into a waterway leading to Lake Michigan. Federal agencies were never informed of the spill, and when a group of law students from University of Chicago discovered the report, not only did they notify the federal agency but also found information showing that U.S. Steel, the company who dumped the chromium, had asked state regulators to keep the information secret.

In Luke, Maryland, a closed paper mill run by Verso Corporation is being sued by the Potomac Riverkeeper Network. Verso had been illegally discharging toxic waste from the mill into the North Branch of the Upper Potomac River. The discharge, which is called black liquor, is a hazardous byproduct from the production of paper almost all paper mills have to dispose of. At the Verso facility, the material is leaking from the site into the waterway. Testing of the discharge showed a variety of toxic chemicals found in black liquor in addition to indications of toxic coal ash. The area waters are heavily used by people to kayak, fish, and swim; in addition, several municipalities draw water for their public drinking water systems from the river. Toxic chemicals such as cyanide, arsenic, boron, and chromium are among those found during testing. As always, it is unknown how long the company let these chemicals leak into the water before they were caught.

2. Sewage Plants

Like industrial plants, sewage treatment plants use our waterways as an extension of the sewer system. Normal discharges from the treatment plant occur after the liquid wastes pass through the tertiary treatment stage. At this point, the liquid waste that makes up about 99% of all waste should have the levels of bacteria, solids, and pathogens safely down. Millions of gallons of this pour into our lakes and rivers everyday around the country.

For those who have a sewage treatment plant nearby, chances are that there are already millions of gallons of waste flowing past. The beautiful area behind a house can be loaded with raw and untreated sewage and industrial waste. It would be extremely unhealthy for children to play in water that is loaded with so much disease-causing bacteria or chemicals. Even when the plant is not operating under overflow conditions and is discharging their normally treated water, it is far from drinking quality, and it still has some levels of bacteria and coliform unacceptable to most people even though it may be acceptable to the government.

There is no way around the fact that a sewage treatment plant must have a place to discharge the treated wastewater at the end of processing. While Tony and I may dispute how clean it is and by whose standards, millions of gallons need a place to go. If there are any sewage treatment plants upstream from where you are located or nearby if you are on a lake or beach, you will be affected by this discharge on a daily basis.

In addition, overflow events at sewage treatment plants are not going to end anytime soon. Discharges of raw sewage after large amounts of precipitation or snowmelt will occur on some regular basis. The treatment plants simply were not built to handle the volume going into the system.

In Buffalo, New York, the sewage system would overflow during periods of heavy snowmelt and rain. This would cause untreated sewage along with toxic materials to be discharged directly into waterways. Upwards of four billion gallons of untreated waste flowed into the Niagara River and its smaller tributaries.

Along the shores of Lake Erie in Cleveland, there was yet another major discharge from the sewer district. Lake waters near the popular Edgewater Beach were the unlucky recipient of over 142,000 gallons of raw sewage and stormwater runoff. Everything appears to be alright, though, as the sewer district issued an advisory for the beachgoers. They said, “people should avoid contact with the water.” Apparently, the way things are is a significant improvement for them as the sewer district made a point of saying it’s much better than back in the 1970s when this would happen 40-50 times a year.

A report was done by Congress for the Great Lakes Basin area on the combined discharge for all the sewage treatment plants using the lakes for their outflow. The report stated during sewer overflow events in 2014, an estimated 22 billion gallons of untreated wastewater was discharged.

At Lake Nokomis, two of the most popular beaches were shut down indefinitely by health authorities on August 14, 2019. An outbreak of E coli was already suspected of causing three children to become sick, and high bacteria levels could make an outbreak of the bacteria cause even more people to become ill.

Even the oceanfront and all its multi-million dollar mansions can’t escape the sewage. Officials are shutting down yet another New Jersey beach due to high bacteria levels. It seems raw sewage from a sewer line leaked into a nearby storm drain, sending tons of human waste into the ocean. The effects of exposure to the bacteria Enterococci are fever, nausea, fatigue, vomiting, and diarrhea. The beaches in the area needed tol be closed until bacteria levels returned to acceptable. This was an annual ritual at the Jersey shore when I lived in the state. Like many beaches along the ocean, sewage is a regular visitor to the waters just like the beach goers. At one point, 47 health advisories were issued throughout New Jersey, effectively placing one third of the beaches at risk of being closed due to high levels of bacteria found in animal and human waste.

In upstate New York, one of the most beautiful lakes, Lake George, is plagued with E. coli contamination. The town was desperately searching for a solution as the elevated levels of E. coli started to affect the lake back in 2016. E. coli again closed down the shining gem of the lake, Million Dollar Beach, numerous times in 2017 and 2018. Million Dollar Beach is operated by the New York State Department of Environmental Conservation. The town tried to find the source of the contamination but even after fixing a broken sewer line, it still had issues. There was talk that some of the million dollar homes along the lakeshore were discharging raw sewage directly into the lake in addition to there being other leaks in the sewer system. Closures lasting the majority of the prime summer months were costing the town lots of tourist dollars in addition to having people coming in contact with the bacteria suffering diarrhea, nausea, and other health issues as a result. While there was no specific information on anyone needing hospitalization due to contact with the water before the beach was closed, many people may have been taken ill with lesser illnesses such as diarrhea. In 2019, Chuck Schumer secured a grant for Lake George to upgrade the sewer treatment after seeing for himself the contamination flowing into the lake.

3. Industrial Accidents

In addition to using our waterways as a disposal facility with the permission of the EPA, industrial operations also have frequent accidents. Many larger waterways have storage facilities for chemicals and petroleum products along them. Leaks and spills are common. Smaller accidents usually go unnoticed and may be unreported. Larger ones, however, affect thousands of people and become a hazard to the population living on the waterway and residents who use drinking water supplied by the local utility.

In Charleston, West Virginia, at least 10,000 gallons of a toxic industrial chemical leaked into the Elk River from a nearby chemical storage company on January 9, 2014. The chemical, 4-methylcyclohexanemethanol, is used by the coal industry in part of their operation. The main concern was that the water utility that used the river for a water supply had an intake pipe just a mile and a half below the spill location. Before the water company was able to issue a “do not use” order, the West Virginia Poison Center was swamped with phone calls. People throughout the area were reporting poisoning and health issues including rashes, nausea, vomiting, diarrhea, and other symptoms. Hospital emergency rooms started to see increases in visits. Over 300,000 people are served by the water company now distributing the contaminated water through the supply lines. Exposure to high levels of the industrial chemical have even caused problems with the liver, kidneys, and brain in laboratory animals.

In August of 2015 an accident at a mine in southwestern Colorado allowed three million gallons of contaminated wastewater to be released into local waterways. The contaminates included lead, copper, aluminum, iron, and several other metals. The spill was so massive that it polluted rivers as far away as New Mexico and Utah. There were sections of water that were turned a shade of orange-yellow from the metals in the water. The EPA apologized for causing the spill. Native Americans who drew their water from the rivers were unable to use them for months after the spill for drinking, crop irrigation, or watering their livestock. Farmers along the waterways polluted by the spill reported losing their crops and in some cases, animals were dying. The EPA claimed there were no immediate health effects to humans as a result of the spill however, people who were unaware of the contamination of the water continued using and ingesting the heavily metal laden water while it was dissipating . There were 540 tons of heavy metals released and a portion of that remains in the sediment along the waterways to be released slowly over time now. Health effects of this may not be seen for another decade.

The beautiful Dan River runs along the border of Virginia and North Carolina. In 2014, just one month after a chemical spill on the Elk River, an estimated 82,000 tons of toxic coal ash poured into the Dan River. Along with the coal ash, a containment pond used by a power plant operated by Duke Energy leaked, sending an estimated 27 million gallons of contaminated water into the Dan River as well. Duke Energy has a reputation for landfills that leak coal ash and has had multiple accidents in the past. The spill contaminated the river with arsenic, chromium, iron, and lead, chemicals frequently found in the toxic coal ash generated by these coal fired power plants. The spill was only one month after the Elk River spill, which left 300,000 people with contaminated water for days. How the coal industry handles their operations raises issues about the dangers to the health of hundreds of thousands of people.

At Bradley International Airport in Connecticut, over 50,000 gallons of foam-containing toxic chemicals known as PFAS went into the Farmington River. These chemicals, which never degrade but stay in the body and accumulate, are carcinogenic. Downstream from the spill, recreational activities on the river are frequent. People living along the river and sportspeople are only told, “Please don’t eat the fish.”

Knowing what may possibly be in the water you are looking at may take away a tiny bit of the enjoyment you have. Since I found out years ago what is in small creeks and at the beach as well, it has made me a bit uncomfortable at times. It also makes me more aware and cautious. When I use the water for recreation, I make certain not to get any in my mouth. Depending where I am, swimming in the water may be out of the question.

Would I live somewhere near the water if I could? Almost definitely, once I checked out the hazards nearby just to be certain what my risk level was at that specific place.

Chapter 39. Dry Cleaners

When I was a child, my mother always used to take clothes to the dry cleaners: my dad's dress shirts and sport coats, my mom's sweaters, and items like that. There was a neighborhood dry cleaner she always used. My parents were friends with the owners, and I was friends with his son at school. I always enjoyed going into the store because it always smelled so good. I loved the smell of the clothes in the bag and the drive home with the windows closed to keep it in the car. After I left and moved away, I heard the store closed because my friend's father had gotten sick. Soon after, he passed away from cancer. When the family went to sell the property, they found out it was contaminated along with the surrounding half block and the water underneath the area. The site sat vacant for almost 20 years until it became too valuable to sit any longer, and it was finally cleaned up and redeveloped. The underground water, however, remains contaminated.

The places that are hazardous to our health do not need to have smokestacks billowing black smoke. There are many everyday places where the chemicals that are routinely used have harmful effects and create hazardous situations. Sometimes, the people using them are aware of this, but other times, they are not. Chemical manufacturers do not always do their best to warn people about the hazards they could be exposed to. The dry cleaning industry is one significant example of this. Living near a dry cleaner or even a closed one may be causing you to breathe toxic fumes in your air and drink hazardous chemicals in your water.

There are a few dirty secrets in the dry cleaning industry from the thousands of hazardous waste sites dry cleaning stores have created to the toxic chemicals used in cleaning our clothes. A close look really takes the clean out of dry cleaning.

1. Toxic Waste Sites at Past and Present Dry Cleaner Locations

The dry cleaning industry created thousands of toxic waste sites. Some of these sites were so bad and caused so much pollution they were declared Superfund sites, the worst of the worst. An Arcadis Geraghty and Miller survey of insurance companies who have to pay for the cleanup of the pollution, estimated more than 70% of dry cleaning operations released toxic chemicals into the groundwater or soil. This number includes both former dry cleaning sites that are now closed and sites that are still operating.

Most dry cleaner owners and their staff do try their best to follow the environmental regulations when dealing with the chemicals they use and the hazardous wastes they generate. Even then, accidental spills in the shop, leaks in storage tanks that hold chemicals, and problems with wastewater lines that lead to the septic or sewer allow gallons of hazardous chemicals to escape. At many of these shops, sloppy housekeeping is more normal than not, considering the volume of cleaning many of them handle.

However, there are also operators who knowingly skirt the rules and regulations. It costs quite a bit of money to properly store and dispose of these dry cleaning chemicals after they have been used. A lot of these sites were contaminated by Perc, the favorite solvent of dry cleaners. Not only is Perc the main chemical used to clean our clothes, it is used as an ingredient in rubber coatings, an insulating fluid for electrical transformers, and a component in paint remover and shoe polish. Despite this cute nickname for perchloroethylene, Perc is a highly toxic chemical. Long-term and even short-term exposure to Perc can have many serious health effects, ranging from respiratory tract irritation, fluid build-up in the lungs, kidney dysfunction, and neurological disorders. Cancer has also been a disease seen in clusters at many of the highly contaminated sites across the country.

Sometimes, operators decided to dispose of Perc in ways that they knew were wrong and in many cases illegal. Some dry cleaning shops directly piped their waste fluids outside to the rear of the property or into the septic tanks where it would flow into the septic field and surrounding soil. Others simply dumped it on the ground and let it soak in. Those who were lucky enough to be hooked into the city sewer system simply poured it down the drain.

These contaminated sites have been created over decades. Many of these shop owners simply did what was allowed at that time, similar to the way disposal of hazardous wastes used to be handled up until the Love Canal disaster. But even after the new environmental laws regarding proper waste disposal, much of the old ways continued in the industry. Moreover, by then, sites were already contaminated.

The big problem with Perc is it only takes a small amount to contaminate a whole lot of water. The waste at these sites often travels through the ground under the shop and into the groundwater below. Once it hits the water, it spreads out from there. The same thing happens when it soaks into the soil. As it saturates the soil under the shop, it starts to spread out, surrounding the contaminated area. Some known waste sites at cleaners have chemical plumes in the underground water stretching hundreds of feet away from the site. Contaminated soil travels the same way, and neighboring properties frequently find their soil needs to be removed along with the dry cleaning site when and if cleanup ever comes.

There is plenty of research that shows the plume of contamination is able to travel underneath the ground and have vapors come up from underneath foundations nearby and get inside homes.

In Ann Arbor, Michigan, the soil and groundwater underneath a still-operating dry cleaners is polluted with toxic fluids used at the cleaners. The underground plume of contaminated groundwater is being monitored as it spreads. One home already needed to be equipped with an air purification system due to levels of toxins seeping in from the soil below. The contamination at this location has been known for over 30 years and still is not cleaned up. A coordinator with the city environmental department said, "It's a site that we've known that low levels of this were getting into a neighboring building."

In Naperville, Illinois, a dry cleaner had a large spill of Perc that contaminated their property. Soil tests showed high levels of Perc had soaked into the surrounding areas. While the dry cleaners as well as the owner of the strip mall knew the site was contaminated, nobody told the nearby residents. For five years the contaminated soil had hazardous Perc fumes evaporating out. One neighbor only found out when an environmental consultant doing work on the spill tested the soil in her yard. Sure enough, the Perc showed up in her backyard in levels, higher than the safe level set by the state. She and her family had been exposed to toxic chemicals for five years while people who knew didn't tell.

The state of North Carolina has about 1,500 dry cleaning sites with contamination of some level needing to be cleaned up. Just in Durham, North Carolina, there were 18 sites showing up on the list as contaminated.

2. Water Pollution

By far the biggest part of the toxic legacy left by dry cleaners is the incredible number of places with water pollution as a result. Cleaning water once it has been contaminated with Perc takes a very long time: 5, 10, even 20 years, depending on the extent and how far it has spread. In many cases the pollution was found before large amounts of people used the water for drinking and bathing. However, there were almost as many cases when it was not. Sad too are the cases where the authorities know the water was polluted but allowed people to continue to drink it and use it. Toxic dry cleaning solvents have been measured in the groundwater at hundreds, even thousands, times higher than state standards.

In Wichita, Kansas, for six years, hundreds of people in a neighborhood used water contaminated from a nearby dry cleaner. For years, they drank it, cooked with it, and showered in it. Even though the city knew about the contamination the dry cleaners had caused in 2011, they did not do any testing in the nearby homes or any of the wells. It wasn't until six years later the residents finally found out they had been drinking known contaminated water all that time.

Unfortunately, the city didn't learn anything from this. In a different neighborhood, another dry cleaner also polluted nearby wells. This time it only took four years of using the polluted water before the people were told.

In Crestwood, Illinois, a small suburban town outside Chicago, residents drank contaminated water for 22 years. The town found out about a large dry cleaner leak that had contaminated the public wells but continued to pump water to 11,000 residents for two decades. There were children who developed cancer and other unexplained diseases during the period the townspeople were slowly being poisoned by the town water supply.

In Billings, Montana, the water was contaminated with Perc from three dry cleaners. A plume of contamination stretched under the ground for four miles. The area affected was estimated at 800 acres. Since Perc can travel through the soil as well as water, the toxic fumes were able to seep into people's homes through the basement and foundation. Residents are currently being told to seal up any small cracks or holes to prevent the fumes from getting in. To have this done by a professional runs about $2,000 for a typical home, an expense the homeowners themselves are responsible for.

In Fox Lake, Illinois, two separate dry cleaning operations, one still running, has contaminated two of the community wells serving water to the area residents. Test of the public water supply showed levels of the contaminants dichloromethane and benzene above safe drinking standards. Town officials say the treatment plant is able to remove the contaminate but are warning many people in the area who use well water to have their water tested for these chemicals and others. Levels of the dry cleaning chemical Perc have been found in the groundwater at one of the cleaners at levels close to 5,000 times federal standards. Another chemical, vinyl chloride, was also discovered at 1,500 times federal standards

In Albuquerque, New Mexico, just north of downtown, dry cleaning fluids from a cleaners leaked into the aquifer and polluted the groundwater. The chemicals are moving in a plume and stretching into new areas. The contaminated aquifer is close to the surface with a high potential for widespread contamination of further water supplies. The plume has grown to over one and a half miles and continues to spread. The state first investigated the contamination over 20 years ago. Ten years ago the state asked the company to put together a clean up plan and submit it. So far nothing at all has been received. The company says they stopped the leak and are working to clean up their site but nothing is being done to address the contamination flowing under all the business and homes. Vapor intrusion into the buildings is a big concern according to state environmental officials and a plan needs to be put in place. The company's lawyer who happens to be the son of a past senator says there is nothing else to do and the contamination will take care of itself. The head of New Mexico Environment Department Groundwater Bureau argues that the extent of the contamination is on par with some of the Superfund she has seen and needs to be addressed. All the foot dragging over the years delayed letting residents know what was underneath their homes even though toxic chemicals measured hundreds and thousands of times safe levels. Health studies have never been done either further delaying any movement. In the meantime, hundreds of people live over the toxic plume and wonder if they will get sick.

3. Air Emissions from Operating Dry Cleaners

There are dry cleaners in almost every town and city. Many people would view having a cleaners next to their building or even on the ground floor of their apartment complex to be a positive thing. It may appear to be a terrific convenience to have a dry cleaner in the building or maybe a door or two up the street in the neighborhood. For those who happen to be living on the second floor above it, though, or up the street close by, every time they open their windows, they smell the cleaners and may be subjecting themselves to continual long-term small exposure to Perc.

The vapors from Perc are able to pass through the walls and floors of dry cleaning operations into neighboring buildings. Once inside, the levels will continue to build up due to the lack of frequent air exchange. These airborne fumes put the families living in these residencies, especially children, at risk. Because children are smaller and breathe many more times than adults, they are especially vulnerable to accumulating the chemicals in their bodies. One of the most common ways air emissions from cleaners are released is during the process of taking the Perc-soaked clothing from the washing equipment and moving it over into the dryers. The even more common way is the exhaust. The air coming from the equipment as it vents from inside the dryer is saturated with Perc vapors. Usually, the vent pipe from the dryer simply opens to an outside wall on the outside. All of the vapors simply blow out into the nearby area.

Studies have shown the levels of dry cleaning chemicals to be many times the safe exposure amount coming from the cleaners. People who live over dry cleaners as well as nearby have reported health issues which are directly related to Perc exposure. A study at the University of Albany, New York showed a correlation between increases in kidney cancer and the density of dry cleaners near those residences.

In New York City, 2.3 million people live within 650 feet of a dry cleaner. New York City, health officials have determined there is an exposure risk high enough to be concerned. Higher levels of Perc are in the air everyone breathes simply because the chemical is widely used. However, even higher levels of Perc are found in the air near dry cleaners. The health department went on to say, "Elevated air levels of Perc have been found in residential buildings with dry-cleaning shops in New York City."

Additional studies have shown the risk from this is so high, new laws are being put in place restricting dry cleaners from being within a certain distance from an apartment building or other residential places. The industry is being required to eliminate changing machines and will also be required by the EPA to install some type of device on the exhaust vents to reduce Perc emissions to the outside. How successful either of these requirements will be is yet to be seen. Hopefully, currently operating dry cleaners will not be exempted from the new laws, which is sadly common with new environmental regulations.

4. Toxic Chemical Residue in Household Dry Cleaning

The health risks from your dry cleaning in your home may not come close to the risks of the industry as a whole, but it is information that needs to be shared.

Despite its name, dry cleaning is anything but a dry process. It involves soaking our clothes in Perc to clean them, and then removing as much of it as possible before the drying process. The majority of traditional cleaners use this method that has been around for many decades, which means they use this undiluted, toxic chemical directly on our clothes, saturating them.

Dry cleaned clothes are appealing. Many people love the smell and how crisply pressed and new-looking they are in their plastic bags. That smell is the residual dry cleaning chemicals still in the clothes. Perc is a colorless, non-flammable liquid with a sweet, ether-like smell. Perhaps that is why we like the odor of freshly dry cleaned clothes. The chemicals in Perc sometimes give our brains a little rush of pleasure when we inhale. The Perc residue remains in our clothes long after the cleaner wraps them in plastic. Most people think nothing of this and leave their clothes wrapped in its plastic in their closet until pulled off to wear.

To limit your exposure, it is highly suggested you remove the plastic when you get the clothes home and hang them in a well-ventilated area for a day until the remaining Perc evaporates. If you bring thick and heavy items to the dry cleaners, such as drapes, heavy blankets, and winter coats, allow them to air outdoors, if possible, for a day or two.

Be very careful and cautious with items such as blankets and pillows, which you will be putting your head on and breathing the fumes from as you spend hours lying underneath.

In summation, people are susceptible to the effects of Perc whether it is through inhalation such as breathing fumes continuously from a nearby dry cleaner or vapors in their houses from fumes emitted out of contaminated soil. Perc in household water evaporates out when residents shower or bathe, and they later inhale these vapors. Consuming Perc in contaminated water has caused harm to thousands of people.

Dry cleaners do represent a serious threat to our environment and health. In many cases the damage has already been done and can't be undone. Thousands of people are living with contaminated water as a result. The unfortunate part is many of those are not aware they are, either because of neglectful government officials or as yet undiscovered contamination.

Every state has a list of known contaminated dry cleaner sites. At some, the toxic contaminants have been removed, and the site is no longer a threat. Many, though, still wait to be addressed. Usually, a lack of money is holding back full cleanup. To cut corners and move on, sometimes the state decides placing an asphalt cover over the site and a couple of monitoring wells around the perimeter is a solution, but that falls far short of a real cleanup. Underground pollution can still migrate off the site, and sometimes, the wells do miss a location while the poisons move towards and under the adjacent properties. It is even worse if there are homes with private wells nearby. Once Perc and other chemicals infiltrate the aquifer, it is very difficult and a long process to get the chemical fully out. Many times, it can't be done, and they settle for a reduction in the level of contamination. Public water supplies that have major wells have needed to shut them down and find other places to draw safe water for their customers once they too find their wells contaminated.

Serious caution needs to be used if you live or plan to live anywhere close to an old dry cleaner or a currently running one. Assurances from the shop owner or real estate agent mean nothing. State environmental officials would be the best source of knowledge on the where and what may be contaminated. A minimum of 1,000 feet away would put you out of most of the fumes from the operation. If, however, there were underground plumes that put you in danger from vapor intrusion, extreme caution must be taken.

Chapter 40. Golf Courses

Golf course living sounds like a dream to many: a nice house with a view and a yard that opens up to a wide expanse of grass, no neighbors behind the house, no houses to block your view. It's just trees, grass, and open space. The only obvious downside is having to put up with an occasional golf ball hitting the house. But is that the only downside?

My collaborator and I have done site searches for many homes over the years, and some of the nicest ones have been golf course properties. Despite their opulence and splendor, these properties contain a few extra environmental hazards besides the sand traps.

1. Pesticides, Herbicides, and Rodenticides

Oh my! This first hazard may be the most obvious. Maintaining a golf course and keeping it in perfect shape for the golfers takes a lot of time and chemicals: pesticides, herbicides, and fertilizers, to name a few. Beginning in the 1980s, there has been much focus on the chemicals golf courses use to maintain their lush and almost perfect fairways and greens. Reports usually like to talk about how many fewer chemicals are being used. However, there are quite a few studies that reveal just how toxic some of these chemicals really are. Not only that, even though many toxic chemicals may no longer be actively applied, their residues and toxic components remain behind.

There was an investigation years ago in New York looking into just how much chemical was applied to golf courses. The people conducting the study reviewed the chemical applications of about four dozen golf courses all in the same state. The final report found that the volume of pesticides these courses used on their properties was seven to eight times the amount a farm used when compared in pounds per acre.

In the report's own words, "The NYS Attorney General Office did a report entitled Toxic Fairways. They looked at 52 golf courses on Long Island and the pesticides used. The report concluded, the 52 golf courses applied 192 different pesticides containing 50 different active ingredients. The total amount applied was more than 50,000 pounds. The report averaged this out over the total acreage and came up with 7 pounds of pesticide per acre. Agricultural land in the U.S. averaged 1.5 pounds per acre. The report also concluded much more than 7 pounds per acre was likely since pesticides were only used on playing areas of the golf courses making the applications even higher." The beautiful fairways and insect- and blemish-free greens are the result of continuous applications of chemicals and fertilizers in the same fashion that makeup is always being applied to an actor to keep the appearance perfect.

Risk assessments of chemical exposure on golf courses never includes the people living along the fairways or a couple of houses away. They are not considered when exposure limits are set to say the golfer who is on the course two or three times a week is perfectly safe at this low level. But the people living along these golf courses are exposed to many times that level. Furthermore, the runoff from the golf courses goes onto nearby property, and chemicals drift off site when sprayed on the course, going into the yards and homes of people living nearby.

Pesticide drift is a significant issue when golf courses spray chemicals. While they are supposed to only be applied during low or no wind conditions, many times this is disregarded to stay on groundskeeping schedules. The risk assessments and scientific research seldom ask if it is safe for people living near these golf courses.

The standard industry defense is that they follow all the directions and guidelines, so it must be safe. Hundreds, if not more, chemicals for which people followed the directions on the label turned out to be highly toxic and poisonous to humans. Also, some of the chemicals golf courses use according to the directions are already banned almost everywhere else for being too toxic. In 2015 the EPA declared there is no safe exposure level for chlorpyrifos, a widely used chemical on golf courses. The EPA proposed a ban on this product, and it is already banned in the European union, New York, California, and Hawaii over health concerns of brain damage from prolonged exposure. The Trump administration opposes the EPA ban. Dow Chemical is the manufacturer of this chemical. Even when studies show that these hazardous chemicals are likely carcinogens, golf courses cite that most of these studies show instances of cancer were in animals and not humans, so there is no evidence of risk to people.

Of course, the chemical companies and golf course industry has a lot at stake in these results. In fact, the chemical industry are major supporters of the sport. Dow Chemical was named the "Official Chemistry Company" of the PGA tour. Dow and the PGA partnered for a four-year period, collaborating with the PGA tour "to bring solutions and products to enhance the sport of golf." Dow Chemical is one of the largest suppliers of chemicals to the golf industry. In 2019, for the second year in a row, AMVAC Chemical Company donated $25,000 to the Environmental Institute for Golf. AMVAC makes herbicides, insecticides, and fungicides. According to the organization Open Secrets, the chemical industry spent close to $45 million in lobbying efforts in 2019. Ten chemical companies made direct political contributions totaling about $2.5 million.

When the EPA drastically curtailed the use of Dursban, it continued to allow application on golf courses, claiming it would not affect anyone there. Children were at high risk for damage from the toxic effects of Dursban. The EPA neglected the children who live near golf courses even though they would have much higher exposure than an adult golfer occasionally using the course. Each time the chemical is sprayed or spread, there is some exposure to people living nearby.

Another horribly toxic compound used on golf courses was methyl bromide, which was banned from use by most places in 2005. The chemical depletes ozone and also affects more than just the target pests. Human exposure to high levels can lead to respiratory issues, nervous system problems, and birth defects in pregnant women exposed. Despite this, there was an exemption for the golf industry to use this poison another eight years through December, 2013.

Long-term exposure to small amounts of these hazardous chemicals has already created a crisis for superintendents of golf courses and clubs. More documentation has come forth about these people having cancer, liver disease, kidney issues, and other medical problems, all of which can only be explained by their long-term exposure to these chemicals. According to Jay Feldman from the group Beyond Pesticides, there is a large body of scientific evidence showing links between these chemicals and diseases. Jay talked about a study of former golf course superintendents that found they had higher levels of four types of cancer: brain, lymphoma, prostate, and large intestine. While the study found the pattern was similar to other occupations using pesticide, it did not say any specific chemicals caused cancer. A recent look at the different chemicals used in golf course maintenance showed that, out of the 30 most used, 19 of them are linked to cancer, 13 of them are linked to birth defects, 21 affect fertility, and 15 of them can damage the nervous system.

The weed killer 2,4-D, known more popularly as Roundup, is by far the favorite to kill weeds even though it also affects humans. Already known as potentially causing cancer, millions of gallons continue to be sprayed all over turf while the EPA does nothing. Essentially, the industry is running over people's health in the name of profit.

Complicating matters, very few golf courses are under any regulations to notify anyone when they are applying chemicals. Some at best will put a notice up in the clubhouse for the golfers. The less the golfers know in this case, the better the courses and clubs think it is. After all, if someone knew they were being continuously exposed to cancer-causing chemicals when they play golf, they may look at playing someplace else instead.

Up until now, this discussion has only explored individual chemicals, but so many various chemicals are applied to golf courses that looking at the problem that way is a dangerous oversimplification. The EPA itself never considers what happens when these chemicals mix. This is a serious hole in the current research literature. Questions are all my collaborator and I have. Do these chemicals form new compounds when they come in contact? Do two toxic chemicals mixing make each one more toxic than before?

2. Building on Golf Courses

From Mission Hills outside of Chicago, to Tamarac, Florida, and out to California, golf courses are being transformed into upscale housing developments. Some courses are being developed for housing while others are modifying existing spaces or downsizing to accommodate housing. This is bringing to light a significant health issue which previously stayed underground--literally.

Many of the chemicals used in the past for golf course maintenance have now been banned due to research and confirmation of links to cancers and many other negative health effects as a result of exposure. These chemicals include arsenic and Dieldrin, which were used in very large amounts to keep golf courses green and pleasing to the players.

Unfortunately, these chemicals do not simply disappear from water application to the soil and grass. They persist and remain in the soil, and in many cases work their way down into the water table, and they even poison wells. Several of these chemicals change their structure as they break down and become new, more toxic versions of the original chemical used. Many chemicals bind themselves to soil particles and are present for decades in toxic quantities.

When the soil is disturbed, especially when construction begins, the chemicals previously trapped in the soil are now free to move around again. Much will be released into the air with the dust stirred up; this material will migrate off site. Dust containing chemicals can drift hundreds and thousands of feet away and get into the neighborhoods and houses surrounding the area. Especially vulnerable are the children who will breathe this dust and also play in and touch the dirt outside.

In Seminole County, Florida, Rolling Hills Golf Club had been around since the 1920s. When the golf club shut down for good in 2014, a developer eyed the property for upscale houses. The neighbors asked the county to purchase the land and save it from being developed. Rather than more houses where the golf club had been for 90 years, the people in Rolling Hills Development wanted a park.

A study of the property was conducted before finalizing the deal, and it found unacceptable levels of two highly toxic chemicals in fairways and greens on the closed course: Dieldrin and arsenic. There were 13 homes with high levels of Dieldrin in their wells as well.

The county moved ahead with plans to purchase the property and turn it into a park and green space; remediation of the contaminated areas still needs to be done. Because many residents live in homes bordering the golf course, some question if they had been repeatedly exposed to these chemicals. Areas of the former golf course were considered hazardous enough to close access to them, but no consideration was made for the backyards and lungs of the people exposed.

Woodmont Country Club in Tamarac, Florida, was going to be transformed into 152 homes, 27 holes of golf, and 28,000 square feet of commercial space. However, arsenic contamination was found in the soil and groundwater much higher than levels considered safe. In addition, the highly toxic pesticide Dieldrin was found in groundwater and the maintenance department. The excavation of the contaminated soil will cause arsenic to be blown around into the surrounding neighborhoods, which would endanger the homes located directly next to the contaminated areas. In fact, it is possible their property may have already been affected from runoff. Many of the homes nearby run the risk of contaminated water. All this contamination was there for years until new development was proposed. The neighbors were likely already exposed to a wide variety of contaminants they never realized.

3. Previous Use

Over the years, there has been a surprising number of golf course projects built on reclaimed lands, also known as old landfills. Land is very expensive in many places, and many landfills used to be built on the outskirts of the cities. As the cities grew, they gradually surrounded and passed the landfill, and now that property has become much more valuable. The problem is converting this land into something that can exploit its value.

Some cities and states have decided turning their closed old landfills into recreational properties is a brilliant way to turn an otherwise useless, hazardous site into a thing of beauty. Governments who own the landfill property encourage developers to design and build something that pays taxes to the town and attracts people to come and spend money there. As a result, golf courses have been one of the special use projects that dozens of old landfills have been transformed into.

Creating a golf course out of a landfill presents unique challenges to the developers because it was recently an enormous pile literally made from trash. To compensate for this, developers will push thousands of tons of dirt on top of the landfill, sculpt some nice hills and slopes, and maybe add a few water holes (being careful not to dig down too far). Presto--a golf course.

Once the magical transformation is finished, the value of the reclaimed land and that surrounding the project suddenly becomes higher and the property desirable. It becomes a prime location for high-end houses and condos. However, the unique problems with transforming a landfill into a golf course also creates unique problems for housing adjacent to the property.

The buried garbage decomposes, creating an excessive amount of methane gas. Methane is a highly explosive gas, and if not properly vented from the landfill, it will find its way to the surface into either the atmosphere or nearby structures. The smell of rotting garbage wafting through windows is not very healthy or pleasing, especially on warmer, humid days.

Below those beautiful green fairways, rotting trash is decomposing at a variety of speeds. As a result, the underlying part continually shifts, rises, and sinks as the trash decomposes at different speeds through the sections. Sometimes there can be a total collapse in a section of the landfill/golf course, exposing old trash and rotting garbage. Even though the golf course was built using fill brought in to cover the old landfill, sinkholes and cracks appear out of nowhere.

There are more than 100 golf courses built on old landfills. Developers sometimes will place houses around the perimeter to be close to the golf course and get a premium price for the lots. All landfills, even old closed landfills, represent a danger of toxic emissions, high methane levels, and vapor seepage through the soil. In no way would my collaborator and I suggest being that close to a redeveloped landfill no matter what the developer tells you.

In Sarasota County, Florida, a dedicated individual led a fight against a developer who wanted to build houses on an old landfill/golf course. The developer stated the residential houses would be in between where all the trash is buried, not on top of it. One of the first attempts to do this back in 2006 was dropped once benzene, chromium, arsenic, lead, and mercury were found in test probes. Unfortunately, another developer smelling not the methane but the huge profits of putting houses on the closed golf course/landfill proposed another development in 2014. Again the plan was to nestle the homes on the spaces between where the rotting mounds of trash were located. One environmental activist tried to halt the construction of the homes knowing the contamination beneath the area was toxic. The response of the developer was to sue the activist for $1 million dollars for telling people about the hazards who then decided not to purchase homes. The suit was eventually dropped.

Outside Los Angeles, a golf course built adjacent to a closed municipal landfill had the ground underneath the neighboring houses open up giant cracks, breaking their water pipes. It seems the houses were built too close to the landfill and the ground was unstable due to the trash decomposing beneath the ground. The development also has methane gas from the landfill seeping up into the residents' homes. These residents are at risk for all the health effects from the gases frequently found in landfills, such as aggravation of asthma symptoms, headaches, nausea, vomiting, and respiratory system issues.

Near Columbus, Ohio, in Grove City, the Phoenix Golf Links course was built on top of a closed municipal landfill. Methane gas from decomposing trash under the course was not supposed to be a problem, and special systems were installed when the course was built to prevent the toxic gas from seeping to the surface. Apparently, there were problems with the methane collection system that was built because over the years the amount of methane gas increased until the levels became so high and the collection system so expensive to fix that the golf course simply shut down. Luckily, no homes were built on top.

4. Toxic Fill Dirt to Create Golf Course

Fill dirt is the materials used to create the undersurface of the course. Many golf courses have large amounts of fill dirt brought in to create the rolling surfaces, the hills and valleys, and other features. This is a significant expense for the developers of the course. Rather than absorb this price as the cost of developing, they have found an inexpensive but dangerous alternative: fly ash.

Toxic coal ash, or fly ash, is the residue left over from coal-burning power plants that produce some of our energy. Fly ash is generated in enormous quantities by these power plants. Because fly ash is the byproduct or waste left over after burning coal, it contains the same cancer causing heavy metals coal has. Coal ash is a cocktail of toxic chemicals, containing the same heavy metals that are found in coal, only in higher concentrations. These metals typically include arsenic, lead, chromium, mercury, and selenium, all of which are known to cause cancer and other health problems. Short-term exposure irritates the respiratory system, may cause breathing issues and shortness of breath, as well as dizziness and vomiting. Long-term exposure, such as possibly living on top of a pile of toxic coal ash, can lead to liver and kidney damage, heart issues, and cancer in body organs including the liver, lung, bladder, and kidneys. Because of the known risks, environmentalists have long fought to have fly ash categorized as a hazardous material, but they have faced stiff resistance from the energy industry, and now the construction industry.

Despite this resistance, fly ash can't legally be dumped as it is too toxic. Disposing of it has been a huge headache for the power plant operators. Much of the millions of pounds of ash sit in storage lagoons, containing hundreds of thousands of pounds of toxic fly ash. Liability for the power plants is huge. However, after industry lobbying, the government decided toxic fly ash can be used if the end use can be considered recycling.

The exact wording of this plan is "beneficial use": take a hazardous product and use it for a beneficial purpose. According to the United States government, fly ash as fill for a golf course is perfectly fine.

In York County, Pennsylvania, a lovely course is proud to feature toxic greens and fairways. I spoke to the mayor of York City who remembered the battle that took place when the project was first announced. The hazardous coal ash was dumped in giant piles and then used to cover the ground before just enough soil was placed over it to grow grass. A company who specialized in helping industry find these creative uses to dispose of their toxic ash was used by the project developers. Over the years, a range of health effects including liver and kidney issues was noted by nearby residents, but none were ever able to be absolutely directly tied to the coal ash as usually is the case with this.

In Chesapeake, Virginia, there are ongoing battles with another golf course featuring toxic greens and fairways as well. The appropriately named Battlefield Golf Club features 1.5 million pounds of fly ash used to create the various hills and slopes. Covering 216 acres of sculpted grounds, the grass hides what lies below. The EPA has also determined the water underneath Battlefield Golf Club is contaminated. While they said the contamination has not yet spread to the wells used by houses nearby, the power company that supplied the toxic fly ash has already paid to hook all the neighbors to the public water supply. That was seen as a preemptive move in case additional people were sickened by consuming contaminated water and wanted to file lawsuits for damages. The EPA noted that connecting these residents who used private water into a public system "mitigated and potential threat to drinking water." Dominion Power put aside up to $6 million to use for connecting residents into public water supplies. This commitment was made just prior to several hundred residents filing lawsuits stating the ash used to build the golf course had contaminated their well water with toxic chemicals.

Incidentally, one of the businesspeople who helped build the course has sued the developer for damages because he developed kidney cancer. In addition to this suit, there are several others including one by 383 residents suing for over $1 billion in damages.

A lovely golf club in St. Louis, Michigan, has the distinction of being perhaps the only golf course on the EPA's Superfund site list in 2010. It takes a lot to get on this list; the site must be one of the most polluted sites in the country and also pose imminent danger to human health and the environment. Right in the middle of an out of bounds area on the golf course sits this Superfund. For over 14 years, Michigan Chemical Corporation used that land to burn and dispose of industrial waste. This waste included highly toxic DDT, polybrominated biphenyls (PBB), and Tris (a banned flame retardant). Estimates are 2,000-3,000 gallons of hazardous waste were disposed of on the site. Approximately 2,500 people live within one mile of the golf course. Around 5,000 use the groundwater for drinking, and six wells for the public water supply are within three miles of the waste dump.

5. Groundwater Contamination

The toxic soup of herbicides, insecticides, fertilizers, methane, and heavy metals created by golf courses eventually seeps its way into groundwater and drinking supplies. And when it does, it creates havoc on the citizens' health.

Decades after it was banned in the United States as a cancer risk, an insecticide called Dieldrin has been discovered in 108 private wells in Central Florida, making the water unsafe to drink. There have been dozens of cases across the country but especially in Florida due to the high water table found throughout most of the state. In the neighborhoods of DeLand's Country Estates, 21 wells were discovered with unsafe levels of the chemical. Records from the state have shown another total of 81 wells in two neighborhoods, Longwood and Sorrento, tested positive for contamination in their wells too. The director of Volusia County Health Department summed up the government position of the poison in the residents' wells by saying, "But it only means there is an increased risk of cancer if there is prolonged exposure over the course of a lifetime." People living here for many years are exactly the example the county health department talks about. Prolonged exposure victims. In addition to the chemical Dieldrin, wells in these areas were also contaminated with the pesticide ethylene dibromide, another toxic chemical banned in the 1980s. All of these neighborhoods are situated next to old golf courses where hundreds of pounds of banned chemicals were routinely applied for years still in the soil and water.

6. Illegal and Banned Chemicals -

Information about possible widespread illegal sales and application of pesticides and other chemicals came out in 2017 after an investigation from the *Journal News*/lohud.com found ongoing issues and gaps in the way the chemical industry sells and handles its products. An incident at a golf course in Rye, New York, in which an unregistered pesticide was used led to the discovery that similar happenings were occurring elsewhere in multiple states as well. Record keeping is extremely lax for everyone from the chemicals' distributors down to the golf courses using them. Many of the illegal sales and use are for the main reason of keeping the courses green and perfect. Unfortunately, most of these chemicals are banned due to the known health effects they have on humans and the potential for poisoning groundwater supplies. Hopefully, the regulators will make some changes in current laws and do more to monitor and track these shipments. However, in the meantime, the question remains not if unregistered and banned chemicals are being used, but where and how much. The issue that started the investigation, the country club in Rye, New York, was fined $500 for misuse of a substance. A paltry amount for a blatant violation that the club covered up. Apparently, nothing really concrete will be done until either piles of complaints are unable to be ignored any longer or until something jeopardizes the health of a community as a whole, such as the poisoning of aquifers supplying water to thousands.

Yes, golf course living can be very enjoyable, but it does come at a price. Even after doing your homework, there still may be unavoidable risks, and you need to decide if you are willing to accept them. Just when you thought you may have found out all there is to know about your golf course backyard neighbor, because of pollution exemptions for golf courses and the insufficient research currently being conducted, there still may be new surprises in store for you. At the same time, because millions of houses are built on golf courses, there will always be people wanting to live in them.

My collaborator and I only want to share what we know about the potential risks you may face by doing so. Would we live on a golf course? I would have to say no because, no matter what, there will be constant applications of chemicals to maintain the course. The cost of doing an all-organic chemical program is still far too expensive for almost all courses, and even then organic also does not necessarily mean safe. Be aware of the potential for chemical contamination and minimize your risk.

Chapter 41. Gas Pipelines

Just as vehicles and trucks move around the country on roadways, gas, oil, and some other products are moved around the country in their own underground highway system. According to statistics kept by the Federal Department of Transportation, as of 2016, there were 1,614,936 miles of gas pipelines in the United States. That's right. There are 1.6 million miles of gas pipelines, mostly buried underground.

All across the country, beneath the ground is a vast network of pipelines that pump gas and oil. These pipelines run from where the oil and gas is found to the area where it will be refined and distributed. These pipelines traverse tens of thousands of miles, pumping millions of gallons and millions of cubic feet of gas and petroleum products. As these pipelines age, they will inevitably begin to leak the contents that they transport.

Giant compressors along the pipelines keep the liquids and gas moving along. These compressor stations also pump out millions of pounds of hazardous chemicals into the surrounding air and neighborhoods. The industry does not factor in the consequences of exposure to these over a long term because they are not required to.

These compressor stations are like a small factory and utilize compressors that continuously run. Not only do they keep these compressors running, but they are constantly emitting gases as well.

There are rules and regulations that are supposed to ensure inspections take place on pipelines at regular, reasonable intervals. Unfortunately, this turned out to be impossible as the number of inspectors is insufficient to even inspect a small number of the total inspections needed. After the massive explosion in San Bruno, California, wiped out part of a neighborhood, the industry had to figure something out to do. With the tremendous amount of pipelines in the network, the best they came up with was a way of prioritizing inspections for the more important pipelines.

The industry decided to look at all the pipelines in the system and place each one into a category depending on how much death and destruction it would cause if it leaked or exploded. The category at the top of the list was called "high consequence areas." This was where the San Bruno one would have been placed, and some people say the explosion would have been prevented if there had been an inspection of the pipeline. High consequence areas have a larger "potential impact circle" from a pipeline incident, as the industry group PHMSA called it. These are areas where there are more than 20 buildings used for residential living, buildings which would be hard to evacuate quickly, such as schools, nursing homes, and buildings usually occupied by 20 or more people a certain number of days per year.

In an effort to prevent further disasters, pipelines with these designations are inspected every seven years if they carry gas and every five years if they carry a hazardous liquid. During the first nine years this inspection program was carried out, there were over 53,000 repairs needed and reported. Sadly, the inspection program only covers about 7% of gas lines and 43% of hazardous liquid lines. There have been many leaks, fires, and explosions that simply go unnoticed other than some paperwork that needs to be submitted.

The pipeline infrastructure was installed decades ago, and while newer sections are being built, many of the older ones in the more densely populated towns and cities are over 40 years old. While newer pipelines are better constructed to prevent corrosion, older pipelines do not have a coating to prevent it.

Most people have absolutely no idea if they are located near a pipeline of even living on top of one. In populated areas, the underground lines frequently are buried under the streets and houses with no signs posted to warn people. In areas where there is open space above the pipelines, there usually are signs warning about underground gas lines. The lack of identifying locations in towns and cities is exactly why utility companies always say, "Call before you dig."

Back in the late 1980s, I dug, but did not call first. At that time I lived in central New Jersey in a rural community and was digging out an area for a garden. Honestly, I figured since I was only going down a foot or two, there was no need to check. Apparently, I was wrong because, as the shovel dug down deeper, I felt it hit something. At first, it felt like a rock, so I pushed the shovel down much harder the next time. Suddenly, there was a whoosh, and gas started shooting out from the hole, blowing dirt everywhere. My entire neighborhood and several others nearby were evacuated to the community center while the fire department had their hoses and equipment trained on the hole. Eventually, the gas company showed up and figured out where in the pipeline the valve was to shut off the flow while they checked out what I did. It turned out there was a heavy plastic pipeline running up my street and up the side of my house from the main road that was not marked on my deed. I certainly should have called. It was very lucky my shovel didn't hit a rock when the pipeline started to leak as I most likely would have been an accident statistic.

1. Pipeline Leaks and Explosions

Many of the miles of underground pipelines are buried underneath housing developments and run through neighborhoods and communities. While the newer pipes are usually made of very high-strength steel to handle the high pressure running through them, older ones were not. Unfortunately, most of the pipes in urban areas are older pipes, and older pipes are showing signs of corrosion and stress from aging. In one 20 year period, federal officials had over 2,840 gas and pipeline accidents. In 992 of these, somebody was killed or needed to go to the hospital, totaling 323 people killed and 1,372 injured.

One big ongoing problem with these pipelines is the difficulty in determining when or where there is leakage or exactly to what extent the leak is because these pipelines are primarily underground. These underground transport lines cost millions of dollars to install. Most of the companies who own the pipelines anticipated that their lines would eventually leak. They built these losses into the price of their product, knowing that it would be next to impossible to locate and fix all of the leaks. The pipeline's contents simply will continue to leak into the ground and spread its toxic cargo out in a plume. The leaking oil and gas may eventually contaminate millions of gallons in underground water supplies. Hazardous chemicals carried through these pipelines will leak as well. While not every pipeline is currently leaking, there is a high possibility along every pipeline that somewhere there is a leak or that in the future a leak of some magnitude may occur.

On September 9, 2010, a large gas pipeline exploded in San Bruno, California, outside San Francisco. Eight people died in the blast, and 58 sustained injuries. In the end the blast totally destroyed 38 homes and did damage to 120 others. Sadly, the pipeline operator was unable to explain what happened.

Multiple explosions outside Midland, Texas, in August, 2018 resulted in five people hospitalized with critical burns. In Kentucky, one person died and five others needed hospitalization when a pipeline exploded in August, 2019. April, 2016, saw a pipeline explosion in Salem, Pennsylvania, blowing up a home and evacuations. The organization, Fractracker Alliance, reviewed data from the federal government the industry sent in and totaled 1,069 incidents resulting in injuries, death, and damages.

The requirements to inform people they are living near a buried, high-pressure pipeline are minimal. Hundreds of thousands of people live close enough to be affected by an accident but have no clue. Real estate disclosure laws do not have a requirement to include information about the proximity of underground pipelines in mandated disclosure rules. While pipeline operators may send notices to current owners about nearby underground pipelines, a buyer may not know about this unless there is an easement on the property deed itself. If there is a right of way on a property, it has to be on the deed as part of it. Most times, though, other than the small sign on a stick that says, “Warning--Underground Gas Pipeline,” no one would never know.

2. Transmission Station Contamination

Gas transmission stations are all along the routes of these pipelines. These stations are usually used for monitoring the flow of product through the pipeline. Some contain equipment that can detect variations in volume between stations, which would indicate any leaks that are occurring. Many of these transmission stations also contain compressors, which are necessary to maintain the pressure and keep the flow of products continuing in the pipeline between stations.

In the past, a material now known to be one of the most toxic and deadliest chemicals was used to lubricate and cool the compressors in these transmission stations: PCBs. Because PCBs work so well to cool these compressors, they were extensively used . When maintenance was done on these compressors, it was a fairly common practice for pipeline companies to clean out the PCBs and dump them there at the transmission stations. The federal government possibly knew about this practice in the early 1980s but kept quiet while dumping continued. The public first learned in 1987 that these highly toxic, cancer-causing chemicals were routinely buried in pits along these gas pipelines by certain pipeline operators.

Many years of using PCBs in pipelines has led to the creation of a large number of toxic waste pits near transmission stations. Even though much of the contamination happened before the use of PCBs was phased out, many sites continue to have levels exceeding safe ranges established for exposure to PCBs. In addition, the PCBs still remain in many of the older pipelines as they were also used as a coating for the interior walls. They continue to be flushed out and disposed of. There are many disposal pits yet to be uncovered that are contaminating wells and water supplies near them. These disposal pits may very well number in the hundreds or even thousands.

After determining there was a high probability gas and oil pipeline operators were routinely dumping hazardous wastes along their pipelines, the EPA opened a probe into 11 of these companies, and discovered several had created dozens of these toxic waste pits stretching thousands of miles.

One such pipeline operator, Eastern Gas Pipeline Co., was caught and forced to clean up 89 sites in 15 states along their pipelines where these PCBs were disposed of. One of the largest pipeline companies, Texas Eastern Gas Pipeline Company, which operates 10,500 miles in 12 states. It cleaned up PCBs and other contaminants at its compressor and meter stations along the pipeline and then illegally disposed of hazardous waste in earthen pits at their compressor locations. PCBs and other hazardous materials were sometimes burned in these pits, and several compressor stations had contamination extending into the neighboring property. At the metering stations used for measuring flow in the pipeline, the company disposed of toxic mercury. Seventy one of these toxic pits were admitted to by Texas Eastern.

In addition to Texas Eastern, other pipeline operators, such as Tennessee Gas Pipeline Company, ANR Pipeline Company and Transcontinental Pipeline Company, also reported the same practice of disposing toxic waste in pits and therefore needed to clean up some of the sites that were discovered to be polluted.

The Tennessee Gas Pipeline is being held responsible for cleaning up PCBs at the compressor stations where they dumped waste along their pipeline between Texas and New England. They had more than 60 waste sites, costing between $20-30 million. While most of the contamination is on their property, there had been toxic contamination onto other properties by runoff over the years. The true extent may never be known, found, or totally cleaned up.

American Natural Resources admitted to creating 29 of these hazardous waste sites in thirteen states. The company spent $10 million to clean up the extensive contamination at those locations.

Transcontinental Pipeline used unlined pits at its compressor stations to dispose of materials including chemicals such as benzene and PCBs, both known cancer-causing substances. The company operated a pipeline network stretching over 10,500 miles across the United States from Texas to New York. Many times the waste was simply burned in these pits and the remains just left there. Some of the compressor stations had waste that spread onto neighboring properties as well. At some of Transcom's metering stations, mercury used in the equipment contaminated the soil and surrounding areas as well. The Justice Department and EPA fined the company $1.4 million and ordered them to clean up hazardous waste at 26 of its compressor stations. It is unknown how many additional locations the company polluted where the waste remains.

The risk of PCB contamination deserves serious consideration when people look at a property that is located near a compressor or transmission station along these rights of way.

3. Blowdowns at Compressor Stations

A blowdown, a typical occurrence for the industry, occurs when the pressure inside the pipeline exceeds the safe level. The blowdown at the compressor station relieves the pressure off the pipe. The typical blowdown can vent several thousand pounds of methane gas directly to the atmosphere in addition to hundreds of other potentially toxic chemicals. These blowdowns can last up to two hours or longer. Due to the tremendous pressure the gas pipelines are kept at, when blowdowns occur the level of noise can be ear-splitting.

Blowdowns are also used when pressure needs to be adjusted in the line for maintenance work. Again, thousands of pounds of chemicals may be emitted, and these releases can last several hours. The industry usually calls these routine emissions, which unfortunately is a true statement. Almost all compressor stations have these routine emissions, filling the air surrounding them with varying levels of toxic chemicals for the neighborhood and residents to breathe.

Methane in particular has a range of effects on the human body. Chronic exposure to methane causes respiratory and neurological problems, problems with the heart, and memory loss.

4. Compressor Station Leaks and Emissions

This is by far the greatest threat to people from these pipelines. The industry technically is in compliance with regulations and frequently claims to be below acceptable levels. Unfortunately, these regulations are based on a single exposure, and many scientists would argue the acceptable levels are set way too high. For many of these chemicals that are emitted, there is no safe level for the human body to be exposed to on a continuous and long-term basis.

The total quantity of the emissions from these compressor stations is not small by any calculation. In New York state, a report done by the Institute for Health and the Environment reviewed emissions data reported to the NYS Department of Conservation and to the EPA. The report stated the total emissions over a seven year period for only 18 of these compressors totaled 40.2 million pounds. Of that total, 7.9 million pounds are known carcinogens. They cause a range of cancers including kidney, liver, prostate, bladder, and lung cancers, as well as several forms of leukemia. In addition, the report stated the amount of chemicals related to health issues of the cardiovascular system was 17 million pounds. The report concluded with a memorable statement: "Without hesitation, we can say not only that the volume and known health effects of these pollutants increase the risk of disease, but that they will result in actual illness."

Leaks occur at these stations at a fairly high rate. Methane leaks are common from the valves and aging gaskets. Corrosion and stress from the enormous pressure weaken the seals and corrode the pipes.

There are also regular emissions that occur continuously. These stations are not just emitting steam; they are a source of harmful chemicals being allowed to pollute the air and slowly poison nearby residents. Compressor stations emit several other toxins in very large amounts, including benzene, nitrous oxide, toluene, sulfuric oxide, and formaldehyde.

On one permit application for a single compressor station, the company stated it would be releasing 46.2 tons of nitrous oxide a year directly to the air. People living within a quarter mile of these compressor stations say they get headaches, nosebleeds, rashes, sore throats, and nausea. Many of the chemicals emitted by the compressors have also been linked to serious health problems such as respiratory issues, cardiovascular illness. Even health problems like cancer and birth defects can occur if there is enough exposure to the chemicals continuously being emitted. Like other industrial facilities, their permits allow them to emit limited amounts of carbon monoxide, volatile organic compounds, and particulates, among other pollutants. Energy companies say the facilities are safe and well-regulated. The people living there and environmentalists say the dangers posed by these emissions are understated and inaccurate, adding that long-term exposure to these chemicals is going to be a great danger to the people.

In Farmville, Virginia, residents reviewed a permit application for one of these compressor stations and saw amounts of pollutants potentially being emitted into the air they would be breathing. The permit application for the Buckingham compressor station proposes an annual release of 70.9 tons of methane, 43.2 tons of particulate matter, 34.3 tons of nitrogen oxides, and 5.3 tons of hazardous air pollutants as part of the 221 tons of total emissions proposed. Exposure to these pollutants can increase the risk of respiratory, cardiovascular, neurological, and developmental diseases. Examples include everything from minor ailments like headaches and eye, nose, and throat irritation to more serious conditions like difficulty breathing, worsening of asthma, high blood pressure, and heart attacks.

The residents mobilized and spent several years fighting the project. They put together research showing the damaging health effects this compressor would have in the surrounding area. In addition, the residents filed legal challenges to the permit based on illegal actions the company had taken. As of May, 2020, a court issued a ruling revoking a critical permit the pipeline needed to continue with the construction. The battle goes on.

In Arlington, Massachusetts, a group called Mothers Out Front searches for compressor stations in town that are leaking gas and other chemicals into the air. When they find one, and there are quite a few, they tag the area with a sign saying, “Fix the Leaks.” People are at least aware these compressor stations are emitting chemicals into the air. The sad part is, they are still forced to breathe those harmful chemicals.

In New York state, a study of compressor stations counted almost 70 chemicals released into the air, including formaldehyde, nitrogen dioxide, carbon monoxide, methane, and particulate matter.. Many of the chemicals emitted at these stations are already linked to different diseases, especially with chronic, long-term exposure, including respiratory and cardiovascular diseases, developmental and neurological diseases, and cancer.

5. Noise from Compressor Stations

When there is a blowdown, which can last for hours and emit noise levels comparable to a jet plane or train locomotive, there is a continuous sound similar to a humming or buzzing that the compressor station emits. People living nearby have said it is impossible to get it out of their heads and not hear it. A constant background noise to an otherwise fairly quiet place could drive you crazy after a while.

Studies examining the effects of this noise on nearby residents has been turning up a host of issues. The noise associated with compressors and these pipelines is a major environmental health concern. Long-term exposure to chronic noise levels higher than usual background noise is associated with hypertension, sleep disorders, and vibroacoustic disease. Additional adverse health effects, such as a range of hearing loss, increased risk of diabetes, and stress on cardiovascular systems, have been reported. In addition, it also leads to issues of poor concentration and lower attention ability, especially in children. Cognitive issues like memory problems, behavioral issues, and decreased performance have also been noted as effects.

A recent study of homes in West Virginia funded by the Maryland Department of Health near a compressor station found that average noise levels for the homes exceeded EPA and WHO recommended limits for both outdoor and indoor noise. Noise is not just a nuisance to many people. It is a health hazard.

6. Rights of Way

The gas company needs to make certain the areas on top of the buried pipelines remain clear for access and to ensure vegetation does not disturb the buried pipes. As is the practice with all industries, the only effective way to do this is to apply large amounts of herbicides. In many areas it is common practice for the line maintenance crews to load fifty five gallon drums of chemical defoliant onto the backs of trucks and drive along the right of way, spraying this chemical agent onto the vegetation. Sweeping their pressure hoe back and forth, they cover everything in their path. Herbicides drifting onto adjacent properties and backyards is unavoidable.

In the past, before being banned, tens of thousands of gallons of these chemicals have been applied, including total vegetative killers, such as 2,4,5,D. Some toxins lie in the ground in high concentrations, leaking into the ground bit by bit over the years. While a large dose of a toxin over a short period of time may be detrimental to your health, so may a small dose of a toxin continuously over a large period of time. Some toxins accumulate in the fatty tissue of the human body and keep accumulating until they reach levels that have a negative effect on the person's health.

Even today thousands of gallons of chemicals continue to be applied on vegetation over these rights of way. While the chemicals that are being used today are considered safe at the level they are being applied, long-term exposure to small amounts may have detrimental effects that are not going to be known for years to come.

If you have a right of way behind you and it is regularly maintained with herbicides, do what you can to find out what they are using and when they spray. Avoid the area during spraying and stay away for several days afterward if possible.

Underground pipelines, if we look at the entire network as a whole, may represent a small risk. Look at specific parts of it, though, and it becomes a different picture. Living near a compressor station or a transmission station places the risk into a much higher category. Exposure to toxic chemicals released in fairly large quantities as part of the operation are a hazard to your health. Proximity to an old transmission station located out in the rural areas may very well still be contaminated with hundreds of pounds of PCBs buried right where the company left them years ago.

The first step towards addressing this potential hazard is finding out if there are any stations near you. There should be a map of them with your local environmental health agency. For what it's worth, I would not live anywhere within a half mile of these places. Noise and chemical emissions make them a risky place to live. If you can stay further away, even better.

Chapter 42. Crematoriums

Any place that burns something in its operation has the potential to create a hazard for the surrounding area. There are a few usual suspects that are obvious to most people, such as chemical factories, trash incinerators, and power plants. Smokestacks sticking out of facilities like these are common and visible. Usually, so is the smoke carrying the emissions away into the surrounding area. There is another type of business that usually operates openly in a neighborhood and does not stick out to the people nearby. People usually don't pay too much attention to these businesses: crematoriums.

Crematoriums, also known as crematories, operate under the radar. Many times, people may not have even known there is one in their neighborhood. Unless they had a reason to know where one was or received a notice for a zoning board or planning board hearing for a new one, they would never know. Crematoriums do not stand out. Because of this low profile, the hazards from crematories were not widely studied. It is now finally being investigated, and experts who understand the negative impact possible are raising a voice.

A crematorium is simply a business where bodies are placed into a high temperature furnace until nothing remains but bone fragments and any metal, such as screws, fillings, implants, or jewelry. The process usually takes about 1.5-2 hours, depending on the weight of the body and the temperature of the furnace.

Once the remains have had time to cool down, the operator removes and separates any metal mixed in with the bone fragments, either manually or with a magnet. The rest of the remains are then pulverized into a fine powder, otherwise known as the ashes. The ashes are then returned to the family of the deceased, if requested, in a vessel or urn of their choice. Some people keep these ashes of loved ones in their houses. Others may choose burial in a cemetery or other place; others spread the ashes to the winds.

In many countries, the practice of cremation has been around for thousands of years. In the United States, it has only been gaining in popularity for maybe the past 50 years. That's part of the reason not much attention was given to these facilities. In the past, they were not widely known and had little visible presence because not many people used cremation to dispose of a body or really understood what it was all about. Back in the 1970s, cremation was a choice for only about 5% of all funerals. In 2015, the percentage of cremations reached an all time high of 53.5% of all funerals. The National Funeral Directors Association expects the trend shifting from burial to cremation to continue over the next 20 years, with the projected rate of cremation reaching 78.8% of deaths by 2035, largely because the costs of cremation are much lower than a burial. That is going to drastically increase the need for many more cremation furnaces and crematories to handle this demand.

In the past crematoriums offering this specific type of body disposal operated totally independently of funeral homes that offered more traditional burials. With the rapidly rising number of people choosing cremation and the decreasing profit margins doing only regular funerals, more funeral homes are adding cremation furnaces, driving a big increase in activity.

Sensing opportunity, many funeral homes that have not had a cremation furnace are looking to add this to their operation. Currently, the estimate is only about 30% of regular funeral homes also have a cremation furnace. Another 9% say they are adding one over the next five years.

Because crematories have gone so long without oversight, it is hard to estimate exactly how many are currently in operation in the United States. and I tried to get an accurate figure by reaching out to the industry group for information. In response to our initial request, they said we did not describe what we wanted well enough. We tried a second time, specifically telling them all we wanted was a fairly accurate number of facilities to quote in an article we were writing. At the time we finished this section, they still had not responded to our request for numbers. With the extra scrutiny the cremation industry is starting to experience, it did not surprise us they were skeptical and chose not to respond. What we do know is that quite a few of the crematories are already operating at capacity trying to keep up with the increasing demand for cremation. Some of the larger ones now operate multiple furnaces or are adding a second or third.

The numbers are hard to track down because, again, crematories have been largely invisible. In the past, there had only been some isolated instances when people in neighborhoods with crematoriums grew concerned over seeing occasional smoke and emissions from the furnaces. Lately, though, that has been creating controversy both with currently operating facilities and proposed crematories opening to meet the increasing demand.

With more investigation into what actually goes on at these facilities, concern is rising among people living near them. Is there a health risk associated with crematories?

The number of studies and research into health risks from crematories at this time is small. However, enough has already been done to form the basis for concerns moving forward.

Part of the concern stems from the stricter regulations that crematories have been able to avoid having to comply with as other similar incinerators in other waste industries already have to. The lobbying group for crematories and funeral facilities may not be as powerful as many others, but they exert influence far beyond their size. Through the National Funeral Directors, crematories have managed to prevent disposal of human bodies from being classified as solid waste. This is a very important distinction because the classification eliminates regulations that otherwise would need to be followed as other industries do. Furthermore, because they are disposing of an entire body, they also have managed to escape regulations governing medical waste disposal facilities as well. There is some oversight, but it is at best minimal and irregular. When oversight does occur, it is almost always done only during optimum operating conditions at the facility. Optimum operating conditions are when a piece of equipment is running absolutely perfectly, ensuring any tests or monitoring done during that period will show nothing but perfection to any observer. In reality, optimum operating conditions are rarely if ever met during regular operation, particularly after any equipment has been used in daily operation.

The difference in classifications allows the operators of the crematorium to not have to meet stringent emission standards specific to the waste disposal industry. Following these regulations would force them to add expensive filtering and monitoring systems to their exhaust stacks. While almost all other industries that dispose of waste products have their emissions regulated by the EPA, the cremation industry does not. Even though the solid waste industry emission standards are inadequate and minimal by some accounts, at least there are some. The cremation industry argues it is not necessary to monitor their emissions because many crematory incinerators operate with double combustion chambers. Of the gases and compounds that escape destruction in the primary chamber, some get destroyed or reduced in the secondary chamber. Then, that air is vented directly to the atmosphere. While the second chamber is efficient at destroying more of what is sent through it, it does not incinerate metals, dioxins, and others. According to the Georgia State Health Department, small amounts of lead, cadmium, hydrochloric acid, nitrogen oxide, sulfur dioxide, and carbon monoxide are released. Those pass through and are vented through the exhaust to the outside air.

In one memorable instance, a project manager for a crematory project applying for a permit to build a new crematorium in Weatherford, Texas told the town's planning and zoning board more pollutants are emitted by a fast food restaurant than the proposed crematorium. This same project manager also told the planning board crematoriums are a heavily regulated industry in spite of that being the complete opposite of the truth. Unfortunately, the town board did not check the validity of these statements on their own but took the word of the applicant. In addition, there were almost no members of the public present at the meeting to refute the crematorium representative because the public stated they were not informed about the meeting in advance. Because of these misrepresentations, the board approved the permit, and the crematorium got a pass. The board did the town a disservice and exposed them to hazardous emissions that were not regulated in any way.

The most common objection people raise to crematoriums are the contaminants known to be in the emissions. As this example shows, the industry downplays their emissions, especially when presenting a request to build a new facility.

Our review of available research has found the following information on health risks related to living near a crematorium:

1. Mercury

This highly toxic metal is one of the top dangers crematories may present. In 2010, the EPA stated that toxins are released into the environment through the cremation of human bodies. The most dangerous of these is mercury. During the cremation process, the mercury in the body becomes vaporized into a colorless, odorless gas. Mercury is especially dangerous in this heated, vaporized form because, when it is inhaled, it is quickly absorbed into the bloodstream and carried to the brain and other organs. As verified by the EPA, even very small amounts of mercury can be dangerous.

The typical response from the cremation industry is stating that the amount of mercury is so small it is not even measurable. However, as with many toxic substances, mercury is one that accumulates in the body. If every third body cremated contains 4-5 amalgam fillings, that adds a tiny amount at a time to the accumulation of inhaled vapors into the lungs of residents nearby.

The EPA does not regulate mercury emissions from crematories like other countries, such as the U.K., do because when the EPA looked at the need to regulate, they depended on information the cremation industry gave claiming they do a good job of regulating themselves so government regulation would be unneeded. They do not mention it would be costly to the operator if they needed to install filters on their equipment and have resisted every effort to require them to do so. The cremation industry lobbying group uses that success on their website as a reason to join their group. To this day there are still no federal regulations on emissions from crematories despite the recognized fact that many pounds of mercury are emitted from crematories every year in every state. The nearby neighbors are continuously exposed to these emissions almost every time a cremation takes place.

There have been attempts from some individual states to place some regulations on the industry to control emissions. In Minnesota, crematories emit almost 5% of the mercury emissions. While the amount is much less than coal fired power plants, it is far too much to ignore. Attempts, though, to enact regulations failed as they have in Maine as well. Colorado estimated crematories emit over 100 pounds of mercury a year but still does not have any regulations to control the emissions.

Tony and I reviewed an application submitted by a new crematory in 2017 in Vancouver, Washington. It listed 11.4 pounds of mercury could be released into the air annually. That was the number the applicant stated based on the applicant's calculations. Given that a few micrograms of mercury can be dangerous, 11.4 pounds is a significantly large number.

Mercury emissions are one of the most dangerous for young children and unborn babies in particular because their brains and nervous systems are still developing. Studies done by researchers at the University of Texas Health Science Center found that close proximity to sources of mercury release were associated with increases in learning disabilities, autism, and rates of special education. Adults can also develop problems with their brains and nervous systems, as well as their lungs, kidneys, and immune system from mercury exposure.

Mercury is not destroyed by combustion. It simply changes forms. The two chamber combustion systems currently used in most crematories are incapable of removing mercury. In addition, government regulations do not require removal of mercury from these facilities, so there is no incentive to add additional equipment to deal with the issue.

The cremation industry is well aware of the mercury emissions issue as it is very widely spoken about at their conferences and organizations. Until the time comes that the lobbyists are unable to stop responsible politicians from protecting the public through regulations with teeth to them, this practice will continue as is.

2. Hazardous and Toxic Air Pollutants

In addition to mercury, there are over a dozen other toxins emitted from crematories. These include arsenic, hydrochloric acid, dioxin, furans, PCBs, naphthalene, and selenium. The list continues. The incineration process does destroy these pollutants, nor are they completely filtered out.

The application for the proposed incinerator in Vancouver, Washington had a total of 20 toxic air pollutants listed as anticipated emissions for a total of 172 pounds per year. Many of these compounds are listed by the EPA as cancer-causing chemicals. Being exposed to small amounts, as may be through inhaling emissions, can potentially threaten the health of nearby residents.

3. Pacemakers

Every year thousands of these devices are implanted into people. While much of the actual device is made of titanium and other harmless metals, pacemakers also contain silicone and polyurethane. Given all the potential for harmful emissions, it is good practice for these devices to be removed from the body before it is placed into the furnace. Unfortunately, this does not always happen. In many cases, the implant is not seen while the mortician prepares the body. Sometimes, the family may tell the crematory about a pacemaker; other times it is not brought up. When the pacemaker is subjected to high temperatures during incineration, it releases toxic compounds and dioxins from the burning plastic.

Also, there have been cases of these devices exploding in the furnace. Pacemakers are usually powered with a lithium iodide battery. When these are subjected to the high temperatures in the furnace, the iodine expands, causing an explosion. A survey was sent out to over 200 directors at crematoriums asking if they had ever experienced an explosion caused by a pacemaker. Almost 47% of the respondents answered they had personally had that happen to them. Several said there was damage to the furnace as a result of the explosion.

4. Replacement Joints

Many people in the past decade or two have received artificial joint replacements. These replacements are usually made of plastic and metal. The plastic most commonly used is polyethylene, and the metal is titanium and stainless steel. The same issues for burning pacemakers exist in burning artificial joints, except they do not explode. The plastic does incinerate and creates dioxins and other compounds that are emitted into the air and not completely filtered out during the process.

5. Radioactive Patients

I added this section to the book when our research uncovered a particularly shocking instance of this happening. In 2017 at a crematory in Arizona, a person was cremated who had recently been given treatment for cancer with radiation. The crematorium, when tested, had levels of radioactive isotopes consistent with what would be found in a lab or medical center working with those substances. Recent radioactive treatments release radioactive isotopes. While the EPA has guidelines in handling bodies who have recently received chemotherapy treatments and other radioactive isotopes, there are instances when that knowledge is either not passed to the crematory or the crematory neglects to ask.

According to the EPA, these radioactive isotopes are hazardous for up to one year after the last treatment was administered to the body being cremated. While the regulations cover how to handle the body and the ashes, it does nothing to regulate or prevent the escape of radioactive particles from the exhaust during the cremation process. The EPA suggests but does not require crematories to take readings of radiation levels in various places in their equipment, building, and around the ashes of the deceased.

A review of all the reputable scientific research and studies that have been done makes it clear that exposure to crematory emissions belong in a category of a substantial health risk. The current practice is to look at emissions data from a single cremation and base safety guidelines on that, but this is insufficient. Continuous exposure over a long period of time has cumulative effects, and it compounds what the people nearby are actually subjected to.

At minimum, crematoriums do not belong near residential areas or schools where people will be exposed to emissions on an ongoing basis. Crematories should be reclassified as industrial operations that need to be regulated like similar operations that emit hazardous chemicals into the air. Until the state and federal government enact real regulations on this industry and truly monitor and limit their toxic emissions, local governments and residents need to protect themselves. We suggest leaving a half mile distance between you and any operating crematorium.

Chapter 43. Department of Public Works and Highway Departments

The Department of Public Works (DPW) is the workhorse department of most local and county governments. In addition, there is the state Department of Transportation. These departments take care of much of the infrastructure that keeps a town, county or state running. From the smooth operation of traffic signals and road signage to the efficient upkeep of roadways, they perform a wide range of activities.

Much of the work they do is out in the community with their fleet of vehicles and equipment. But they also need a base with work facilities where the town, city, or county can carry out other needed activities. These facilities can range from a simple office area to fill out reports and have meetings to a place to store extra drainpipe and fencing. Typically, there is also a maintenance yard or two for their fleet of cars, trucks, and heavy equipment. Other typical uses for department facilities could be using the area for recycling, composting, temporary land filling, and hazardous waste collection. Sometimes the yards are locations for old and unused equipment that just lays there and rusts.

There are many toxic and dangerous chemicals used and stored at DPW yards and facilities. Some of them are used on a daily basis. While most of the activities carried out every day do not present any environmental issues by themselves, problems can occur simply because of the materials and activities at the location. Problems may also occur if one or more individuals working there disregard their actions' effects on the environment. Improperly disposed motor oil could be poured into the ground or drains. Chemicals such as cleaning compounds, degreasers, and machine lubricants could easily be spilled out behind or around the buildings. When vehicles and large machinery are hosed down or repaired, hazardous fluids and oils could leak on the ground.

My past job took me to many DPW yards on a weekly basis. Because I am interested in the environment, when I saw something, I would ask about it. Over the years I made friends with many of the DPW workers, and they would share stories with me. In speaking with these people, sometimes when an accident occurred, the people that are involved in the incident prefer to clean it up themselves because they are afraid of getting in trouble and losing their jobs.

In one case, shared by a camping buddy who works for a DPW, a forklift driver accidently hit a pallet, rupturing a drum of industrial degreasing solvent. The contents of the drum spilled out all over the floor. The involved parties just washed the solvents down the drain and disposed of the drum out back into the dumpster. If the spill were handled properly, the local HazMat team would have had to get involved, as well as the proper government environmental agencies.

Once the HazMat team responded to the incident, paperwork would need to be completed, state agencies would be notified of the incident, and possibly the local news may have picked up the story. The entire incident may have been turned into a big issue with potential negative ramifications for the workers involved and a very expensive cleanup. It seems, then, that the system discourages people from properly dealing with these situations. It is preferable to sweep the problem down the drain, so to speak, than to deal with it the right way.

Sometimes the spills are not accidents, but negligence. Over decades, continuous oil leaking from trucks, residues dripping from machines, and fluids seeping from equipment get into the ground bit by bit and build up to potentially hazardous levels. It is also important to remember that it was not too long ago that dumping these potentially hazardous materials down the drain, onto the ground, or into the local landfill was an acceptable means of disposal. It was only recently that we have begun to understand the consequences that our past actions are having on our environment and, consequently, on our health. Even when intentional contamination has not occurred, just the routine everyday operations at these facilities presents a very strong potential for a hazardous situation to occur.

I have a good friend who worked for a DPW in a central New Jersey city for 23 years. He would tell me stories about washing down vehicles and heavy equipment with hazardous substances right on the bare ground. There were drain pipes inside the building where they worked on the heavy equipment, and all of the hazardous materials went down the drain, possibly even contaminating the large creek behind the building. Old motors and other machinery were thrown into a pile out back and covered with dirt. The road scrubbers would return from their rounds and empty their tanks of dirt out back as well. Gallons of oils, gas, paint, degreasers, and other materials were handled sloppily. He said it was common practice for most of the crew to work in this manner. Sometimes, I would visit him during lunchtime at the yard. I knew several other workers from camping trips, and they would share some of their stories as well. Especially since their employment with the department went back so far, they knew about past practices that were no longer being done, but the consequences of which remained in the soil and buried out back.

I can hear the moans and yells from operators who take great pains to do their job correctly and meet environmental standards, but our research unfortunately bares out that even though there are totally responsible individuals, totally responsible DPWs are the exception.

1. Underground Storage Tanks

Most DPW yards have on-site storage tanks that hold bulk oil, gasoline, or other substances. The majority of DPWs have been at the same location for decades. There are many instances of the original storage tanks having never been removed and leaking their contents into the soil and aquifer. Towns with a public water supply usually draw their water from wells that may be impacted by chemicals leaked from the DPW site in the vicinity. There are hundreds of DPWs that have old, leaky tanks. In 1984, the EPA mandated underground storage tanks be replaced with special lined tanks to protect against leaks. There was a 10 year phase during which tanks were to be switched to the new ones. Well over a million old tanks, primarily made of metal and buried since the '50s and '60s, were already in the ground. All the fluids and sludge in the old tanks first needed to be removed and disposed of at special facilities. The costs associated with digging up and removing the old tanks were high.

In Bolton Center, Massachusetts, the site of the highway department was determined to be the source for an ongoing leak from underground storage tanks onsite. Gasoline leaking from tanks created a plume of contamination that impacted a nearby water supply well. It is unknown how long the well had been polluted, but all the while, people used the water.

In Fairfield, Connecticut, the DPW yard never removed their old tanks when new ones were put in. It was discovered that the tanks had leaked diesel fuel and oil. There was contamination 300 feet away from the spill area. While 300 feet may not seem like a big distance, based on the high water table in the area and potential for extensive groundwater contamination, concerns are high. The cleanup, which was once thought to be an easy quick one, would now take five to ten years based on the extent of the contamination. While so far nothing is being said about the chemicals having traveled offsite, contaminating adjacent properties, neighbors are watching anxiously and should have their own testing done.

In Dumont, New Jersey, the DPW was cited by the Department of Environmental Protection for contamination. Groundwater at the site contained petroleum products that had leaked from underground storage tanks. Some of the groundwater and soil contamination has been there since the mid-'80s, seeping through the soil and creating a chemical plume in the aquifer.

2. Degreasers, Cleaners, and Solvents

Maintenance yards for DPWs are very busy places. All of the equipment the department uses, from weed wackers and lawnmowers to earth-moving and digging equipment, is stored and maintained at the yard. Working on the machines and making sure they keep running takes a lot of maintenance. Changing the oil, lubing the parts, replacing worn parts, and doing repairs are all part of what's done at the yard. After it has been repaired or used, the equipment needs to be hosed down to wash it. Bulldozers, backhoes, paving equipment, and dump trucks all get cleaned. Many present-day DPW yards have a wash bay where the equipment can get pressure washed and cleaned. The waste water goes into the floor drains. If the floor drains are properly set up for a wash bay, the contaminated water is captured in special equipment and disposed of properly. If the water is simply sent into the town sewerage system or, worse, drains into a nearby creek, that presents a much larger problem.

However, there are many DPW yards that still wash the equipment outside. Heavy duty degreasers, solvents, and whatever else is used run onto the pavement or dirt and go into the ground. The runoff goes into nearby ditches, eventually making its way off the property. Years of accumulation of polluted wash water build up in the surrounding soils.

3. Past Dumping Practices

Depending on how long the yard in question has been used by the department, past activities almost always included waste disposal. For instance, the DPW yard in New Brunswick, New Jersey, was found to have contamination from an unknown source. The property tested positive for toxins usually found in landfills potentially chemicals. Old landfills and informal dumps were common at older DPW yards. While many are no longer used, the threat they pose may still exist. It is entirely possible toxic chemicals deposited at the old landfill have contaminated the soil in the area putting neighbors at risk.

So far, the city is not admitting to anything. City officials did say the current DPW yard could have been the site of a landfill or ash dump 75 years or so ago as some of the contaminants found indicate coal ash and a landfill were there. If indeed that is the case, unlined dumps were usually used for everything from construction debris to industrial waste disposal.

Past practices of disposing materials that were not considered hazardous in the past but are now known to be so, are commonly found in older yards, mostly from the 1990s and earlier. In many cases they still remain where they were disposed of years ago until a problem needing to be investigated occurs.

4. Particulate Matter and Noise

Frequently, the town may have a section of the maintenance yard where the locals bring all the organic waste, such as tree trimmings, grass, leaves, and other types of waste. In many towns they may grind up the waste and let it compost into mulch. The town may use it for their own projects or sell it for a small fee to the residents to use.

While recycling the wastes into compost and mulch is environmentally friendly, it creates a situation where small particles from the debris are continuously being swept up in the air and blown off the site. Many of these particles are inhaled by neighbors and lead to irritated respiratory systems and increases in asthma.

Even if the DPW yard does not have a mulching operation, they do operate trucks, heavy equipment, and diesel-powered generators. Diesel fumes are already a known carcinogen, and the fine particulate matter these engines emit gets trapped in the air sacs of the lungs, irritating the entire system. The exhaust-filled air, even if only for several hours per day, will have a detrimental effect of people living nearby. The fumes alone from all the diesel powered equipment they use would make for some very exhaust-filled air for a while.

The noise generated from all the trucks, vehicles, and equipment presents a hazard of its own. Noise has been proven to have a detrimental effect on people's nervous system when they are subjected to continuous loud noises. It has even more of an effect on children, resulting in decreased ability to focus and concentrate.

Early in the morning, the first diesel truck cranks up and starts idling to get ready for the workday. Multiply that times the many trucks, heavy equipment, and other noise from the yard, and you have a very loud neighbor. Sure, maybe they don't start until 6 in the morning, but to have that every day to start off with does get to be quite a strain on your body's nervous system.

5. Fill Dirt from Outside Places

Many DPW yards have piles of dirt they use for projects around town. Some of the dirt and fill is from excavations they did at other projects. Sometimes, they bring in fill from outside contractors that they allow to dispose at their site.

At the Fairfield, Connecticut, DPW yard, a pile of demolition debris and fill was found to contain levels of PCBs and lead by an unknown source thought to be an employee with knowledge of the problem. An outside environmental consultant was brought in to test the fill and soil at several town locations. By that time, unfortunately, contaminated soil was used in many local projects.

When investigators requested some of the files related to this issue for review, it appears some of them files related to this issue were missing when requested for review. The files may have likely had information about where the contaminated soil came from and who authorized it being dumped at the DPW yard. Recently, it was discovered an official of the town stole the files possibly in a coverup of who knew what and when. They were quickly returned to investigators by the official once it was known he removed them.

This turned out to be a major scandal involving kickbacks, bribes, and illegal dumping of hazardous wastes. The co-owner of a construction company was accused of paying kickbacks to the superintendent of public works to allow the construction company to dump hundreds of truckloads of soil containing toxic levels of lead and PCBs on property belonging to the town of Fairfield, property adjacent to the town's public works garage. The construction company co-owner was charged with first-degree larceny, first and second-degree forgery, paying kickbacks, and illegal dumping.

While the blame fell on the outside contractor managing the fill pile, the fact remains that PCBs are in the DPW yard next to neighbors. The uncovered pile is subject to erosion and seepage into the ground from the weather. The DPW of Fairfield didn't know that a fill pile on their own facility was contaminated until it was too late. Schools, parks, and other public lands wound up with contaminated fill pulled from the pile by the DPW. Lead, PCBs, and other toxins are also now being cleaned up around town.

In Teaneck, New Jersey, a former DPW maintenance facility used since the 1960s has been discovered to be contaminated. The property was being considered to build housing for senior citizens. As of the writing of this book, the town is currently investigating to see the extent of the contamination and if the site would be able to be cleaned up enough to place housing on it. Because Tony and I have experience and knowledge of what DPWs used back in the 1960s and '70s and how disposal in general was handled, we would not be surprised to find out the property has much more contamination yet to be found on it than hoped for.

Not every DPW yard is a definite hazard. However, there are usually a range of work activities that increase the possibility that it may be. Since many DPW maintenance yards have been used by the department for years, the hazards could have happened long ago and may not have been discovered or made public.

If I moved into a new place or just now realized there is a DPW facility nearby, I would check into certain facts. How long have they been there? What have they been used for in the past? What are they currently being used for? What might it be used for in the future? Ask the state environmental office if there have been any issues in the past that were investigated or suspected.

You can easily check with the town and the state to see if there had been old underground tanks that were removed and replaced. You can also see if there are large piles of fill dirt or mulch that may be a source of contamination or of much dust blowing around on windy days. Don't take what they tell you at face value, either. Do some investigating on your own by reading through old news articles on the internet and reviewing old town meeting minutes.

Looking at records of town meetings gives an open look into what problems the town has faced in the past and what, if any, solutions they took to resolve issues. Tony and I both agree living at least one half mile is a good idea.

Chapter 44. Tire Dumps

It is estimated by industry groups such as the U.S. Tire Manufacturers, that every year 250-280 million tires are disposed of in the United States. If we add up the total thrown away since the year 2000, we have over five billion tires thrown away. The amount to be disposed of is tremendous. Where do all these tires end up? They have gone to an assortment of places, from simply dumping them in landfills to burning millions of them in industrial furnaces.

In general, many tire dealers and wholesalers disposed of their tires by paying a small fee to whoever would take them away the cheapest. Many of these tire haulers had deals with farmers and landowners outside of the cities who would allow them to dump the tires on their property for a few dollars. Some tire dumps contain millions of these old tires. States began to collect fees on disposed tires when new ones were purchased starting back in the mid-1980s. Today, 37 states collect fees ranging from 25¢ to over $3.00 to turn in old tires when a new one is bought. While the fees are supposed to be sent into the state to fund specific programs, the money is not really tracked, such as in the state of Texas. If a tire dealer can get $2 from a consumer for tire disposal and pay a hauler $1 to take it away, the tire dealers have a nice little way to make some extra money. The tire haulers also do, especially if they dispose of the tires illegally or perhaps cross state lines and dispose of them in a nearby state charging 50¢ a tire. One tire dumper in Georgia was estimated to have earned over $100,000 before being caught illegally dumping the tires.

The largest tire dump in the world was in the state of Colorado. Known as Tire Mountain, it was estimated to hold over 80 million used tires. Work to reduce the numbers through a shredding program funded by state tire disposal fees is still ongoing, but millions of tires remain.

Tire dumps do not need to have thousands of tires in them to pose a risk. While, obviously, the larger the dump, the bigger the potential is, small dumps also are an issue. Many local service stations and tire shops have a pile of tires out back. The larger the volume of tire sales they do, the more used tires they take in.

Landfills, which used to be the place to dispose of old tires, no longer want them. In many cases, they will no longer accept them. The reasons are legitimate. Old tires take hundreds of years to decompose, and during that time they kill off much of the beneficial bacteria the landfill uses for decomposition of organic waste. In addition, tires usually work their way back to the top of the landfill. Moreover, because they are bulky and hollow, they take up more space in the landfill, and available landfill space is rapidly decreasing. Unless the tires are cut into pieces or shredded into strips, they are no longer welcome to be disposed of at landfills.

With all the tires needing disposal, there need to be options for reusing at least some of them. The tire industry has promoted recycling of old tires for years, but the uses and markets for that are limited. Companies who take the tires and grind them into crumb rubber can market the crumb for use in playground surfaces and partial fill for new roadway surfaces. Some innovative companies use the crumb to make new products, such as being turned into mulch for playgrounds or sandals, added to asphalt for roadway construction, and powdered into dust for industry, but that still only puts a tiny dent in the overall quantity disposed of each year. Recycling only uses a quarter of all old tires, or about 70 million. About 40 million old tires a year are still disposed of. The recycling operations that do exist also have the same issue with keeping a giant pile of tires onsite for their use.

Currently, the largest use of old tires is burning them for fuel in furnaces, such as those in the cement and paper industries, as well as several others. Almost half of the annual waste flow of tires gets burned. While the proponents of this disposal method claim it is recycling at its finest, opponents say that burning tires is not recycling. Rather, the facilities they are burned in are polluting the air and subjecting communities to toxic chemicals, an issue discussed at further length in the chapters on cement plants, solid waste incinerators, and industrial operations in this very book.

Despite these uses, many old tires are simply dumped. Tire dumps exist all across the country, especially illegal ones. For every one that is licensed, there are several that operate without permits while they accumulate thousands of old tires. As Tony and I have travelled, we have seen more than our share of piles of tires dumped off the back of a truck sitting on the side of a road. In our national forests some of the dirt roads have become dumping grounds for disposing of tires. On culs de sac in the country, trucks from the nearby city come at night to drop their tires.

Even individual people have created significant tire pollution. In Clayton County, Georgia, a man who was dubbed a serial tire dumper was recently arrested. He was charged with renting U-Hauls and illegally dumping over 100,000 old tires at places all over the county.

Sadly, there are dozens of people like this lured by the fast dollars and almost unlimited places to dump old tires without being seen. In the town of Florissant, Missouri, business owners are finding their area a constant target for midnight tire dumping. One day it was behind a closed restaurant, the next, in a vacant lot. Old tires are being dumped on a regular basis, and the culprit is still on the loose.

Clearly, tire dumps are a national issue, and they pose several environmental and health risks.

1. Fire and Smoke

The most dangerous hazard a large tire dump has is a massive fire in the dump. Not only will there be impact from the smoke but also from the potential for soil and water contamination. Nearby properties have been contaminated with toxic chemicals from the runoff of burning tires and chemical firefighting foam used. Buildings and nearby homes are unusable while the stench of smoke and chemicals are cleaned out or dissipates. Once a pile of tires starts to burn, they are very difficult to extinguish. Many tire fires in the past burned for days. Anyone who ever tried to put out a fire from a burning piece of rubber has experienced just how difficult it is to extinguish.

The EPA recognizes the extreme danger a fire would create in a massive pile of tires. Big piles of tires have everything fire needs. It has fuel since tires are made from oil, giving the fire a continuous source of material to burn. It also has plenty of oxygen since the hollow places in tires allow for air to circulate. Depending on the size of the tire pile, it could burn possibly for weeks before it is finally extinguished.

Once the tires start burning, they create thick black clouds of smoke containing toxic chemicals. People who have lived through a tire fire, even a small one, know how toxic the smoke is and how it permeates everything inside their homes, turning it all into a smelly, smoky mess. When burned, tires release oil and heavy metals into the environment. The site of the fire may be completely contaminated with toxic materials and become a toxic waste site needing clean up. In fact, some tire fire sites have been declared Superfund sites due to the incredible amount of hazardous material left.

The city of Phoenix, Arizona, says tire fires are a threat to the environment and health of individuals in the same ways an oil tanker accident or a railroad car carrying hazardous substances are. Many emergency response teams are aware these fires create potential hazards due to exposure to volatile organic compounds, carbon monoxide, heavy metals, and polycyclic aromatic hydrocarbons (PAHS). Exposure to the toxins in the water, smoke, or soils is hazardous to the respiratory system, cardiovascular system, and may cause liver and kidney damage.

One of the largest tire fires ever was in the town of Mountain Falls, Virginia. Approximately seven million tires went up in smoke, and the fire burned for about nine months unable to be extinguished. The smoke from the fire went 3,000 feet high and spread across four states. During the fire, thousands of gallons of oil was released into the ground and area from the melted tires. A nearby stream was contaminated with the oily tar that flowed into it. The EPA emergency response team was able to work on containment of runoff from the site and managed to collect over 800,000 gallons of oily wastes that would have otherwise made it into the surface waters nearby. People in the path of the toxic smoke had difficulty breathing in some cases, and many had nose and throat irritations from inhaling the smoke that remained in the air for weeks. The site was considered a Superfund because of the extensive contamination, and it took almost 20 years to be cleaned up enough to be removed from the list.

In Waterford, New York, a massive tire dump holding an estimated five million tires caught on fire, burning about 750,000 of them before being able to partially extinguish the blaze. A nearby General Electric plant that manufactured silicone needed to be evacuated. The tire dump had been closed down several years earlier due to concerns over the number of tires stored and the risk of a potential fire. It took firefighters nine days to extinguish the blaze, which was fueled by 750,000 tires. Nearby residents remained inside their homes to avoid the toxic smoke.

2. Groundwater and Soil Contamination

Tires themselves are not considered hazardous waste; however, when a tire burns, it breaks down into its hazardous components. Toxic gases such as benzene, phenols, butadiene, and polycyclic aromatic hydrocarbons are released from the burning rubber and oil. There are heavy metals used in the construction of the tire that are released, and each tire contains about two gallons oil, so during the fire, runoff from thousands of gallons of contaminated water is a potential hazard to groundwater

A tire fire back in West Rockhill, Pennsylvania, back in 1986 saw 70,000 tires go up in flames in a massive blaze. The tires were a disposal site that a 30 unit tire shop used for dumping old tires at. It took more than 30 fire companies and almost 24 hours to get the fire under control. At the time of the fire, crews from nearby naval stations joined the battle, using firefighting foams that are now known to contain PFOS, a carcinogen. Almost three million gallons of water was used fighting the blaze, and recently, the Pennsylvania Department of Environmental Health has been finding that dozens of private wells in the area may have been contaminated with PFOS and PFOA. Local residents who over the years suffered illnesses such as cancer are now wondering if the chemical contamination in the water supply was the cause.

In Traverse City, Michigan, a huge tire fire consumed 750,000 tires and burned for 22 days back in 1996. The groundwater was contaminated with pyrolytic oil, and the EPA stepped in to clean up most of the contamination. In 2019, over 20 years since the fire, wells in some of the households near the fire site are testing positive for PFAS in the water. While the health effects of the chemical are still being studied, the EPA has quickly set guidelines for levels in drinking water. The residents, however, are not taking chances with consuming their contaminated well water and switched to using bottled water instead.

In addition to fires causing groundwater contamination, as tires sit and decay, some of the materials that went into making them begin to break down. While the tire itself may take decades to crumble away, chemicals such as lead, zinc, and a host of others will slowly leach out into the soil. Depending on factors such as how long the tires sit, how much precipitation there is, and the type of soil, these chemicals can cause the groundwater to become contaminated. The EPA recognizes the danger that decaying old tires may have as they leach toxic chemicals into the soil and potentially the groundwater where the tires are disposed of. Tests of soils under where some of the larger tire dumps have been cleaned up have shown levels of volatile organic compounds, zinc, and lead. While the levels were not high enough to warrant remediation beyond the normal clean-up procedures, they were enough of a concern to test the groundwater for the contaminants.

We were unable to find examples of communities affected directly in this way, but the EPA recognizes the threat to soil and groundwater that decaying tires may have as well as the potential to affect the health of nearby people dependent on the drinking water source.

3. Insect and Rodent Breeding Grounds

If there was a perfect place for insects to breed, it would be the tire. The round shape and the hollow inside create a perfect environment. The shape of the tire allows rainwater to be easily collected during a storm. There is always some shade inside to slow the evaporation of the water and also keep it cool. At the same time, the rubber material that tires are partly made of holds the heat so the temperature is warm and steady inside, almost like an insect incubator. Debris, such as leaves, accumulates inside, providing a perfect source of food for adult insects and larvae as well. Combining all these factors creates insect heaven, the perfect place for insects to lay their eggs. The CDC says many of the mosquitoes and other insects that spread disease such as West Nile Virus and encephalitis come from tire breeding grounds.

Even small collections of tires pose potential hazards. During the summer months especially, that small pile can potentially breed thousands of disease-carrying insects. Every year the CDC and local governments ask people to make sure they empty all standing water to minimize potential breeding grounds. Even places as small as an abandoned pet water bowl can become a perfect place for insect larvae. In a pile of a hundred tires, millions of insects can breed.

In Moncks Corners, South Carolina, a subdivision next to a now closed tire recycling company had a big problem. The development was plagued with swarms of big, bloodsucking mosquitoes. One resident said they kept coming in big waves and would bite anything that was exposed on their bodies. It was so bad it was difficult to be outside. In Anderson, South Carolina, near another tire dump the same company operated, neighbors were also having huge issues with uncontrolled populations of mosquitoes swarming the houses nearby. One resident remembers, "They were eating us alive," as he put it. Nearby residents also had thick black soot coating their vehicles and houses which came off the dry rotting tires. More than just the nuisance of a swarm of bloodsucking insects--already a concern in itself--residents of both of these communities were at a heightened risk of diseases spread by insects.

In Simms, Texas, an illegal tire dump in a wooded area has grown from an estimated 20,000 tires to almost 50,000. The tires were disposed of over at least a two-year period on private property that nobody was actively using during the time. The county was hoping to have the tire dump cleaned up quickly to avoid a potential mass infestation of mosquitoes and reduce the hazard a fire in the dump would create. The Bowie County Deputy Fire Chief was concerned that if the tire pile caught on fire the resulting blaze and chemicals released from the tires could leave the area a Superfund site. The health inspector, meanwhile, worries about the mosquitoes spreading disease in addition to the rats, snakes, and other animals making the pile a new home.

Tire dumps are not always easy to spot. Especially in rural areas or undeveloped properties outside towns and cities, there are unlimited places tires can be disposed of. The lure of cash money in exchange for allowing someone to dump a few trailer loads of old tires on it is tempting to many people. It doesn't take many truckloads of old tires until there are thousands in a pile. In cities, especially in more run-down areas with vacant lots and abandoned buildings, piles of tires may find their way there and quickly multiply. The state environmental agency sometimes has a list of any registered tire dumps and also may even have an additional list of illegal ones they know about.

Because of the difficulty in finding tire dumps, the best you can do is to be observant if you see a pile of tires near you. Find out about having the pile removed before it starts to grow and become a hazard. That would be the best course of action. Also, watch any neighborhood businesses that regularly have a pile of tires due to their business to see how often it is taken away and how big it gets before that happens. Many towns also have regulations limiting the number of tires that can be stored at any one time before they need to be taken away.

If the tire dump appears to be a regularly used area for disposing of tires, legal or otherwise, chances are good it will grow larger. Giving at least a half mile distance between yourself and the dump should take you out of the highest insect zone. Since groundwater contamination is a possibility, if you are dependent on well water, a mile would be a safer buffer. We also strongly urge testing your water on an annual basis for metals and volatile organic compounds. Remember, public water supplies may also be affected if the well is in the area of contamination, so be aware your public water supply may have issues needing to be watched.

Chapter 45. Cell Towers

The debate over whether cell towers are hazardous to our health has raged on for decades. Neither side gives in, but both claim they are right. According to Cell Tower Lease Experts, as of 2016 there were 307,626 cell towers in the United States. Including all the sites on rooftops, flagpoles, and other places, it adds up to an astounding 964,444 locations. Besides the fact they are really ugly, what risks do cell towers pose exactly, and what are we exposed to?

The answer to that really depends on what you are looking to hear. New studies continue to come out. Regardless of your position, you can find studies to support your view. Tony and I do not claim to be scientists, but one fact is certain beyond doubt: some people are more sensitive to the effects of cell transmission waves than others as we shall show later. With the proliferation of locations with cell transmitters, avoiding exposure is extremely difficult. They appear in a wide variety of places: the tops and sides of buildings, mounted on flagpoles, churches, schools, parks, etc. We can't escape.

What harm could they be doing to our bodies? The government and the cell phone industry use the tired old phrase, "The level of radiation is below federal standards." The standards used to measure exposure are for short-term exposure only. We are only beginning to find out the hard way what the effects are for long-term exposure. With our current best understanding, there are a number of factors that influence how hazardous cell towers can be.

1. Proximity to source of EMF

Data from the National Institute of Health and other organizations show that the intensity of EMF waves decreases rapidly as you move away from the source. The research and studies also show the level of EMF waves is far less at ground level under a tower. What these studies do not address is the long-term effects of continuous, low-level exposure to electromagnetic rays.

In order to accommodate the rapid expansion of the transmitters needed by the cell industry, many other locations are being sought as potential locations to place new transmitters. Telephone poles, streetlights, sides of buildings, and billboards are only a few of the creative places the industry seeks to use. Many of these places get paid a hosting fee for allowing the transmitter to be located there. This opens up many places that previously would never have been used, creating new and potentially hazardous locations.

One of the biggest dangers we are now facing is instead of all the towers being located very high off the ground, lower locations are shifting EMF waves closer to the ground and exposing many more people to them.

Studies have come out that show evidence of increased risk of miscarriage due to high level EMF exposure during pregnancy. Soon after, another study, funded by the NIEHS (National Institute of Environmental Health Science) also found an association between high EMF exposure during pregnancy and miscarriage. In addition, there were at least four more studies published in widely recognized and respected journals during the past 15 years that examined the relationship between high EMF exposure and the risk of miscarriage. The studies measured the overall EMFs the pregnant women were exposed to, and not just specifically those from a source as a cell tower. There are other sources of EMFs in our homes and workplaces; however, none emit such a constant and steady stream passing through the tissues of the people in the path they travel.

An elementary school in the town of Ripon outside Sacramento, California, had a low cell tower placed on school grounds near the classrooms children use. Over the last few years, four elementary school children have been diagnosed with cancer. The school, when asked about the possible link to the cell tower, said, "The EMF emissions are below federal standard." The school refused to take down the cell tower while further investigations are carried out. The school also receives $2,000 a month to host the tower on the property. Eventually, the school gave in after the parents continued to insist the tower was the possible cause of a cancer cluster among elementary school kids. While no conclusive tests have definitely shown the tower was the cause, the tower was removed and relocated because of the strong correlation.

In Mill Valley, California, the town banned new cell transmitters in residential neighborhoods. Their decision was based on studies showing negative health effects from the radioactive waves. Of course, the industry contested the decision. How it plays out in the courts is yet to be seen, both the industry is spending many millions of dollars on lobbying efforts across the country.

In Sandy Springs, Georgia, residents are getting together to stop Verizon from installing new transmitters across town. A class action lawsuit against Verizon claiming the transmitters cause cancer and lower property values is headed for the courts. One resident said the new transmitter would be 100 feet from his bedroom window. If the residents are successful in stopping Verizon in town, this will have ramifications across the country. Most of all, it would allow people to choose not to be subjected to the EMFs that nobody is certain will not turn out to be a health risk.

2. Length and frequency of exposure

When the government establishes exposure regulations, they are almost always based on short-term exposure. Unfortunately, this does not address the people who live, work, or spend many hours day after day exposed to low levels. Similar to regulations governing toxic chemicals, long-term cumulative effects of exposure are neglected to be considered in the exposure.

There is a tremendous difference when a person has a single exposure to something toxic or harmful and when a person has continuous exposure to the same thing. Tiny bits add up and become huge problems. For example, most people go to the dentist once or twice a year. At one visit each year, the dentist takes an X-ray to look for any problems. That one X-ray per year would not have any long term negative effects. But, the technician giving the X-ray leaves the room when the machine is turned on. They do not want to be exposed to something that is safe in the short-term but harmful in the long-term. They know the cumulative effects of exposure will cause problems and be harmful. Until then, a cautious approach to potential effects of cumulative exposure of these EMFs would make sense.

3. Sensitivity to EMF

There are people who claim they have a sensitivity to EMF exposure, usually referred to as EHS (electromagnetic hypersensitivity). While mainstream science has not officially recognized this as a disorder, it does appear to be real and not simply a manifested reaction.

In the words of the World Health Organization, "EHS is characterized by a variety of non-specific symptoms, which afflicted individuals attribute to exposure to EMF. The symptoms most commonly experienced include dermatological symptoms (redness, tingling, and burning sensations) as well as neurasthenic and vegetative symptoms (fatigue, tiredness, concentration difficulties, dizziness, nausea, heart palpitation, and digestive disturbances). The collection of symptoms is not part of any recognized syndrome.". Both WHO and the National Institute of Health acknowledge the existence of EHS. Until officially recognized as a disease, however, it remains unofficial.

When I lived in upstate New York, there was a family who lived directly in the path between two cell tower transmitters. They had been living there for years before the cell towers were installed. The town held public hearings prior to the approval, and all the neighbors nearby attended. In spite of concerns of potential health issues, the town gave their approval for the towers. One family in particular raised the issue that the EMF waves would travel between the towers and directly through their house. Once the transmitter tower was turned on and operating, all three people living in the home started to have headaches and strange sensations in their bodies. When they were not in the house and away, their headaches went away as did the strange sensations. As soon as they returned and were in the house again for a few hours, all the negative symptoms returned.

In spite of there being no definitive medical proof showing a direct causation between the exposure to EMF waves and their health effects, to them it was as real as anything could be. Cases like theirs occur with regularity, and the medical profession says there still is no proof the two are related. However, the current research that does exist has not explained these people's symptoms. These are real people who are really suffering. Currently, the best explanation is that EMF waves are causing or exacerbating these health problems. Rather than believe these people and look for a solution, the current research suggests they are imagining or faking their symptoms. More research needs to be done to make sure that we are not putting further people at risk.

4. Effects Yet Unknown

Tony and I acknowledge that the information in this chapter is more difficult for some people to accept because the research is not finished yet. We feel it is our duty, however, to point out that there is a clearly established historical pattern surrounding pollution like this. Certain chemicals that were once used widely by industry and even wider by an accepting public, such as DDT and non-stick coatings on products, were found later to have been responsible for causing many people's illnesses, newborns' birth defects, and thousands of cancer deaths. They too were at first thought to be miracle products, making our lives better, with no drawbacks.

The chemical industry gave us Alar to spray on our apples and Paraquat to kill weeds and unwanted pests with. We blindly believed them when we were told they were safe. Millions of pounds were used without so much of an afterthought to our safety. Such was our blind trust in the government and the belief they would never let us use unsafe products. Blind trust in anything will only get us in a bad situation.

The most noteworthy example of this is DDT. We were told to go ahead, spray it and kill what we need to with it. They never told us the deadly effects it would wreak on humans and wildlife. Those millions of pounds of DDT is now coming back to haunt us.

This is not the only example. Flame retardants such as Tris were used extensively for years on infant's and children's clothes; this chemical is now known to cause cancer. Teflon-coated cookware we used in our kitchens has given us diseases. Certain chemicals such as BPA in our plastic containers harm us and are considered potential carcinogens. The pattern is clearly established. We must trust these people who are reporting the early symptoms so as to not cause further, more dire, consequences later.

Radiation that passes through a body beyond a certain intensity heats up the cells and causes damage. There is growing concern about the potential non-thermal levels of radiation causing damage too. Now, most of us do not have a cell tower a couple hundred feet from our home sending electromagnetic waves through our homes and bodies whenever we are there. But for those of us who are located directly in the path of these transmitting waves, there definitely should be a level of concern about what the cell phone companies have not told us or what they have yet to find out.

The research is beginning to bare out the dangers cell towers place. There are already studies pointing to brain tumors, miscarriages, Alzheimer's disease, and more. An Australian study even found that children living near cell towers had substantially higher rates of leukemia compared to those who did not.

To reiterate, for every study saying cell towers do not cause harm, we can find matching ones that say they do. Moreover, many of the guidelines and regulations dealing with the EMFs from cell towers were written over 25 years ago and are based on what was the standard back then.

Incredible advances in cell phone communications have been made, and tens of thousands of new transmitters have been added, all sending waves much faster than ever before. Using outdated studies on technology no longer in use does not protect the public. Numerous pushes to update the regulations using the standards and outputs of today's towers have been fought vigorously by the industry. Will the wireless cellular industry become like the tobacco industry?

With the expansion of the wireless industry into 5G, scientists say we will be blanketed in more radioactive waves than ever before. Recently, WHO has come out in support of slowing down this expansion until more is learned about the health effects. Hopefully, the effects of these transmission waves would have many more in-depth studies done by independent health organizations with no agenda or stake in the results. Instead, the battle lines have been drawn, and both sides are firmly dug in, saying they are correct.

The course of action that makes the most sense to somebody in the direct pathway of a cell tower transmitter would be to avoid it. The small body of conclusive evidence raises the possibility of there being effects beyond what we are able to detect and see damage from at this point. Until the science is settled, we advise caution. We suggest if there are transmitters in the area you check out the exact location of the path the waves will take.

Chapter 46. Electric Power Lines

Electricity surrounds us all the time. Throughout our homes, we have many electric appliances and machines we use daily. Take a moment to think about all of the things we take for granted that are always there when we want them. Television, radio, hair dryers, coffee makers, vacuums, electric blankets, refrigerators, clothes dryers, water heaters, can openers--this list can go on seemingly forever. All of these appliances and daily conveniences that we have come to see as necessities are run by electricity. Very few of us have solar panels on our roofs or windmills in our backyards to make our own electricity to power these things. The vast majority of America is hooked into the enormous grid work of electrical lines that connect virtually every home in America to each other. We are all wired together, for better or worse.

It was back in the mid-1930s when President Roosevelt decided all of the country needed to have access to electricity. The Rural Electrification Agency was established, and, with that, electric power lines began to connect far off rural areas to the grid. Over mountains and through valleys, thousands upon thousands of miles of power lines were run.

The EPA tells us the U.S. power grid has 160,000 miles of high voltage power lines crisscrossing the country. In addition to that, there are millions of miles of low voltage lines like the ones we see connecting our houses. The ones we are focused on here are the high voltage lines. Those are the ones carried from one giant metal stand to another, stretching as far as you can see. They look like giant monsters with arms stretched out wide.

There are people who see potential risks to their health from various aspects of this electrical grid. Some are looking at research on the effects of exposure to the electromagnetic fields the high voltage line produces and are questioning the effects they have on us. Others study the use of chemicals used to keep all these miles of power lines cleared and accessible and see a problem to the people who live along a right of way. The research is still being done, and conclusive evidence to support those decisions is still forthcoming, but when Tony and I are faced with deciding between raising our concerns or not bringing them up at all, we choose discussion. If in the future it turns out without a doubt that this is harmful, we would feel we were negligent in not including it. However, part, but not all, of what we see as a risk for this particular topic is well-documented.

Because the research isn't completely finished, Tony and I needed to see for ourselves the effect high voltage power lines had on our human bodies. We have been places to stand underneath a vast array of these towers and wires, and we felt our skin tingling. The hair on our arms stood up and moved as the magnetic field passed over and through our body. Honestly, it felt creepy, so we left the area quickly. This brief experience gave us extra reason to question the risks associated with living near these high voltage lines.

1. Exposure to Constant Levels of Electromagnetic Fields

Electric power lines produce electromagnetic fields (EMFs) as electricity passes through them 24 hours a day, every day. Since the 1970s, there has been an ongoing debate about whether these high voltage lines and the EMFs they give off have any effects on the health of people exposed to them. In a handful of studies conducted in the late 1980s, the scientific community demonstrated an increase in children's leukemia rates in children who were exposed to very high levels of EMFs in their homes. In those 1980s studies, some scientists said the percentage of people who were being exposed to these very high levels of EMFs was only about 1% or less of the population. Even at 1%, that is an awful lot of people exposed. More studies were done as better methods of looking for a direct link to high voltage lines, and scientists were unable to establish a definite link to a large population group. However, these additional studies over the decades up to the present continue to show potential risks in small percentages of people exposed.

In the meantime, the International Agency for Research on Cancer (IARC) labeled extreme low frequency magnetic fields as a possible carcinogen in humans. They based this on the results they saw of several childhood leukemia studies and high exposures to EMFs. The National Institute of Environmental Health Studies (NIEHS) weighed in on the subject by stating, "EMF exposures cannot be recognized as entirely safe because of weak scientific evidence that exposure may cause a leukemia hazard." Cancer.gov says the studies were reviewed, and there is no consistent evidence for an association with the exception of children exposed to high levels, which is, again, 1% of those studied.

While, admittedly, no consistent evidence of a link has been established, there is evidence that stands on its own that has been. High voltage lines are confirmed to emit electromagnetic waves as the electricity passes through them. The risks associated with long-term human exposure to these waves depends on who you ask about them. The overwhelming majority of agencies, studies, and articles support the conclusion that the waves do not cause negative health effects to people living near them. In spite of this, the World Health Organization made the statement that there is a small body of evidence that does suggest negative health effects could possibly result due to exposure. In their own words, "low-frequency electric fields influence the human body just as they influence any other material made up of charged particles. When electric fields act on conductive materials, they influence the distribution of electric charges at their surface. They cause current to flow through the body to the ground." Therefore, depending on how strong the EMFs are, they do have an effect on the body. Usually, these low frequency magnetic fields send electric signals to the nerves and muscles. If the field is strong enough, they could cause stimulation to the nerves and muscles.

After we have carefully sifted through and reviewed all the evidence and studies, there still remain several that do indicate some existence of a relationship between EMFs and increased health risks. Granted, these studies were done on a fairly narrow, focused group of test subjects. However, they were also done by highly respected agencies and groups whose results, however alarming, can't be simply dismissed as isolated instances. They are in the minority but stand on their own.

There have been studies completed that initially pointed to the possible increased risk of a miscarriage if pregnant women were exposed to high levels of EMFs during pregnancy. Based on some of the initial findings of these studies, another study funded by the NIEHS also found additional evidence supporting this correlation. Their study added to the mounting evidence of associating miscarriages with exposure to high EMF levels. In total, there were more than eight studies that looked at EMF levels from several sources and were not done for the purposes of linking a specific source and proving a single-sided view. Ultimately, though, all eight found some level of evidence pointing to negative outcomes to EMF exposure.

There is also some evidence that long-term exposure to EMFs may lower the efficiency of our bodies' immune systems or even cause cancer. The same situation as the other studies arises here. A small amount of different research shows a possible link to exposure. Again, we have those words such as "possible," "weak," "potential," and "small."

It must be strongly emphasized that there are no large bodies of conclusive studies yet that positively link EMFs to these problems. Most of this chapter is based on a small group of studies, hypotheses, and possibilities. Some scientists assert that the facts will bear this out in the near future. They are confident that there is a genuine need to be concerned over this now and take measures against potentially harmful exposure.

One scientifically accepted fact is that EMFs cause whatever they are passing through to heat up, and EMFs can pass through humans. The amount of heat produced depends on the strength of the waves. Microwave ovens work on this principal. Studies say the levels of EMFs from the high voltage lines is much lower than would be needed to produce large amounts of heat in the body because government standards presently are set to keep the EMF levels below that point, likely to not set off alarms that could result in a public outcry for a real and in-depth study. That said, there still is no good answer if long-term exposure to low levels of EMFs eventually causes a negative health response in the person exposed.

2. Chemical Applications Along Power Line Rights of Way

One thing nobody disputes is that these lines and towers are really ugly. Another thing nobody disputes about these really ugly power lines is that every year, thousands of gallons of herbicides are sprayed under them to kill all the vegetation in the area. The utility companies like to use the most powerful herbicide possible as it is much easier and cheaper for them to simply kill everything along a right of way and keep it all dead for as long as possible. Many times no notice is given to the people living along a right of way. Because the power company owns the property under the transmission lines or has the right of way deeded to them, they can do what they like when they want.

Over the years, there have been millions of pounds of herbicides sprayed all across the United States along power lines, gas transmission lines, railroad tracks, highways, and such. Back in the '60s and '70s, the primary herbicide that was applied was 2,4,5-T, also known as 2,4,5-Trichlorophenoxyacetic acid. This was a basic ingredient used in Agent Orange. Yes, the same powerful, highly toxic, and deadly defoliant used in gallons by the thousands during the Vietnam War was also used in similar proportions along every right of way across the country. After many years of freely spraying thousands of tons of this highly toxic chemical along each right of way, many of which bordered people's backyards, the EPA finally stepped in and slowly looked into banning the use of the substance because of how toxic it was. Much of the push for the ban, as usual, came from the American public. The industry was extremely unwilling to ban this product as they were making considerable amounts of money off it.

One of the reasons for the high toxicity of the product was the fact dioxin compounds are one of the by-products contained as a result of the manufacturing process. Nobody disputes the fact that dioxins are one of the most deadly compounds known to people.

Initially, the defoliant was only banned for use in household products. In spite of the evidence the chemical caused birth defects, the EPA continued to allow the power industry to apply the poison in every right of way. Dow Chemical, the maker of the toxic chemical now known to cause birth defects, fought the ban of the substance. When it finally was obvious the EPA had no choice but to submit to public pressure and the furious outcry, Dow stopped trying to keep the product in use. However, after supplying millions of gallons of this toxic chemical for 30 years to kill everything, Dow still insisted it was perfectly safe for use.

What is the best thing for the industry to do when they find out they have a problem with a chemical? Usually, their response is to create another one that works as well or better, especially one they can get to market quickly. Once the product is on the market and widely used, it takes years before enough people are affected by the toxic side effects to get any action to regulate or ban the new substance. Over the years, that's exactly what they have done. The latest example of this is the chemical glyphosate, the active ingredient in products like Roundup and Rodeo.

In a string of towns in western Massachusetts, the power company Eversource maintains its rights of way under the power lines with the herbicide Rodeo concentrate. Glyphosate has recently been called a probable carcinogen by the World Health Organization. Many European countries have already banned its use. This highly toxic product, however, is still widely used and sold in the U.S. The use of this product, which kills all vegetation it comes in contact with, is part of the ongoing spray program Eversource operates. They plan to continue the use of the herbicide in spite of being declared a probable carcinogen and in spite of a court case in which the maker of glyphosate, Monsanto Chemical, was ordered to pay millions of dollars to a person who contracted non-Hodgkin's Lymphoma after using Roundup regularly.

In Claiborne and Union Counties in Tennessee, residents along the power line right of way have been subjected to unannounced spraying by the utility company and its contractors. Sometimes, they could be out in their yards enjoying their property only to find people randomly appearing. In several of the towns that were sprayed, people began to note a variety of issues related to the spraying. Crops in some areas were withering away, and a large number of bees were turning up dead. Several residents reported their pets had died. People who came in contact with sprayed plants or ate crops near the sprayed areas reported rashes and high blood pressure issues. Some were hospitalized for kidney issues. One resident who went swimming with his dog in a lake nearby was hospitalized for kidney problems, and his dog needed to be put down due to kidney failure.

Several new chemicals have been used by the power industry, who insists spraying herbicides to keep a right of way cleared of vegetation is necessary and safe. The data, however, on the new group of chemicals they are using doesn't sound like that is the case. Aqufact, one of the chemicals, is listed as causing birth defects. Another, Garlon 3A. may damage kidneys with prolonged and repeated exposure. Yet, the industry claims they are following the directions for use and they are safe.

The toxic chemicals used to keep the right of way property cleared for the high voltage power lines does present a hazard to people in the area. Luckily, the majority of developments do not have high voltage lines running through their backyard. Unfortunately, though, there is a large enough number that do make this an issue needing to be addressed. Due to the inadequacy of research on effects of EMFs, for now, many people are convinced there isn't enough proof. But, on the subject of the toxic herbicides being sprayed, there is no reasonable debate.

If you happen to live next to these high voltage power transmission lines, it is best not to get excited and panic about EMFs. Again, this emerging evidence is too limited and inconclusive to draw definite conclusions at this point. The best possible course of action an individual can take would be one of extreme caution. Also, if you are looking at property, you may also choose not to locate next to existing power transmission lines. If you already live next to these lines, as the scientists say, look at the emerging body of research and pay attention to it.

At the same time, the issue of the herbicide applications is something you may have some control over. In some cases, the utility companies or their contractors doing the spraying agreed to give each property owner living along a right of way advance notice when their property would be sprayed. While property owners can't control what is being sprayed and when, knowing in advance at least gives you an opportunity to avoid the spray days and potential drift and exposure to you, your family, and your pets. Contacting the utility and explaining who you are, where you live, and what your concerns are may get you some cooperation.

Chapter 47. Cemeteries

There are quite a few places we would never really think could possibly present a hazard to us. Many of these are places we drive by all the time or already live next to. Sometimes, though, in the strangest of places there are environmental issues we need to be concerned about. The word "cemetery" does not make pollution come to mind. Cemeteries as a source of pollution are often overlooked. Tony and I asked a group of people who live next to a cemetery what they thought about their neighbor. Almost all of them talked about having a quiet neighbor and never having to worry if the land will be developed. On occasion, several of them joked that maybe they think a little bit about the ghosts that live there.

In all of the years we have been doing environmental site searches, looking at cemeteries only came up when we were looking at adjacent land uses or zoning. We too never thought to look into any possible environmental issues. It was only after a visit to see some friends in Florida that the thought of a cemetery having negative health potential became an issue. My friends knew I was interested in pollution and had come across a news article talking about groundwater contamination near old cemeteries up in Virginia and Pennsylvania. Intrigued by this, I started to gather some research and found that, indeed, there had been a few studies and tests on source pollution caused by cemeteries. I also found out quickly that while the amount of research into this area was small, all the conclusions pointed to evidence confirming a high probability of potential negative impact.

There has historically been a lack of research and data partially because almost nobody thought there was a reason to look at the possibility of cemeteries and groundwater pollution. Any adjacent properties to existing cemeteries never gave an extra thought to any potential health issues. Tests usually conducted for groundwater contamination covered the standard list required by the government and local agencies, but nothing else. Up until recently, that standard testing for water quality did not include most of the chemicals and compounds used in embalming and burial. Obviously, they can't find things they aren't looking for. Even the most common chemicals found in cemeteries are not included in public water supply tests or private well analysis.

Once I started to follow the new information, I started noticing some very disturbing health issues that went overlooked by almost everyone. Indeed, there has been a link made from cemetery pollution to contamination of groundwater supplies: liquids drain out of a landfill into the groundwater and similar drainage which occurs in cemeteries.

One surprising fact about cemeteries is the sheer number of them. The actual number of cemeteries and graveyards in the U.S. according to cartographer Joshua Stevens is 144,847. Stevens used data compiled from the Bureau of Labor and Statistics and the US Census Bureau. These cemeteries date from as far back as the Civil War, and some are as recent as the present. Cemeteries range in size from the massive three million grave Calvary Cemetery in Queens, New York, to the single-grave cemetery containing Dr. Francis Farley in Tewksbury Township, New Jersey.

Biologists and other scientists started to closely study cemeteries in the early 21st century to determine if they could be a source of pollution. What they found was quite astounding. There was a high potential for various types of pollution due to the materials placed in the ground. This included the decomposing bodies, the fluids used to embalm the bodies, and the types of materials used to bury the deceased in.

Now knowing cemeteries have the potential to pollute, the scientists took a close look at what specifically the risks were to the environment and the people living near a cemetery.

1. Groundwater Pollution

There are quite a few ways water can be contaminated from a cemetery. Embalming fluids to preserve the bodies until burial are hazardous. It takes about two or three gallons of embalming fluid to embalm one body, which does not seem like much. However, multiply that by the number of bodies buried each year at a cemetery; that is thousands of gallons put into the ground every year. Depending on what container the body was in when buried, it could be 10-20 years until decomposition of the casket allows the fluids and the decaying body inside to leach into the ground. Because many cemeteries have been burying people for 50, 60, or 80 years, many of the buried caskets have already disintegrated, exposing the bodies, long decomposed and rotted.

Older cemeteries and graveyards present a unique problem because of arsenic. Up until the early 1900s, arsenic was the primary fluid used to preserve bodies. Arsenic, while naturally occurring, is also used in quite a few industries. The waste is handled as hazardous and disposed of accordingly. However, when arsenic was used for embalming, each body embalmed would have about two to four gallons of arsenic at the time of burial. As the body began to decompose, the arsenic was released into the soil and sometimes made its way down to the water aquifer below.

An article posted on Smithsonian.com discussed the link between arsenic in old cemeteries and high levels of the poison found in nearby water sources. The article directs readers to a study done by the U.S. Geological Service, which tested a potentially contaminated water supply near an old cemetery in Iowa. The government agency found arsenic level three times higher than the safe levels set by the EPA for exposure in water consumed by people. To reiterate, that was three times the levels considered the threshold for safe.

When arsenic was finally phased out due to its toxicity, it was replaced with formaldehyde blended with glutaraldehyde, methanol, and other solvents. Like arsenic, formaldehyde is considered a carcinogen by the EPA. In addition, because of its potential hazardous effects on the environment, it is regulated as a hazardous material. If I poured several gallons of this down the drain or onto the ground, there would be large fines and penalties for illegal disposal of a hazardous substance. The funeral industry, however, is allowed to bury about three gallons of this toxic material each time they put another embalmed body in the ground. Because some cemeteries have thousands of gravesites and hundreds of burials each year, this adds up to hundreds of thousands of gallons of embalming fluid. One estimate of the total volume of embalming fluid placed annually into the ground comes in at a whopping 3 million gallons based on burial rates. Another estimate has 1.3 million gallons. No matter what the exact number is, the amount placed into the ground is tremendous.

Until very recently, the funeral industry has been allowed to bury these toxic chemicals without any oversight or scrutiny. Finally, though, studies seem to be forcing the hand of the regulatory agencies responsible for keeping water supplies safe. The EPA has finally become more than a little concerned about embalming fluids being placed into the ground concentrated in cemeteries. Now, the EPA acknowledges the threat it poses to water supplies. In Europe, the concern of the toxicity of formaldehyde and the potential for negative health effects resulted in the substance being phased out over the next several years. Knowing the U.S. too will be forced to take action, the funeral industry is quickly trying to find a less toxic alternative.

Scientists estimate that the average body will take 5-12 years to decompose inside a coffin. The actual coffin containing the body could take upwards of 25 years to decompose, depending on the materials used in its construction. With wooden coffins, the preservatives used on its surface can add to its decomposition time. Coffins made of metal will obviously take much longer before they start to leak their contents into the soil. Moreover, the actual time it takes until leakage is also affected by factors such as the types of soil in the cemetery, the amount of rain and snow in the area, and the overall quality and integrity of the coffin.

Precipitation percolates through the ground, and the chemicals released from the decomposed bodies and rotted coffins slowly drain through the soil into the aquifer below. Once there, they spread out in a plume headed in the direction that the water flows.

Besides the toxic chemicals used during the embalming process, there is also documented concern about the fillings in teeth containing mercury. The average body may contain only three fillings, so this may seem already hazardous, especially considering the tiny amount of mercury it takes to reach dangerous levels. Adding hundreds and thousands of small sources together creates a large issue. Mercury has been detected in water on and near cemeteries in levels higher than acceptable.

Even the caskets themselves are a hazard to the groundwater. Casket manufacturers are listed as generators of hazardous waste because of the toxic by-products they make when coating and applying the finish to the caskets. Once in the ground, toxic chemicals used to preserve the wood and stain and varnish the coffin also degrade into a toxic stew. Before environmental rules were made stricter, paints frequently contained heavy metals such as lead, copper, and chromium. Many older cemeteries still have sections where those chemicals were widely used on the coffins buried there.

2. Pesticides and Chemical Treatment

Cemeteries have large, open expanses of green lawns. Like golf courses, it takes a lot of maintenance and upkeep to have them stay this way. Many cemeteries use continuous applications of herbicides and fertilizers to maintain the lush green. In an effort to keep the expense of lawn maintenance low, heavier applications of weed and pest control chemicals are applied as needed. Chemicals drifting from cemeteries is the same as for farms or golf courses. Nowhere near 100% of the chemicals applied stay only on the cemetery property. The adjoining properties are exposed to whatever has been applied.

In addition, because many cemeteries have been around for decades, there were many toxic chemicals used in large quantities before they were phased out and taken off the market. Testing on cemeteries shows much of the same level of contamination as golf courses. Because the cemetery is not being developed into a housing complex, this hazard goes unnoticed, untested, and unreported. Only the people living along the perimeter of the cemetery and nearby using wells would ever know about this, and only then if they checked their water or soil for specific chemicals

3. Bacteria and Viruses

This hazard gets a little creepy, but based on concerns and tests from well-respected organizations such as the World Health Organization, its inclusion is necessary. The human body contains enormous amounts of bacteria and viruses. These do not die when the body itself does. These pathogens might survive on the decaying body for years, eventually leaking into groundwater. The level of the water table has a lot to do with the potential for transmission of germs. In drier climates with lower water tables, precipitation is the only method for bacteria and viruses to get into the groundwater. However, in areas with high water tables, poor drainage, and frequent precipitation, it is easy for bacteria and viruses to get into the groundwater and travel. In places like Florida where the water table is very close to the surface, this could be a perfect situation for the transmission of bacteria into the groundwater.

This issue is a concern to many towns, and many of them have tried to address the issue with local laws. These towns have local regulations with regards to the depth a body can be buried in relation to the water table. There also may or may not be regulations dictating distances from the cemetery boundaries to the neighboring properties' wells. That is something the homeowners would need to look at.

The U.S. Department of Interior and the U.S. Geological Survey Agency published the results of a study in 2018 that shows cemeteries may be a potential source of both chemical and microbial contamination to groundwater and surface water in the area. More studies should be undertaken based on this.

Wells, both private and municipal, can be at risk from contamination. The larger and older the cemetery is, the greater the possibility of some level of pathogens in water supplies are. People living near cemeteries should monitor wells and test the water quality frequently for specific chemicals. In addition, water tests should be done for arsenic and mercury if the cemetery is an older one, and formaldehyde, lead, zinc, copper, and iron for all others. Nitrates, pesticides, and herbicides should also be included in the water testing.

Having a cemetery as a neighbor can be an advantage for those who prefer the quiet atmosphere it offers. The potential for future development is minimal. The only real development is above-ground vaults for burials, some of which can be 30 feet or higher, right behind people's backyards. This would resolve the issue of decomposing bodies in the ground, but the trade-off is the visual effect. Moreover, there is the issue of the toxic materials in the ground from years of spraying being disturbed during construction of these vaults. Once in the air, they would drift into neighboring properties. If you live near a cemetery and they are looking at above ground burial buildings, we suggest you get the town to include soil testing for specific chemicals like ziram, dicamba, and 2,4-D or commonly used pesticides and herbicides that are carcinogens or now banned before construction disturbs the area.

Living near a cemetery takes a certain type of person who is not bothered by the concept. The health risks can be checked and tested for occasionally to see if anything changed. The value of the property near cemeteries is definitely lower compared to similar properties far away from cemeteries. If those factors do not bother you, then this may well be the place you have been looking for. Just remember to test your water and be careful when the cemetery applies chemicals to its green areas.

Chapter 48. Zoning

Zoning describes the specific purposes a parcel of land is allowed to be used for. The zoning designation of a particular piece of property is what allows a chemical factory to be there but not a housing development, or allows there to be a pizza shop, but not a supermarket. One of the single most important places Tony and I look when doing a search for possible health risks is a thorough review of the zoning for the community. This tells us many of the obvious places where to look for potential uses that may affect locals' health at the present time. While it is only an overlay of the community, the zoning in a town or city may have much more to do with its residents than they realize. The basic designations of zoning and how zoning is used in a community will impact the residents much more than most other laws and rules will.

When looking at the zoning of a community, it is also important to look at the surrounding communities' zoning. Pollution and toxic emissions do not stop at the edge of town because someone drew a border on a paper map; they contaminate the areas nearby. Looking next door at adjacent communities will also give information that is as valuable as the zoning of the town in question.

An excellent way to get an overall idea of how things are laid out for the community is to start with a review of the town's master plan. The master plan also shows the town's vision for the future and the type of a community they want to be. Many towns and cities are already fully developed with very few open spaces for new construction to occur. Others may still have hundreds or thousands of acres open and waiting for the office buildings, strip malls, and houses. Depending on how much of the existing available land is already developed, that vision may vary drastically from one place to another. In new communities that are still growing, the zoning is viewed as a crucial part of the master plan for what the community will look like when developed.

The planning board in a community government reviews new construction and new uses in existing construction to ensure the zoning laws are followed. This gives the local government control over how a town or city will grow and where residential, commercial, industrial, and agricultural operations are allowed to be. The zoning laws usually apply to both the present and future and allow control over what limits are placed on the differently zoned areas. Non-conforming uses that already exist usually will be grandfathered, meaning they will be allowed to stay because they were built and operating prior to the zoning changes. However, if they are sold or closed, that use sometimes is no longer allowed without the new owner receiving a special variance from the town

In almost all discussions of zoning, there are four main land uses that are zoned for: residential, commercial, industrial, and agricultural. Usually, the town will have a general description for each of those four categories. Then, there is usually a further breakdown within the main category for similar or subsection uses. To understand how zoning works requires an understanding of each of these categories.

1. Residential

Any place in the town that is designated as residential would be allowed to have some type of housing built there. However, within that category, there are many variables that the town most likely will have specified for the specific area and parcel of property. There are all types of different residential housing from single-family structures to multi-family and condominium complexes. Add in apartment buildings, the density of housing units allowed per acre, and other factors, you have over a dozen different possible projects all considered to be residential uses.

A typical example of how residential zoning could be laid out would be to have an area designated for high-density residential use like condos or townhouses. Another section would be dedicated to only single- or multi-family housing. Residentially zoned lands may also have restrictions on how much land is needed to build a single-family house. Some residentially zoned areas may require building lots to be a minimum of one acre or higher. The town could restrict multi-family houses, or it could choose to allow rental apartment complexes. These regulations run the entire range. For example, a community might have a minimum residential lot size of five acres for any single family house. Others may allow high-density housing of 20-25 units per acre. Towns that require high acreage many times only have a very small and specific area zoned for business to serve the local population.

2. Commercial

A commercial zone is the area in the community where various types of businesses are allowed. The potential range of different businesses is wide, so similar to the way residential zoning has categories and sub-zones within it, commercial does as well. For example, a sub-zone could be an area designated for retail only or small office building development only. There may be variations in what one community considers a commercial operation another may call light industrial. There are many similarities, but different interpretations do exist between towns, which need to be examined closely. A commercial zone in many communities is for the places where much of the support services the community uses are located. Examples of the usual types of business operations allowed in commercial zones include such places as restaurants, bars, office buildings, retail operations, car dealerships, hotels, daycare centers, and similar, non-polluting, low-impact uses to the community. There may be large amounts of commercial property for lots of different businesses, or it may only be large enough to accommodate a few stores and that is all.

When I lived in Clarksburg, Millstone Township, New Jersey, the town had a couple of acres set aside in a town center for businesses like a general store, a pizza shop, and a video store. If we wanted to go grocery shopping or needed a hardware store, we drove 20 minutes to the nearest town that had one. There was no shopping center with big box stores, no Wal-Mart, and no movie theater. That is a decision the town and people living there made to be able to preserve the rural character and charm of the area. It is a choice to have a non-business environment. We chose this specific community because of the type of zoning it had.

3. Industrial

Industrial zones are sections of land where companies who manufacture products, warehouse and distribute products, or do research and development can set up their operations. It often includes office buildings, operations doing repair work, trucking companies, and other types of uses. This category presents the most potential hazards because of the large number of different types of potential businesses that may be allowed to operate. These industrial zones are usually sectioned into two categories: light industrial or heavy industrial. Examples of light industrial uses could be moving and storage companies, window and door manufacturing, vehicle repair businesses, office buildings, warehousing and distribution centers, and other types of clean, low-impact operations. A heavy industrial zone may be needed for operations such as a cement plant, paper mill, metal manufacturing, chemical company, and others that would potentially impact the environment and health of the community. Each community establishes its own interpretation of what they feel is light or heavy industrial uses. and many have a very specific list stating the uses allowable in each. If a use not specifically allowed is proposed, the board may review the request and decide if it is a conforming use and allowable, a non-conforming use and issue an exemption, or deny the use all together.

This is the category that requires the highest level of caution. As residents of a community, or prospective residents if looking for a place, it falls on us to very carefully go through the specific uses to see what potentially could be opened in addition to what already exists. A use that is allowed under the zoning regulations is very difficult to shut down even when it turns out to be the largest polluter in the town.

4. Agricultural

Agricultural zoning is for areas with farms or open spaces the community wishes to keep that way. This way, these properties would have protections limiting complaints from neighbors and prevent encroachment from commercial developments and residential properties. People who move to these areas because they like being near the farms and open lands sometimes suddenly realize that animals smell. The agricultural zoning protects these farms from any actions being taken against them as a result of these complaints. Sometimes, as communities grow, the agricultural zones may be changed to allow some development on some of the previously restricted lands. As this occurs, and as farms sell off due to increasing pressure from developers and rising property values, the grandfather law is especially needed to preserve remaining farms from being forced to close due to changes around them in previously agriculturally zoned lands.

The same zoning that allows for a beautiful horse farm with green rolling hills as a neighbor may also make it equally possible to have a 3,000 head pig farm that daily spreads tons of fresh, pungent manure on the same property. There are not usually distinctions unless the zoning laws have very specific limitations and guidelines incorporated into them. Unless there are size restrictions on the types and numbers of animals, then anything is possible. I will add, however, when an animal operation is over a certain size, there are usually state regulations regarding air and water emissions that require permits. This has kept factory farms out of certain areas where they otherwise would have set up.

In part because of all of these complications, most zoning maps of a town resemble a jigsaw puzzle. They are laid out with various sized pieces placed all over. There may be a dozen different areas for different types of residential zoning sections around the town. For example, industrial operations may have only two or three areas near the main highways. That land may be eyed for warehouses and industrial zoning due to access to roads.

Commercially zoned property may be two very small areas at the main intersections or crossroads but not elsewhere in town.

While there are rarely two towns that have any similarities in zoning other than the categories, some reasonable conclusions can be drawn. Chief among them, if an area is zoned to allow for a specific use, many more of the same type will eventually follow. Intersections near major roads and highways usually develop with commercial zoning to allow space for stores, gas stations, and such. Large acreage zoned for industry will turn into sprawling industrial parks, hosting whatever type of manufacturing facilities want to do business there. Some towns choose to have few or no areas in which industry or commercial operations are allowed to locate and operate. Other towns may have entire areas zoned for industry that is "shovel ready" for a company to propose a project and move in. The range of zoning rules and regulations allowing some projects but not others spans across every possible project.

While the general categories of zoning may be the same for most towns, how they choose to enforce the laws varies. Individual towns also have different criteria for offering variances and granting special permits. These permits can allow development in a zone intended for other purposes. Because of the mix of business, money, and politics, it gets especially complicated.

The variation in permitted projects and where they are zoned to be tells the story of a town. Those zones show how the town developed and gives us an idea of areas we may want to live in and others we may want to stay away from. Even without access to formal zoning maps, simply driving around a town and doing a visual check will give a strong indication of the present situation. It is extremely important to keep in mind the present situation is subject to future change. Future politicians, political connections, and industry lobbyists are always on the lookout for a new place to exploit. To avoid your community becoming one of these exploited places, public pressure that continues to prod and make sure officials know where the majority will vote helps.

Unfortunately, some town boards may not always have the best interest of the citizens at the forefront of their decisions. They may even attempt to change zoning in ways that do not benefit the residents. The promise of additional tax revenues and jobs have swayed many town boards into allowing zoning changes for incompatible uses. Once a new business is in place or a change in the zoning laws happens, a precedent is set, and soon the community may find an industrial park or a sea of houses where farmland had been just a few years ago.

Hopefully, the city planning commission in Bowling Green, Kentucky, will have the residents' best interest. A trucking company is trying to get the town to change the zoning on a property currently used as a farm to a light industrial classification that can be used to operate its business. The residents in the area are fighting the change as the new business would change the character of the area. The town board is reviewing the request from the trucking company and states they will listen to both sides before making a decision.

However, some members of town boards do act in the best interests of their citizens, such as in Martinsburg, West Virginia. A request to change zoning was denied by the town. A developer wanted to build more than 300 apartments on a property next to a heavy industrial zone that includes a cement plant. The board expressed strong concern because the cement plant had a track record of air emissions violations and had even been fined for non-compliance. The board also heard from current residents about how the cement plant negatively impacted their lives, arguing that more people should not be subjected to the pollution and health risks they are. The vote was close though at 4-3 with three members thinking putting apartments directly next to a cement plant was an acceptable risk to take for people who would live there.

A community's zoning gives you some indication of the general type of community it is. You are able to carefully review the areas in the community that could currently have operations with potential to harm the health of you and your family. The current regulations also give you an indication where future detrimental operations could locate.

Chapter 49. Proposed Projects

Proposed projects are as important to know about before moving to an area as they are to keep track of once you are already living in the area. The impact a potential proposed project may have on people's health and housing investment can't be understated. Even if you do not own the place you live, you certainly owe it to you and your family's health to be aware of what could happen and what you could be exposed to if the proposed facility becomes a reality.

Finding out about a negative proposed facility as soon as possible is critical. If you approach the situation while it is going through the process of becoming a reality, you have more options than if you didn't start until the project was already underway. Once you have already moved to a town or after a project is already approved and preparing to be built down your street, you can be put in a position you really do not want to find yourself.

Tony and I have known people who found out a sand and rock quarry was coming in across the street from them only after they moved into their home. Another friend of ours called us up hysterical about a proposed trash incinerator the county wanted to build less than a half mile from their house. It was in the works for 18 months before our friend knew about it because they never looked. Yes, keeping an eye on proposed projects does take a little effort, but the small amount of time will pay off.

There are only two ways to stop a highly destructive, dangerous, and poisonous project. The first is to close an operating project down, which is a very long and hard road to go down. Success, if achieved, comes at a great cost of personal sacrifice in both time and health. The other way is stopping a proposed project before they ever get the chance to put a shovel in the ground. This is more possible than most people realize.

Proposed projects come in a range of sizes and from a variety of sources in both public agencies and private companies. Some of the projects are even proposed and backed by the federal, state, or county government. There may be a proposal by the state to build a new highway or extend or expand an existing one. It could be a county trying to expand an existing landfill. It also may be a private company who wants to build a cement plant or open a gravel pit in the town.

When evaluating proposed projects, I always ask myself three questions:

- What are the potential health risks to people in the area of the project?
- What is the environmental impact the proposed project will have to the region in general?
- Is this project one that is suitable for anyone's backyard?

Every person deserves the same consideration not to be subjected to the negative impacts the project will have wherever it goes. I am part of the NIABY movement--not in anyone's backyard. If a proposed project is not good to be in my backyard, it is no good to be in anybody else's.

Usually, the majority of proposed projects are simple, harmless ones. Others, however, will leave a lasting impact on the town and people, and not a good one. To be clear, there are three types of proposed projects.

1. Some are highly desirable ones such as when the first White Castle hamburger shop announced it was opening here in the Phoenix area. People waited and watched for months once the project was first announced. When it finally opened, people

mobbed the place, sometimes waiting in line for three hours for a bag of hamburgers. A new Krispy Kreme or a Costco may be another example of a development with a lot of expectations and people being really excited about a new project being built and opening.

2. Others are the regular, everyday projects like a new convenience store or drug store coming in. Maybe the town is adding a park someplace, or a developer is building a new office building or a housing development. Regular, everyday projects usually do not create a whole lot of excitement either way, and if there are any negatives, they usually only involve complaints about some extra traffic or making certain the project blends in with the existing area.
3. The worst kind of projects are the ones that the majority of the neighborhood or town view as highly undesirable and unwanted. There is a big difference between a project that is undesirable because it just isn't wanted by the people living there and a project that is undesirable because it is a threat. This type of project is undesirable because it's a hazard to locals' health and the environment. The impact of this type of project is going to affect much more than the small area it will be constructed on. It is going to affect the entire area surrounding it.

For this dangerous type of undesired project, there are usually documentable, negative consequences to people's health from similar projects already in operation in other places. The safety of the water supply for the community could be impacted. Pollutants and chemical emissions from the proposed facility could be highly toxic and contribute large amounts of unhealthy particulate matter to the air. It may also be that the project itself is not the problem, but the company proposing it is. They may have such a horrendous track record at other operations they run as to make this one a potential threat. Many times Tony and I have seen shady and incompetent operators forced out of one town or state only to try to open the same facility someplace else where they are unknown.

Many times the project developer or the town itself will try to downplay the impact of the project by calling it a small operation or by comparing it to non-threatening everyday examples to soften or disguise the true impact it is really going to have. However, the size of the project is absolutely no definite indicator of the potential negative effects it may have on the surrounding area and community as a whole. There is no predetermined size to what constitutes one of these highly undesirable projects or what makes one worse than another. The real factors to consider are the actual impact the project will have once it is completed and running. These highly undesirable types of projects may be very large in size and impact a huge area in the community, but sometimes they could be much smaller in actual size and still have an impact on a big area. It all depends on what the specific project is.

A small project with a very large impact would include a medical waste incinerator. The space the building itself would occupy may only be a single acre of land, but the toxic pollution it would put out would impact the area for miles around.

In Reno, Nevada, Stericycle is trying to construct a medical waste incinerator. The company incinerator was finally chased out of North Salt Lake City, Utah after years of polluting the neighborhood and being cited for numerous violations of hazardous emissions and non-compliance with EPA regulations. An attempt to locate in North Las Vegas met defeat in 2019 with tremendous community opposition. Now the Reno proposal is working through the permit phases as we write this chapter. The entire area will be subjected to emissions of mercury, cadmium, lead and many more toxic chemicals. Hundreds of trucks of medical waste from the western U.S. and Canada would roll through the area bringing tons of waste to dispose. The town is rising up in opposition to try to halt this project before it is approved.

A large-scale project in this category has sometimes changed the entire character of a community forever. One only needs to look at some of the huge mining operations that left a toxic legacy behind for residents to deal with once they closed for decades. The mining company operators sold the projects as creating needed jobs for the community and supporting the local economy, but the opposite happened. The jobs eventually left, and the economy collapsed unless another equally undesirable place came in. The town was left with the devastation the mine created in addition to thousands of acres polluted with mine tailings and acid waste which will last decades.

In the rural southeast section of Ohio, farms, horses and the forest are the peaceful life people in the area choose for themselves to live. Now in Perry County that way of life appears to be threatened with a proposed 545 strip mine for coal that will destroy the state forest it is to be located in. The area was already devastated once by a strip mine that never did the land reclamation and left the forest barren and destroyed after taking what they wanted. Now the scenario may be replaying with another company coming in. Using jobs as a way to get support in the high unemployment area and the claim the land will be better in the end it just may happen again unless the people can stop the project.

Because of the damage and impact one of these highly undesirable projects may have, immediately letting the local government know the community will not accept and allow this to become a reality is the best response once the proposed project is known about. Much of the time the government agency or company proposing a project is well aware in advance of what the reaction the community will be when word gets out about the proposal. They try to feel out an area and see in advance how much resistance there may be before they officially propose. They try to locate key politicians and community influencers. There are professional people who make a career out of selling undesirable projects to communities and deflating resistance. Tony and I have gone up against them and seen the tactics they use to attack opponents to their projects. We have seen them refer to an extremely educated and knowledgeable woman talking about the negative impacts of the project as a "hysterical housewife." They don't pull punches. If your community establishes itself early on that it will put up a big fight, the project developers may reevaluate their decision once they see how strong, willing, and knowledgeable the project opponents are going to be. Protracted fights by communities, especially once they involve using legal tactics, are very costly to project developers. Looking elsewhere for another location is easier. If your community establishes itself as one that will put up a big fight, they likely won't see the project as worth the effort.

Luckily, there are people who dedicate themselves to telling the truth about the negative impacts many of these undesirable projects will have on the community. Many places have already seen more than their share of destruction and do not want anymore. Plenty of times the proponents of undesirable projects pick a community that is already host to a variety of nasty places and try to force another one in.

In Muncie, Indiana, residents luckily found out a metal recycling company was trying to get a permit to build an operation that would recycle steel dust. The company already has one facility in Alabama, a facility which is the second largest emitter of the toxic chemical mercury into the air. The plant in Muncie, based on the permit application, is likely to become number one. The residents are arming themselves with facts and have environmental consultants working for their side. While the company has its permit application going through processes at the state level, residents are poking holes in the environmental track record of the company and the claims of all the jobs and money it will bring to the community. The consultants are gathering the needed facts to back up why the permit should be denied.

In late 2019, officials in Anchorage, Alaska, announced they were commencing a study for the purpose of building a huge trash incinerator. Not only were they considering building it to burn trash but also as a possible disposal method for the sewage sludge from the waste treatment plant. Authorities also awarded a sole source contract to one company instead of using an open bidding process, citing the need to speed the study up. The project is garnering massive opposition from residents who already have to put up with some of the most polluted air in the nation A study by the American Lung Association shows Anchorage has the 14th most polluted air in the country. The addition of the pollution from a trash incinerator would increase the level of particulate matter in addition to greenhouses gases to almost intolerable levels. While authorities say the actual construction may take seven to ten years, citizens are not taking this lightly.

In Montana, a group of people have become fed up with the mining industry, especially as yet another proposal for more of the same is trying to be pushed through. In the words of the Montana Conservation Elders: "The proposed Black Butte Copper Project is the latest in the long list of Montana mine proposals that sell the project to the public as a job creator with no downside. History, something near and dear to our group, demonstrates the true stories of mining in Montana. For a few jobs upfront, Montana's environment is irreversibly damaged, and at the end of so many mining operations the mining corporations declare bankruptcy — American bankruptcy laws are for corporations, not for citizens — and these corporations have insufficient bonding to cover the costs of cleanup leaving taxpayers with the expense, and subsidizing these temporary mining jobs. Ghost towns and polluted streams across Montana speak to this boom, then bust cycle."

Montana is not unique in this practice. Too many times one of these highly negative proposed projects seemingly comes out of nowhere and, depending on what it is, may totally turn a neighborhood or a town upside down--or worse. Projects such as opening low-level radioactive waste facilities can impact more than a town. Sometimes these projects have such a large impact they may even turn the entire county to do battle with a developer of a proposed project.

Examples of these battles show up in our news feeds almost weekly. Many times, quiet, clean, rural communities suddenly found themselves as a proposed host for landfills, incinerators, and sludge dumps. Places such as Boggs Township, Pennsylvania, with a population of 1,751 is targeted for a huge landfill. Gonzales, California, in the Salinas Valley is a proposed site of a trash incinerator. In Andrews, Texas, with 19,000 people is the dumping ground for all the mercury waste in the nation and currently proposed for the high level nuclear waste as well. Rural communities are frequently targeted by companies proposing waste disposal operations due to their low populations to offer resistance and, most times, lack of political clout in the inner circles of state-level politics. Companies often employ bullying tactics of lawsuits against the communities knowing the legal costs of fighting these projects is in excess of many of these towns' annual budgets.

Choosing rural areas for the most undesirable projects is not an isolated incident but one that happens fairly frequently. Tony and I have seen this happen personally in the mid-1980s in central New Jersey, witnessing the enormous impact a proposed facility like this could have been like a bomb going off. Everyone's life changes almost overnight. People suddenly become activists, and researchers trying to learn all the facts about why the proposed facility should not be constructed. We spent four full years becoming citizen-experts in the operation and failures of hazardous waste incineration and disposal methods. After four long years of battling, our community managed to halt the construction of the facility. That victory came at a price: the fight became my life, superseding my job, and many times I would rather have been with my family.

Recently in 2015, there was another proposal just like this on the Pennsylvania side of the Delaware River, directly across from some towns on the New Jersey side. In Falls Township, Pennsylvania, Elcon Recycling Services wants to build a hazardous waste incinerator capable of burning 200,000 tons of industrial wastes per year. The list of chemicals includes lead, cadmium, mercury vinyl chloride, and over 250 other known toxic compounds. The plant would emit hazardous substances 23 hours per day according to their proposal. The company is promoting the 55 jobs created by the toxic waste incinerator as an incentive for the town to welcome it in. The battle against the incinerator is in its fifth year now, and neither side is backing down.

Other examples include the current battles to create a temporary storage site for all the high-level nuclear waste in the country. Several sites have been proposed by private companies seeking to build this nuclear waste storage facility, but the resistance to this at some sites is not just coming from citizens, but the higher levels of government as well. In New Mexico, the governor announced he is opposed to the project and does not find the proposed location acceptable. The opposition from the highest office in the state deals a potential setback and high hurdle to the private company proposing the interim nuclear waste storage facility.

The possible list of proposed projects covers a lot of areas, many covered in other chapters. There are many more that are not specifically mentioned, but you know them when you see them. All you need to do is look on the internet and search for "community fights proposed project." The results will really open your eyes.

For example, in the Finger Lakes in upstate New York, there is a proposal for building one of the largest trash incinerators in the country, right in the middle of the region's wine country. Seemingly out of nowhere, a company with minimal ties to the waste disposal business decided to build this huge incinerator. The town is fighting back. The town and the region also have the governor of the state opposing the project as well.

In Cumberland County, Virginia, the county is proposing to open what residents are calling a "mega dump" for trash by expanding a current operation. The residents say they first heard about the proposal in June 2018, and by August that year, the County Board of Supervisors jammed it through and approved it. The neighboring county is against the project as it is close to the border of the two counties. The residents are using some legal tactics in addition to making their opposition and facts about the impacts of the proposal known. The landfill expansion recently announced it was reducing the physical size of the project by half in an effort to appease the opposition. That turned out to be a Trojan horse as the tactic was to still double the trash but just pile it wider and higher on one giant pile instead of two.

Up in Washington state south of the town of Newport, a company from Canada that already has the support of local and state agencies is seeking to build a huge smelter for processing silicone. Smelters in particular are high on the list of projects with health impacts far greater than the size of the operation itself. Any type of operation that processes quartz creates micro-particles of silica dust, which is one of the most harmful particulates there are.

Residents in the town are organizing, using facts and research of similar projects to demonstrate how harmful this project will be to the health of the residents and damaging it will be to the pristine environment that people live there to enjoy. The local and state officials, meanwhile, argue the smelter will provide dollars and jobs in an attempt to offset the negative impact of the smelter. The locals remain undeterred.

Battles over proposed projects sometimes drag on for years as the developers try to wear down the resistance. We have seen incinerator projects that took over five years from the time it was announced until construction started. We have also seen projects that, after ten years of community activists waging war, have finally been pulled from consideration due to varying factors.

In the small rural community of Grant Township, Pennsylvania, the people enjoyed a quiet life. Away from the industrialized cities, the community was located in an area complete with forests, rolling green hills, and freshwater streams. Then, suddenly, they found themselves pulled into a battle with a waste company seeking to drill and operate an injection well for toxic and radioactive fracking wastes. For seven years, the citizens battled the waste company, employing a variety of legal challenges. The small rural town even rewrote the town charter and adopted new local laws. The town banned the disposal of fracking wastes and injection wells on the grounds that it violated the people's rights to clean air and water granted to them by the state constitution. Interestingly, the town also gave the same rights to the animals and plants as well. The state environmental agency revoked the permit given to the injection well company. This tactic may be a new one for the activist playbook to deal with other similar issues elsewhere.

Finding out about proposed projects can sometimes be a bit difficult, especially the negative ones. Even doing searches on the internet sometimes may not turn up anything even though important events may be happening. As a result, sometimes these projects seem to come out of nowhere, but that usually isn't the case. The reality of it is that the people behind the proposal usually are quite aware that the public will not welcome them with open arms. Sometimes they meet one on one with a county or town official so there is no record of the meeting or what was discussed. Whether it is the government looking to build a new landfill or a company that handles hazardous waste that wants to build an incinerator, these projects try to keep a very low profile as long as possible. They know once the word gets out that things will get a bit riled up.

The best source to look for information are the minutes and meeting agendas from town and county planning boards and zoning boards. Keeping an eye on those will let you know when something may be on the horizon. Better yet, get on the board to get an inside view. I served on both the zoning and planning board in my rural community. Quite a few times there would be a company representative or an attorney feeling out the town about a potential project they were looking to site. It always surprised me how few people attended the regular meetings, but when word got out about a potential dangerous or destructive project, the place was packed. Many times when a company sees a high level of resistance from the beginning, they go elsewhere and try again. That is your community's best line of defense.

The Search Begins

Now that you have read through the hazards in this book and have a better idea how to see behind some of the everyday places, we have accomplished a big part of our mission. Just being able to realize hazards may be found in many places and the only way to protect yourself is with knowledge clears a significant hurdle. If you want to start to do a search of your specific area on your own, we offer some general suggestions on how to proceed throughout this next section.

With the new information you now have about hidden hazards, you are now ready and prepared to conduct your own search. We have completed some searches in as little as 5-6 hours depending how many hazards needed extra checking. Most people have finished their search in a few days, working on it several hours a day when they have the time.

Conducting your own search and environmental assessment for hazards located either where you currently live or a place you are considering, is not very difficult. All it takes is a computer, a phone, and some time for researching on the internet, making phone calls, and for driving around the area you are checking out. While it is not difficult, it does take some effort to research and compile your information. Some hazards will be much faster than others to check out, but the time spent will be worthwhile if you want to do a thorough job and have peace of mind. If you are already living near one or more of these hazards and the actual health effects and risks are too great, perhaps you may find out that now that you can see the risks, you wish to move to a safer place when an opportunity arises.

There are some companies who claim to do environmental assessments for you and will charge you hundreds of dollars to generate a report on your neighborhood. Most of the information they will give you is from the same free sources and online databases you would use yourself. We are also certain that the scope of hazards they check is not even close to all the ones that should be checked out. There are many real hazards they simply pass over and dismiss as not worth looking at. Maybe they just do not realize how harmful these hazards can be or maybe they just want to take the easy way out. If we felt this was a good investment and was a complete assessment and review, we would say so.

To start your own review, the first step would be to go through the list of all the hazards we talk about and determine which of those definitely are not in the area you want to check. We use a simple checklist that you can print for yourself at www.toxictom/checklist. We also included a page titled "The Hazards" in the resources section in this book you can use for this purpose.

As with most of our site searches, there will be quite a few you can cross off the list before the real work even starts. All you need to do for this is go through the list, look at the hazard, and ask yourself, "Are there any of these located where I am looking?" Some will be as easy as looking out your back window and seeing there are no high voltage electric lines, trains, or gas transmission lines running through. Some will be eliminated just knowing that there is not an airport or a cemetery in your neighborhood.

At this stage, it is now a good time to take your first ride around the area looking for any hazards. Some of these hazards will be fairly obvious if they exist near you. Easy ones to look for as you drive are places such as gas stations, large farms, or golf courses. Depending on where you live, you may be able to eliminate more than quite a few off the list in the first two steps. Just take your time and see how many you can cross off the list and don't need to look any further for.

Keep in mind we talked about past uses as being potential hazards. You will address that subject a little later in your search.

Once you have gone through your list and done your initial elimination, look at the ones you have left. Mark any of them you would in no way want to live near if they existed where you are living or looking to move now that you know the risks you would be exposed to. In other words, the purpose of this is to create a group of hazards that would be deal breakers for you. If your marked list included, for example, trash incinerators, landfills, and cement plants, you would now be able to eliminate specific locations and areas from consideration if you were looking to move. If you find these hazards near where you currently live, there may be some steps you can take to limit your exposure and risk if you are unable to relocate. Understanding the risks you are exposed to from certain hazards and deciding how acceptable this is to you is a personal decision. Nobody else can decide for you.

The hazards left on your list you can now group for the way you need to search for them. The federal Environmental Protection Agency will be a main source for hazards such as Superfund sites, hazardous waste disposal facilities, locations of nuclear reactors, etc. Your state environmental agency should be the main resource to locate hazards such as brownfields, hazardous waste sites in the state, mines, sand pits, and several others.

There is a list of websites we put together to help in some of the searches for some of the specific hazards. We also have all the main websites for every state's environmental agency. You can access the pages and live web links at www.toxictom/resources, or at the end of this_section in the book. To aid you in your search, the Resources section of this book has a collection of websites that work very well to help you locate some of the hazards. We do try to give you sources to use as starting points for some of the searches. You will need to dig to find some of the others. If you think of it as a scavenger hunt, it can almost be twisted fun trying to find the next hazard near you. Or, better yet, hopefully you can scratch it off the list when you don't.

The information is out there and accessible. When we started to try finding a resource list of all the hazardous waste sites in

Connecticut, for example, we started at the EPA.gov site and needed to dig down almost 10 layers to finally get to the complete list of sites. As we were digging, we found other links that took us to other hazards we wanted to check, so it was almost like a path with several ways to get to the end. Over the years, we have spent thousands of hours navigating the EPA site searching for information. It is not always the easiest, but almost everything is there to be found.

Usually, one of your best resources will be your own state environmental agency. Almost always, the people at your state environmental agency are very helpful. Give them a call and briefly explain to them what you're trying to find. Once they understand the information you are looking for, they may be able to direct you to a specific department and person to speak with at the agency. Other times, we have had people walk us through the website while we were on the phone with them as they showed us where to get the information we wanted.

One of the best aids we have found for helping with our searches is using Google Maps with the satellite view. We have discovered hidden junkyards, larger populations of animals on farms nearby than expected to be there, and some landfills (both operating and formerly closed) using this tool. Start with the closest view you can get on the address you are searching and carefully look for anything out of the ordinary. Slowly enlarge the area one click at a time, carefully looking for anything that may need closer examination.

Using the computer to search for a subject on the internet will bring up much of the information you are looking for. If you start a search for a specific hazard near you by asking, for example, "where are the nuclear reactors located," you will find a list of what you need.

One of the biggest issues we had doing searches was the volume of information out there. We would see stories about other places and people we found interesting and would spend time looking at those and not focusing on our own search. Try to avoid going down that road if you can for now. There will be time later if you want to read about places people are fighting against and dealing with some of these hazards.

In many cases you will find speaking with people in the community nearby where you are looking may be excellent sources of information. Many may have lived in the area for years and would love to share stories about what happened in the past with you. We have learned about quite a few hazards in this way. Asking them who else you could talk to could lead to finding out about other issues as well.

With all the new technology available, we have even used small inexpensive drones to conduct real time searches around the neighborhoods. If you have access to this new tool, it is fantastic, especially in the suburbs and rural areas. You can see all sorts of places you may have missed using Google Earth.

At this point, hopefully you have a good idea of what you need to do and agree many of these hazards present health dangers to you and your family.

One final piece of advice while you are doing your search. Do not trust anyone who has any amount of vested interest in keeping these hazards hidden. That includes politicians, real estate people, developers, and industry representatives.

We have listed all of the places and things we consider environmental hazards in our book below in alphabetical order for easy reference. Keep in mind that, in the search for these hazards, many will overlap in certain areas, so do not consider any hazard other than the hazard you are working on completed. You should still do a complete search for each individual one on its own.

1. **Agricultural Operations** - These include small farms both crops and animals, orchards with fruit and nut trees, and also large nursery operations growing for resale. A lot of these are easy enough to see by doing a simple drive around the area you live or are looking at. The zoning map you have from the town, and neighboring town if close to the border, will indicate if any of the property is zoned for agriculture. Keep in mind that just because it may not specifically be set aside and zoned

that way, using it for that purpose may be allowed by the town. Many agricultural operations will be using chemicals on their property for various purposes. They also may be using manure or sewage sludge for fertilizer. You need to determine if this is the case with the one you are looking at. Call the county environmental department and check if the operation has any permits for applying sludge on the property. Also, check with the environmental department to find out if there have been any past violations and what they were. I always suggest talking to the farm itself and asking them about when they spray and if they are willing to notify you in advance. Always do a water test to look for nitrates and chemicals such as pesticides and herbicides from the spraying. A soil test would not be out of the question as many chemicals that are currently banned were used in the past in large quantities, and those chemicals stay in the soil for years.

2. **Airports** - A check on the internet under the heading of airports should give you a list of the main ones in the area. You do need to do a separate check to see if there are smaller ones that primarily service corporate and private aircraft. We also have seen small housing communities where a private airstrip and hangars are part of the development. Keep an eye on the news and watch if there are any plans for runway expansions. Remember, these are governed at a federal level, so you may not even know about plans unless you actively watch for them. The FAA main website https://www.faa.gov/airports/ has links where you are able to get information on specific airports and see what is happening.

3. **Brownfields** - Most states maintain a database of all properties listed in the brownfields program. The agency in charge of the environment would be your connection for information. In addition, many sites currently classified as hazardous waste sites may be eligible to become brownfield sites. All they need to do to qualify is meet certain criteria. Keeping that in mind means the list you review of current brownfields may change and have others added to it at any time in the future. Just because a site is now considered a brownfield doesn't make it any safer than the hazardous waste

site it was previously. This also holds true especially after a brownfield site has been developed and now has another use on it. There still may be significant contamination at the site. Checking the remediation records from either your state environmental agency or the EPA will show what was found on the site, what was done to clean it up and, what is known to still remain that was allowed to be left there. With brownfield sites, it is entirely possible additional waste will be discovered at a later time that may be even more dangerous and toxic than the materials already removed. Dangers exist at many of these sites forever.

4. **Cell Towers** - It used to be these were usually visible and stood out like a sore thumb in any neighborhood. It was hard to miss the tall metal telephone poles with a multitude of transmitters decorating the top. Then, the industry started to get a bit creative. Traditional cell towers on tall steel poles are also being installed to imitate native trees in the area. There are many other locations where cell towers are using non-traditional locations. The transmitters are being placed onto any object high enough to serve the purpose. We see them on the sides of buildings, hiding on billboards, on telephone poles, and other places. They do need to be registered with the town, and a town database should be available to check in your area. In addition, there are groups who also operate databases that track all the known locations of these and open up their information to anyone interested. Applications for new towers and additional transmitters on existing ones usually need approval by the town in which they are going to be installed. The planning and zoning board would almost always need to have a public hearing on these applications prior to them moving ahead. Keeping an eye on the meetings and applications could prevent you from having a cell tower 150 feet from your bedroom window. People have found transmitters in places near their homes where the waves are actually being sent through their houses to the next tower.

5. **Cement Plants** - Usually another easy place to spot as many tend to stick out with their multiple tower silos and smokestacks. Concrete plants should not be confused with

batching plants that simply blend the ingredients together to make the concrete. Concrete plants have furnaces that make the raw product. There are some that may not be as obvious, though, so doing a search for cement plants on the internet would be a good way to find almost all of them. Many cement plants use alternative fuels such as hazardous wastes and old tires. Because they do need a special permit to be able to burn these, there are databases where you can locate all the cement plants in the nation and also see that ones are approved to burn alternative fuel. The local planning board agenda should always be watched for any potential operation. Almost every cement plant needs an air emissions permit from the state environmental agency. Securing a copy from them and reviewing it would show some of the toxic emissions and the quantities they expect to emit. These permits do not cover all the possible chemicals emitted and do not account for larger quantities than those permitted for. Backroom deals have been made in the past to bring in tires, plastics, and other waste. Many have been done very quietly; it was not known until the final meeting when it went public. There is always some sort of trail, however, and careful searching and watching the meetings will give you a clue if something is happening.

6. **Cemeteries** - Looking around the area will work most of the time. There are smaller cemeteries and sometimes old and almost abandoned ones that may be hard to spot. Checking with the town usually will turn up several you missed. Churches, especially in older rural areas that are now growing larger, may have had small cemeteries on their property that you would not be able to see while driving past. We have walked the grounds at many churches and have never been asked to leave. Always be respectful where you walk, and remember, the older the cemetery, the greater the chance for toxic chemicals to have leached into the soil and migrated into the groundwater. The eastern part of the country, especially the Northeast, tends to have quite a few older cemeteries from the time when arsenic was heavily used.

7. **Chemical Factories** - These blend in with almost any other industrial facility in the area. Their range of uses include

fertilizer factories, drug manufacturers, petrochemical plants, and more. An internet search through these individual categories may help, but it will be a bit more time-consuming. While it is by no means complete, we usually access a list of businesses in a town or city from the local Chamber of Commerce. Getting all the names in town gives a list to go through. Many will be obvious as to what they are. Others that you think may be chemical factories you can search one by one on the internet to see what they do and what that specific location is. Also, the town emergency services almost always maintains a list of factories with quantities of dangerous chemicals in case of an emergency or accident there. Ask if they will tell you or share if one of these is near you. The EPA Toxic Release Inventory will show you what the company reported to the EPA, at least what they claim was emitted. Based on many of our searches, the information the companies report is frequently underestimated and has many omissions. There are websites that track chemical companies and list the violations they have had. Use this resource and get that information as well. Chemical plants can be especially dangerous, emitting large quantities of highly toxic chemicals. Plants that use or manufacture a very limited range of chemicals can be some of the most dangerous, like Sterigenics with its emissions of ethylene oxide.

8. **Coal Ash Dumps** - Most coal ash dumps are in the vicinity of a coal-fired power plant. It is unusual to have a dump more than ten miles away. That does not mean there aren't any, but it's not usual. Once you know there is a power plant in your area, contact the environmental agency in your state. Ask for the locations of any coal ash disposal sites the company has been known to operate in the past or present. There are also some online resources of groups who have tracked these dumps. They will help locate specific locations as well. The larger issue is that, with hundreds of thousands of tons to dispose of, the industry has dumped these in many places over the years. Piles of coal ash have been found in places ranging from canyons and old mines to closed quarries and sand pits. Lax and loose regulations have made this a major environmental health issue, and secret and illegal dumps will

continue to be discovered. Especially be careful if you know you have a power plant within 10 miles of where you are looking. The dump is not always going to be self-apparent. There is farmland covered in coal ash, parks created from piles of coal ash, golf courses built with coal ash as fill, and many other places that allowed this toxic substance to be dumped and disposed of. Be wary of any place that has an appearance of being artificially created. If there is a power plant nearby, chances are very high it is a coal ash dump.

9. **Contaminated Fill** - This hazard is a difficult one to do a thorough and conclusive search for. We will say that the vast majority of projects do not contain fill that is contaminated with toxic or hazardous materials. However, there are so many projects and developments that used fill as a part of the construction, hidden contamination is always a small possibility. Making things more complicated, many of the companies that were used to supply the fill for projects may have honestly been unaware the fill they were dumping contained hazardous substances. Unless the contamination has already been discovered, there may be a chance of a future discovery. If there is existing knowledge that, for example, coal ash was used at a site or that mine tailings were used as fill, you will have some information of possibilities. Checking with people in the nearby area may give you some idea of what may have happened in the past. We have spoken to neighbors in areas we are searching who have told us about seeing massive amounts of demolition debris being trucked in and covered over with dirt quickly. Perhaps there are high levels of asbestos contained in the debris. We highly recommend that if you are unsure at all to get a simple soil test done. A check for contaminants associated with petroleum and a small range of toxic chemicals and metals should suffice. There are millions to be made in disposing of contaminated fill, so this is an ongoing concern. There may also be a paper trail of where the fill came from that you can access from your local building department by asking for the project information. Whether testing your water or soil, let the company know what

you are looking for and they will be able to target the test for that group of chemicals.

10. **Crematoriums** - Blending in with most neighborhoods, these places do not usually stand out. A quick search on the internet should turn up any already in operation nearby. Checking funeral homes as well is advised. The downturn in traditional burials has many funeral homes adding cremation furnaces at their existing facilities. Be cautious and watch for any applications before the town planning or zoning board for expansions or new faculties requesting approval. Some states require a minimum air emissions permit for crematories; in that case, they will maintain a list you can check. Remember, these air emission permits may have the list of all the toxic materials they will be releasing even if the state has no requirements to monitor for poisons such as mercury and heavy metals. Do not discount the effect of exposure to even a small amount of mercury on your health if you are close to any of these places.

11. **DPW Yards and Highway Departments** - The town hall can usually give you information on where any of the community DPW facilities are located. The county DPW and the state Department of Transportation would be the source for these locations. Let them know you would like all the locations, even if they are small field offices or storage areas. Also ask if there have been any over the years that were closed and either still are vacant or have been turned into something else. The older yards many times turn out to be the biggest risk. Sometimes, when they were repurposed or sold, the contamination was not remediated and remains at the site. Old locations and dumps require sifting through the internet and looking at old news from the area you are checking.

12. **Dry Cleaners** - Presently operating facilities are easy to find. Some cleaners are actually doing the work at that site and using chemicals there, and others do not. What you should determine first, though, is if the location actually is a site they do cleaning at or if it is only used for drop off and pick up of clothes and items. If there is cleaning done onsite, there is likely contamination from solvents and perhaps even Perc. If there is no cleaning done at all on the site, the only hazard

is the cleaned clothes, and that is minimal. Even if there is no cleaning presently being done on-site, find out how long the specific cleaners has been at the location. If there was actual processing there in the past, the chances of chemical contamination are quite high. The use of toxic chemicals is not only a hazard in the past; it currently continues to be a hazard. If the cleaners is older than five or ten years, it is quite possible some extent of contamination may already exist. Also ask about any buildings that housed cleaners in the past. Remember, many old cleaners are hazardous waste sites that were supposedly cleaned under the brownfields rules. There may be quite extensive soil and water contamination left that was not cleaned up.

13. **Electric Power Lines** - These are one of the simplest potential hazards to see. Remember, we are only talking about the high voltage lines that are always the ones hung from huge metal structures. The structures tend to be no more than a quarter mile apart with the wire hanging between them. The ground underneath the wire is fairly bare perhaps with only a very low covering of vegetation. These back up right behind many housing developments and even cut through neighborhoods built around them. Check and see if your property is at the right of way line where the spraying and chemical maintenance will be taking place. Depending on your level of comfort, a simple soil test near the rear of your property for certain herbicides to check present levels would be an option. Power companies contract out the spraying services in many areas. If you can find out from your power company who they use, it is possible to ask the contractor for notice when they will spray and what they plan on using.

14. **Factory Farms** - One of the top hazards, factory farms affect areas much further away than the immediate location. Driving around the area, you will possibly be able to spot these; however, you have no way of knowing the size of any operation. We also have seen smaller farms routinely increase the size of flocks and herds many times. There are some private groups who monitor factory farms, and contacting them could be a source of information for your area. They may even

help you to find ones near you. They can usually be found on the internet as groups protecting the local waterways and watersheds such as shorerivers.org, waterkeeper.org, and many others across the country. Just search for the river or waterways near you. And the local group should come up. Many times these factory farms cluster in the same area, making an entire region a miserable place. Looking at the local news and doing internet searches for factory farms in certain regions turns up a lot of information. Examples like the lower end of Delaware are one huge factory farm of chickens. Many other places have incredible concentrations of hog farms. Be very careful when looking in any area with factory farms. The level of risk is extremely high for you in many areas of exposure. Factory farms are supposed to have air emission permits and water discharge permits as well. Your state environmental office will have records you should be able to access. Depending on the state you are in, the level of cooperation you get may vary. Some states are friendly to factory farms, and the political system in favor of them. The current political climate at the time I wrote this in 2020 is in favor of loosening restrictions on the factory farms and removing protections that were put in place for the health and safety of people living nearby. Already the requirement of reporting emissions has been removed, taking away vital information you need.

15. **Fracking** - Places where fracking is taking place are widespread, and not much land is needed to set up a well or a cluster of wells. Based on the very low level of regulation in both siting and monitoring, the hazard level is fairly high, and what specific hazards to be on the lookout for are hard to discuss as a group. The waste from fracking is not limited to the immediate areas near the well, although some are. Waterways in the area carry away contaminated soil and toxic water to other locations downstream. Piles of waste rock and drill cuttings may simply be piled haphazardly nearby, creating hazards from toxic chemicals and radioactive wastes. There are states and counties that have bans on fracking, but many of those bans are currently being challenged in the courts. This hazard may turn out to be one you have limited control

over and may wind up living near regardless of how hard you check. If you live near a fracking site, the one thing Tony and I can definitely recommend is vigilance. Keeping an eye on the community you live in and the regulations over fracking may give you some level of safety.

16. **Gas Pipelines** - Existing gas pipelines are part of the transcontinental pipeline network. There are sources online that show their locations. Some of them show the sizes of the pipes as well. With the threat of terrorism, some sites have been taken down, making it a little more difficult than before. If you live along a right of way, you can see the line where vegetation changes due to the spraying of herbicides. Gas pipelines that run under neighborhoods and developments similar to the one that exploded in San Bruno, California, are usually mapped as well. The town you live in should also be able to give you this information if you explain why you want to know. The new boom in oil and gas development is creating a need for additional pipelines to connect to existing ones. Like the years-long fight over the construction of the Keystone XL Pipeline, many other proposals are also under consideration, so proposed pipelines and any possible right of way needs to be checked. In areas with high levels of gas development, many pipelines are being built to connect with the main infrastructure. Very few of these lines are being monitored for leaks and may not be inspected for years to come. If you live in the suburbs or rural areas, underground pipelines are usually marked with a sign on a post about 3-5 feet off the ground notifying you of the buried pipeline. They are usually placed about a quarter mile or so apart directly over the pipeline. Walk the areas near your property where you see open space and look for any indication of these signs.

17. **Gas Stations** - Both closed gas stations and currently running ones present hazards. The ones that are closed are limited to the potential pollution of the soil and water from leaking underground storage tanks in the past. Some of the contaminated sites may be listed on brownfield sites or another database maintained by your state environmental agency. There's a separate category for just petroleum brownfields. It

is also necessary to check if any gas stations that had been closed in the past have a new use on the property. We have seen old gas stations converted into everything from breweries to office buildings without any testing done on the soil or any old tanks removed. You are looking to ensure there was both testing and cleanup done. Existing gas stations, as we explained in that chapter, have an entire list of hazards and risks associated with them. The positive part is they are able to easily be seen. Keep an eye on your local planning board to watch for new facilities being proposed near you. We have seen locations that were near residential neighborhoods have huge truck stops proposed and then constructed against the protest of the people living there.

18. **Gas and Oil Wells** - The number of both traditional and fracking oil and gas wells is up to almost a million. There are some states that have a database available to check your local area for active wells. If your area is in oil country, there may be nearby oil fields being tapped or about to be tapped. In addition, the same department that monitored operating wells should have a list of the so-called orphan wells no longer active. There may be hundreds still sitting nearby that you would never know about. Call your state environmental agency and ask about a list to see. Orphan wells tend to be discovered in many developed areas. Just because you live in an area where the homes are 20, 30, or more years old, that does not mean there may not be dozens of old, unsealed, and leaking wells nearby. If there is no list, you can only use what information you can gather about the likelihood of the area having been an active well area in the past or not.

19. **Golf Courses** - These are one of the easiest potential hazards to spot when doing a simple drive through the area you're researching. Checking the age of the specific golf course will give you an indication of what chemicals may be found on the property from past use. Many of the most toxic chemicals that were once heavily used in the past have since been banned, but they stay in the soil and remain toxic for decades. Many golf courses are very approachable and sensitive to the neighbors' concerns over chemical drift when they are

spraying. Some will be happy to notify you so you can choose to stay indoors or leave during spraying time. Testing your water for pesticides and herbicides is strongly recommended. Make certain the area where your home is was not previously part of a golf course. Large multi-course facilities have sold off nine or eighteen holes for developments and kept the rest active. We have done internet searches for specific developments next to golf courses and found they were indeed built on old parts of the course.

20. **Hazardous Waste Incinerators** - If they are commercial facilities, these are registered and permitted by the EPA. In addition the state environmental agency also has information on who holds permits to dispose of hazardous waste in this method. However, we have found the EPA website not to be complete on several occasions with active incinerators missing from their list, even incinerators now responsible for waste containing PFOS and others. An internet search under the heading hazardous waste incinerator and the geographical area you are looking in should bring up any. Keep in mind many other places are also allowed to burn hazardous wastes as fuel such as cement plants, paper mills, and a few others that have legal exemptions. While they are not officially hazardous waste incinerators, they are in fact more dangerous, so check as well. Additionally, individual companies may operate their own onsite incinerator for their own wastes. These are much harder to find information about or track down the locations of. At the same time, most companies that would generate so much hazardous waste as to have their own incinerator are already likely to be on the list of places you have eliminated living near.

21. **Hazardous Waste Treatment, Storage, and Disposal Facilities** - All types of disposal facilities and companies involved with hazardous waste must have registrations with the EPA. In addition, many states also require these companies to file reports with the state environmental agency. Looking at the EPA list of registered companies handling hazardous waste is a great place to get most of what you are looking for. All trucking companies hauling hazardous waste need special

licenses registered by the state. The state environmental department should have a list; if not, the state department of transportation should. Even after you check this out, you will still need to take a look at companies in the area where trucks that are used for hauling the waste may be cleaned or stored. There are many hazardous waste treatment and storage companies who store large amounts of waste onsite near one of these disposal facilities. Be aware of the potential for an accident. Companies currently handling one type of waste may expand in the future and also may look to bring in much larger volumes in addition to potential new waste stream of many times more toxic materials.

22. **Hazardous Waste Sites** - These sites number in the thousands in many states. Once all the leaking underground tanks, old dry cleaners, vacant industrial factories, and others are added up, the numbers are astounding. The federal government tracks these sites in several different places, so a complete and easily accessible list is difficult to get. However, using the main EPA website, you can go through links to reach each individual state's database of every site known and listed. If you contact the state environmental agency, they may be able to direct you to their own site quicker than looking through the EPA will take. Internet searches for waste sites in specific towns may turn up some interesting and helpful information. Never depend on the state hazardous waste site for complete information. While it is a good source, it is not thorough and exhaustive; there may be hundreds of other known sites yet to be placed on the list for one reason or another. Internet articles about possible sites and ongoing investigations will add to your list of places to avoid.

23. **Highways** - Busy roadways near you are usually easy to determine. If you want more specific information on traffic counts, the department of transportation in your city or state usually has accurate figures of vehicle traffic. Remember, it is not only the highways you are looking for. The key term is "busy roadways." Check all the roads within a half mile as the ones with more traffic will have higher concentrations of particulates and chemicals further away. Intersections with

traffic lights mean more vehicles at a complete stop emitting more fumes. If the busy roads near you have a lot of stop-and-go traffic during busy hours, consider that to also have higher levels of pollution in the air. Tony and I use the figure of 10,000 vehicles per 24 hour period as the beginning of what we consider "busy." At this level there are accumulations of particulate matter and exhaust chemicals that begin to affect the respiratory system. While this may seem high to you, many main roads in larger towns and cities carry 40-60 thousand vehicles per day, and major highways can rack up hundreds of thousands of vehicles. Roadways hemmed in on either side by buildings or walls increase the level of pollutants by at least 50% because of the lack of air movement. The particulates remain trapped for much longer. Consider this as well when you are checking roadways.

24. **Industrial Operations** - These places can be a single building off by itself or a tremendous industrial park miles wide with dozens of factories in it with all types of businesses. Regardless of the number of places, it is important to get the names of the operations and check them on the internet to see what they do at that location. This is also where the tool maintained by the EPA called "Toxic Release Inventory" becomes highly valuable. Searching companies by name and city will quickly tell you a specific list of the chemicals emitted and in what quantity. Keep in mind, though, the major shortcoming: all the emissions are not monitored for. Additionally, there are many emissions not disclosed by these operations. On top of that, cumulative effects of many polluters in the same area magnifies the risks greatly. Consider the density of industrial operations when assessing the risks. If you look at the actual industrial operations yourself, do not be deceived by appearances. Small smokestacks and vents do not equal small emissions. Some of the most poisonous chemicals are emitted from small vent pipes. If you do not see lots of smoke, don't assume that means no emissions. Many toxic pollutants are not black smoke, nor are they always

visible when they are emitted. It is what the operation makes and emits that matters.

25. **Injection Wells** - This is another type of facility the federal government tracks because permits need to be issued to operate one. However, there are many injection wells that were never officially set up with the EPA. There is a big loophole with regards to a way to dispose of waste. What this means is there are old oil and gas wells that the government allows the industry to inject waste materials into if the intent is to get more yield from the well. Many oil and gas companies are taking advantage of this loophole and using it to inject millions of gallons of what is actually hazardous waste from the drilling operation into the old wells. There are also large companies that have their own injection well onsite and pump millions of gallons of hazardous waste into the ground on the property. Checking on the Toxic Release Inventory, you can see how much waste is disposed of onsite at a specific company's injection well.

26. **Junkyards** - In most states there is a permit needed to be able to operate a junkyard, whether for automobiles or recycling purposes. The state environmental agency should have a searchable database with the locations of all these places. Unfortunately, many more do not follow their state's requirement to register, and the lack of enforcement leads to the existence of many illegal junkyards operating. Take a look through the list of known junkyards from the state and also ask if they have additional ones they are aware of. Any that already have been designated as hazardous waste sites will be on that database. The town offices are a great source for finding others the state may not know about. The zoning department may know of several it has been trying to enforce local regulations on and can share with you where they are. All junkyards have the potential risks of soil, air, and groundwater pollution. Doing a drive around should turn up many of the larger junkyards. They are in business to make money, so they usually have a sign at their location to advertise. It is not uncommon in many rural areas to find houses with a dozen or so cars in various states of completeness just sitting around.

While this is not a junkyard, leaking fluids are a risk to the water supply, so be aware of these places too.

27. **Landfills** - This is one of the more difficult hazards to get a complete list of partly because so many are sitting still undiscovered or long-forgotten. Operating landfills usually do have permits from the state environmental agency and would be on a list. Hundreds of other landfills that are already closed, however, may not. If the closing of the landfill was properly done, there will be ongoing records and reports of the monitoring well results and about the leachate being taken out. Some landfills are similar to other hazards in that the ones closed long ago are only found once they become a hazardous waste site; in other words, they aren't found until it is already too late. Careful checking in the area you want to search and speaking with some people there may get you interesting information of what occurred in the past. Construction debris landfills, old town landfills from back in the '60s and '70s, and other tidbits may be found this way. Ask the people in the town hall about long-closed quarries and landfills they can remember. Doing a search on the internet for stories in the area from the 1990s or earlier can also yield some old landfill stories long ago forgotten. So many forgotten landfills and dumps, both legal and illegal, over decades number into the tens of thousands. Our search method has been very effective at finding them, at least as much as possible considering the unknowns of some of these landfills.

28. **Medical Waste Incinerators** - These facilities are somewhat regulated by the federal government and the Environmental Protection Agency. There is a list available from the EPA of currently operating commercial facilities. Check on their site for these specific locations. When you do locate one, check the age of the incinerator as there are less stringent emissions standards for older facilities. Also, quite a few of these operations are expanding capacity by adding additional incinerators at the same location or increasing the capacity of existing ones. Many hospitals have operated their own onsite incinerators over the years, and some may still be doing so. Check with your state environmental agency to see if any

hospitals near you are doing so. In addition, ask also if there are any other facilities being allowed to dispose of the waste. Trash incinerators are becoming the new medical waste incinerator in many places.

29. **Military Bases and Government Installations** - These number in the thousands and cover millions of acres. Usually, you can get a list of military bases and government installations in your state by searching the internet. Another good way we have used is to look over current maps, and the federal and military properties are always clearly marked. The full story about the extent of contamination at any of the installations is never really known. There are so many nuclear waste dumps, hazardous waste pits, contaminated underground water sources, and other hazards. Even knowing where all these locations are, you will never know that certain locations are dangerous. Assume all of them are. We hate to be this pessimistic, but military bases and installations that don't have nuclear waste issues and hazardous waste sites are rare. Hundreds have been found to have contamination from PFOS and PFOA across the country; there will be an exceptional few that remain no threat to the health of people on the base and nearby. However, since the military is unlikely to be forthcoming with the chemicals and disposal practices used on their bases, it is practically impossible to distinguish these few safe bases from the vast majority of unsafe ones.

30. **Mines** - Each state has a department that handles mining. There should be an accessible database listing all of the mines currently operating. If you are interested, access the mine's specific permit from the regulatory agency and see how much air pollution and water pollution they are claiming they will emit annually. Quite a bit of the pollutants emitted through the air and discharged in the water may not be reflected in any documents you are able to see. Always check on violations tracker websites too. Sadly, just because you do not find anything, that doesn't mean there are no violations. There are hazards from all types of mining that makes living within three miles a high level risk many people find unacceptable. Just a

reminder that permits are able to be renewed or modified, so be aware nothing is permanent.

31. **Mine Tailings** - Disposal areas for the tailings covers hundreds of acres at some mines and may be piled 50 feet or higher at times. Even if you are at a two mile distance from the active mining, there may be tailing disposal sites nearby. Finding all the places the tailings were dumped over the years takes some effort. A drive through the area may find some active sites, but older sites that are no longer in use may not be so obvious any longer. Speak with the town officials if they have a list of locations. Search through the state hazardous waste site list for any locations that are already listed. Like coal ash, tailings were disposed of in canyons, old mines, unlined landfills, and any place the company could dump and leave them. Especially be wary of areas that look artificially filled in. Over the years it was common practice in some places to use mine tailings as fill for construction and for other building projects.

32. **Nuclear Reactors** - The Nuclear Regulatory Agency has a list with every operating nuclear reactor: https://www.nrc.gov/reactors/operating/list-power-reactor-units.html. You also should check to confirm there is not a closed one near you as they will still have high-level radioactive wastes in addition to the thousands of tons of low-level waste stored there as well. The only plus of a closed facility is there should be limited accidental releases. This is one hazard you will easily be able to find: every currently running and shuttered facility across the country is clearly documented. Check the NRC website to see what types of releases and violations are likely at the reactor in question: https://www.nrc.gov/about-nrc/regulatory/enforcement/current.html. With nuclear reactors, you are going to have to do some extra research about specific areas of possible concern surrounding the facility you are looking into. For instance, leaks of tritium have been very common at nuclear plants, and groundwater contamination is an issue you need to specifically check for. Violations and reports of leaks are also available at the NRC website for individual facilities. Doing internet searches for stories about these incidents

once you have the dates and place may turn up even more additional information.

33. **Past Uses** - This is sometimes one of the more difficult ones to find all the facts for. The town hall is an excellent resource, especially the people in the planning or zoning departments. They may recall being directly involved in something or have reliable second-hand knowledge of any changes to certain properties that happened over the years. Looking through the old tax records for the property that are available from the town or county may also reveal names of other places that existed on the property you are researching. It is important to look back as far as possible because some property may have had three or four different places on the site at one time or another. Industrial parks have large turnover, and there could have been highly polluting operations long gone that left behind hidden waste and contamination.

34. **Power Plants** - Generating stations of any type should be looked for and avoided. While most of the concern is with the coal-fired ones, gas-powered are dangerous in their own way with many of the same emissions and others specific to themselves. Locating a power plant generally is as easy as looking to your state agencies and finding who regulates the utility industry. A call to them or a visit to their website should give you access to a list of all the generating stations in your state. In addition, they should have data on the size of the facility, the type of fuels it uses, and how old the plant is. Check with area residents regarding past issues with waste disposal and pollution. There will be issues no matter what power plant it is. Whether it is the tons of toxic emissions and chemicals the plants pour into your air or the disposal of toxic coal ash into your water, threats will be present. Even closed plants will be a threat to the environment and community through the unlined pits leaking their toxic ash for decades to come.

35. **Proposed Facilities** - An ongoing search of the local news and meetings is the only way you will be able to stay on top of any proposal in the area. If you are first moving to a new area, review all of the minutes for the past 6-12 months of both

the planning and zoning boards. That will get you up to date if anything is already in the works and starting to go through the process already. Do remember to check surrounding communities as well. With internet access to the minutes of public meetings from the county and town, you can stay up to date on what is happening to ensure nothing has an opportunity to slip in quietly.

36. **Public Water Supplies** - With over half of public water supplies having contamination from chemicals and pollutants, there are no real assurances your water is safe at this point. That said, having a sample of your water sent out for testing is the only way to determine what is in it. Get testing done for a complete range of chemicals, including heavy metals and contaminants commonly found from waste sites and industrial pollution. The latest hazardous chemicals of concern are the potentially cancer-causing PFOA and PFOS. These are being found in thousands of public water supplies. Check out how your water company did overall on the Environmental Working Group (EWG) public water database: https://www.ewg.org/tapwater. Just enter your zip code or state. It will be a real eye-opener to see what you are drinking in your supposedly safe water. EWG is one of the most outstanding environmental groups making sure corporations and polluters are exposed and held accountable for their actions. Many water companies send out an annual state of the water summary showing the test results they report. If you receive a similar report from yours, consider it to be in no way a true representation of the safety of your water. I may sound paranoid, but the report only has an extremely limited handful of potentially harmful chemicals tested for that are required. This is miniscule compared to the hundreds and thousands frequently found in public water supplies and hidden from the public. The EWG database is a comprehensive, realistic picture of the poison in your water. Use it with confidence that it shows a real picture of public water supplies. Still, get your individual water tested.

37. **Radioactive Waste** - This has become a much more widespread hazard to check for over the last ten years. In the past, most sources of radioactive waste were limited to

a handful of disposal sites and generators. Now, because of how many new generators of low-level radioactive waste are involved in fracking, the dangers of being exposed has increased tremendously. Disposal of radioactive waste is being done in ways that do not protect people and the environment. If you are in an area where there is fracking activity, you are at increased risk from illegal disposal. The NRC website has data on the operating facilities for low-level radioactive waste. Unfortunately, this list is no longer complete. You will now also need to check with your state environmental agency and see if there are locations where fracking waste is being disposed of. It is only going to get worse, and you need to be vigilant and thorough, making sure there is not a place near you accepting radioactive waste. The internet is a good tool for finding places already caught that you should avoid.

38. **Radon** - There is only one way to check this out: test for it. Call the people at the state environmental agency and ask if the area you are in is considered a high risk or low risk for radon. Even if you are in a low risk area, there are hot spots that may create a situation in a small area. If you're in that area, you need to know and can only find out from a test. The test kits are so inexpensive and accurate that it makes total sense to test regardless of what the normal radon level where you are is. It is worth repeating again. No matter where you live, test for radon. Every person who lives in a place from the second level down to the basement needs to do a test for radon.

39. **Railroads** - Tracks carrying trains run through quite a few places cars and trucks are not permitted. Since you are only looking for those that run near where you live, it should be a fairly quick search. Keep in mind, though that old tracks that are no longer being used may one day be reopened. The cost of building new ones and acquiring land is so expensive that Tony and I have seen many old routes started up again. If you have active tracks nearby, look especially hard to see if there are areas of side tracks where train cars are frequently parked or left for a while. We know of many places where chemical cars are dropped off by one train and sit waiting for another train to hook up to them. Hundreds of thousands of gallons of

toxic chemicals sitting and waiting near you may be a ticking time bomb. Also, check for train tracks where passenger train engines frequently gather and idle overnight to keep the engines warm before running their routes again in the morning. These are frequently near populated areas, so make sure to check for them.

40. **Sand and Gravel Pits** - In most states there is a department that handles the permits for and monitoring of these operations. Usually, it is the Bureau of Mines and Reclamation, but it has other names. Almost always, mines and any operation that removes something from the ground needs to have a permit and an operation plan filed with the office. You will be able to see if there are any currently operating near your area. If you find one, the operating permit would have information about what they are removing, how much they are removing, and how long they currently are allowed to mine. Many are able to expand their operations and extend their permits as well. A date of expiration on the permit does not mean that that is when they will absolutely be closing. Also, checking if explosives will be used as part of the operation is smart. The closer you are to the pit when explosions take place, the higher the chance of damage to your house structure and your nervous system.

41. **Sewage Sludge** - All states have regulations in place regarding land disposal of sewage sludge from treatment plants. Contact the environmental agency in your state and ask who handles sewage sludge. Ask to get a copy of the state regulations and see if there are places near you that may be potential locations. If there are places near you where sludge disposal is permitted, you will be able to know. Don't let anyone minimize the potential impacts of sewage sludge disposal to you. Dumping thousands of gallons of treated sewage waste in your area can't be dressed up in any way that still changes the facts. It is still feces. We have had people show us nice bags of rich-looking, light and fluffy soil and tell us this was what it looked like. Perhaps it does at least look better, but only after it completes decomposition and all the issues surrounding that process. However, it still contains

toxic chemicals such as PFOS and others never tested for. Other places sewage sludge is applied in huge amounts are in state and national forests, county and municipal parks, and other public areas. If you border one of these, check. The land application of sludge, or "biosolids" as the industry prefers to call it, is dangerous. Some cities and towns have properties where the sludge is regularly brought. Many states regulate where they can be applied. Because the public and agencies that regulate these hazards are finally starting to realize the danger from chemicals and metals in the sludge, many places are beginning to ban sludge applications. However, what chemicals in the sludge that has already been applied will remain there for a long time.

42. **Sewage Treatment Facilities** - The town or city usually has a wastewater treatment department you can contact to find out where any of the treatment facilities are located. Larger areas may have several treatment plants of varying size around the area. In addition to knowing where the plants are located, I suggest also asking where the wastewater is being discharged to and where the sludge is being disposed of. While the immediate area around the treatment plant is what we are concerned with, places where the sludge may be stored or the treated wastes will be piped out are also part of this hazard needing to be looked into. Every sewage treatment plant has to pipe out the millions of gallons of liquids and dispose of the sludge left. Do not let these facilities tell you that there are no places the treated effluent goes as that is impossible. In towns where fracking waste and industrial waste are disposed of at the landfill, there is another danger now. The leachate from the landfill is usually brought to the sewage treatment plant for disposal along with the radioactive materials and toxic chemicals the sewage treatment plant is usually not equipped to remove before sending the treated waste into the nearest waterway.

43. **Superfund Sites** - Another of the easier hazards to locate in spite of this being one of the deadliest. There are many databases with a complete and current list of these sites. You can find out information about clean-up plans, the work

that has already been done, and the chemicals and materials contained in the dumps. While other parts of the EPA website are lacking, the database specifically related to Superfund sites is user-friendly with fairly thorough, fairly reliable information. Do not underestimate the dangers from any of these sites. They are the sites that have been determined to have the highest levels of immediate risk to human health and the environment. The government likes to label these waste sites as "controlled" and "uncontrolled," depending on the level of exposure to the deadly toxic waste and chemicals present. Even if the site is considered controlled, that does not reduce the dangers of what is there. In addition, many so-called controlled sites became uncontrolled again. These waste sites are among the deadliest in the entire country. The EPA's list does update with new sites being added frequently, so check periodically for new listings.

44. **Tire Dumps** - There are so many places piles of waste tires are found. The environmental agency in your state may only know about a handful of legal tire dumps and perhaps has a short list of illegal ones needing to be cleaned up. With so many tire shops, repair shops, and junkyards operating, the only real way to know if there is one nearby is to actually drive around the area, looking for a place where there may potentially be tires stored. Look behind nearby tire shops and car mechanics. There have been cases where there were several thousand tires piled up behind these places. This is one hazard that you really may never be totally certain of, but you can at least do your best to check.

45. **Trash Incinerators** - Trash incinerators, resource recovery facilities, or whatever you want to call them are regulated by the state environment agency where you live. Locating these plants should be as easy as a phone call to the agency asking where they are located. Usually, a search on the internet will also bring up every trash incinerator operating in the state. Due to the size of these plants, they are very easy to see, especially their large smokestacks sticking up. In many of their locations, they stick out like a big smoking thumb. If they are in a heavily industrialized area, though, they blend in a little more,

and finding the exact location would be necessary. Keeping in mind these trash incinerators are also starting to burn other types of waste such as medical waste, check with your state agency and get a copy of their permit to review. Air emission permits are required of these facilities. They are available from the state environmental agency. Like many of the other hazards discussed, the quantities and chemicals listed on the permit is only a fraction of all the ones emitted.

46. **Trash Transfer Stations** - These facilities are usually a part of the solid waste program in the town or city they are located in. That usually is also true in rural areas as well. Call the department in your town that handles the solid waste and get a list of where they are. Check what services are available at each of those locations so you are aware if there are any recycling, composting, or hazardous waste collection operations as well. Most places, even the transfer station itself, do have a sign near the facility pointing the direction. Many towns have multiple private companies doing trash service, so even if you use one service, check into the others as well. Trash transfer stations even pop up in places where the company operating the facility may not pick up trash but use the location as a transfer point on the way to the landfill or incinerator for the other routes they operate.

47. **Waterways** - It is simple to see if there are any bodies of water in your area. Whether they are small creeks, large rivers, or huge lakes, you will want to know what is upstream and using the waterway to dispose of wastewater or other materials from their operation. Tony and I have done Google satellite searches on the computer looking along the waterway. When we see a potential operation along the water, we find out who it is and what they may be discharging. Also, once you know the names of the companies, you are able to do a search on the toxic release inventory the EPA operates to get more specific information. Many companies are located along the water for the sole purpose of having a toilet to dispose of their wastes. We frequently find many companies have large storage tanks of chemicals close to the water; these are another concern to look for. Remember, whether you are along a river, at the

oceanfront, or on a lake, the day or two after a heavy rain is almost always when the coliform levels in the water skyrocket from overflow events at the sewage treatment plants.

48. **Well Water** - Anyone who uses well water for drinking and household purposes has the possibility of the water source being contaminated. The only way to determine for certain is to do a test of the water. Depending on the location and what possible hazards are in the area, the specific tests should be done focusing on those potential contaminants and pollutants. Speak with the people at the testing company and tell them what your concerns are. If you are near farms, an industrial operation, or military bases, it is important they know that. By letting them know some of the potential sources of pollution, they can focus on the chemicals to look for. The more specific you can be, the easier it is to look for certain groups of chemicals. Once contaminants are in the water, they travel and change in levels. Testing once and thinking you are safe is not enough, so retesting and asking your neighbors about any issues they may have would be advisable.

49. **Zoning** - The zoning department in the town or area you live or are looking to move to will be a great resource. Start with asking for a map of the town showing all of the different zones along with the definition of what each one allows. Remember to also check the zoning of neighboring towns as they may have an industrial zone in their town up to the border of a residential zone in your town. The zoning map is a guide to every place you are going to be looking for hazards that exist at the present time. Except for pre-existing uses still operating under the grandfather clause, what you see is what you get. Remember, zoning may change in the future, and variances are issued at times.

Now with the information you have, you are able to make an educated decision about not only what risks you are exposed to but how large a risk you are willing to accept for those you are unable to minimize or mitigate. We need to know when something has the potential to harm us instead of being unaware it is there. These hazards aren't going away, so, for the meantime, we have to arm ourselves with knowledge so we can navigate these hazards. Then, at least, we get to make a choice and act on our choices. When assessing these risks, there are important questions we have to ask ourselves.

1. **How far away should you be from the hazard?** With some hazards, the risk decreases rapidly as you move further away. Some, like high voltage power lines and gas pipelines rights of way, only affect the area immediately nearby. Others, such as sewage treatment plants and chemical factories, may have an impact up to several miles or further from the actual hazard. You need to examine the individual situation and set a zone of risk you want to be outside of.

2. **How high is the danger level from the hazard?** There is a significant difference between the risk presented by a tire dump and a Superfund site. While many comparisons are merely generalizations, you can draw some reasonably safe conclusions based on the facts.

3. **How high is the future impact from the hazard?** A leaking landfill known to have already contaminated the groundwater presents more of a threat of becoming worse in the future. The continued exposure to the pollution from a trash incinerator will have cumulative effects on your health in the future. Almost all of these hazards will only get worse as time passes. The point is to decide how much of a hazard it is and how quickly the negative consequences increase and accumulate the effects on you and your health.

4. **Can exposure to the hazard be minimized or mitigated?** In some cases, like radon in your house, the answer is yes. Contaminated water caused by any of the hazards in this book can sometimes be addressed with special filters and equipment to substantially reduce and almost eliminate the risk factor. However, sometimes a water filter won't be enough. Additionally, other hazards are totally out of your control. You never will be able to reduce the effects of toxic chemicals from a hazardous waste incinerator. The same would be true if there was a railroad right behind your house. Nothing in your control could change the impact on you. At the same time, if you're

financially able to move, you can control how close to these hazards you live. And if you're not financially able to move, you can at least be aware of the risks so you can take appropriate action sooner rather than later.

5. **Is the cost of reducing exposure reasonable?** Installing air filters and exhaust equipment in your home for radon is reasonable as is installing special filters for water problems. Both definitely are not cheap, but reasonable and long-lasting. Putting in a new well or adding special purification systems to remove certain highly toxic chemicals may cost tens of thousands of dollars.

6. **Is short-term exposure a risk?** Some hazards only need to affect you for a small period of time to cause harm to your health. Emissions from a hazardous waste incinerator during upset conditions could release a cloud of poisonous chemical gas that would immediately affect you. The same would hold true if there was an accident on a railroad track where tanks of ammonia ruptured and leaked. As these instances really do happen, make sure you're aware of any of the short-term risks so you can minimize your exposure to them.

7. **Is long-term exposure risk?** With long-term exposure, some of these hazards build up in your body and begin to have a compounded negative effect. If your well water was polluted with chemicals from a leaking underground gasoline tank or fracking chemical waste in the aquifer, poisons would build up in your organs and eventually begin to cause diseases. The same is true of the emissions from many industrial operations. In many people living near some of these hazards, it took several years before noticeable decreases in health and signs something was wrong started to show up, but the end result was cancer. If you are unfortunate enough to live near one of these risks, make sure to be vigilant about minimizing exposure and learning the early signs of serious medical conditions so you can stay on top of them.

8. **Does the existence of this hazard increase the potential for other hazards to move into your community?** Undesirable places have an uncanny way of being in close

proximity to each other, like Cancer Alley in Louisiana and The Armpit in New Jersey. It all had to start with the first one. Anytime one of these hazards gets a foothold in an area, it makes it much easier for the next one to come in as well. If you notice, many hazards on this list are usually located near each other even though they may be totally different types. The short answer to this question is yes. Especially if you are moving, keep that in mind. You might be comfortable with the risks of living near a window factory, but that window factory makes it easier for a cement plant to move in.

9. **What is the possibility the hazard will be removed or closed in the future?** This is only part of a larger question. Just because something may be closed does not mean it is no longer causing harm. A closed landfill really never goes away. Everything in it stays for decades, and as the EPA acknowledges all landfills eventually leak. The same is true of a Superfund site. There have been many instances where more chemicals were found that were missed the first time. This happens enough to be a real concern. Thus, you have to be aware of what sorts of operations remain dangerous long after they're closed so you can take appropriate measures.

10. **How many hazards are in the area?** There are some operations, like a power plant, that are hazards even if they are the only one in the area. Other times, there may be so many different hazards in a small area that the total effect is definitely very risky. In the Phoenix area where I live, about 10 miles from my house there is a new housing development being built. It is located next to a very busy highway along a railroad track hauling freight cars and along the side of a sand and gravel pit. I personally can't imagine why anyone would build homes right next to all three of those high-level hazards. My question was answered when I saw the large number of people moving into the community there now.

These developers are profiting by preying on people's lack of knowledge.

Focus on the information that is available to you and make educated decisions that are best for you and your based on the facts.

RESOURCES

Distance From Hazard Chart

The safe distance chart of the hazards in the following pages is the guide we have developed based on the most accurate information we have. Some people may argue there is no safe distance to be from some of these hazards. Other people argue there is no risk at all presented by the hazard to be concerned about. We offer our chart based on the records of health issues, accidents, potential for risk and other factors. Use the chart as you see fit but please do not take the word of industry or politicians in deciding the true level of risk. Base that instead on factual information from independent scientists and from people who experienced the reality of living near these hazards and the effects they caused.

DISTANCE FROM HAZARD CHART

HAZARD	BUFFER ZONE NEEDED	YES	NO	MAYBE
SECTION ONE HAZARD REVIEW				
SUPERFUND SITES	2 miles			
TRASH INCINERATORS	2 miles			
HAZARDOUS WASTE SITES	2 miles			
POWER PLANTS	4 miles			
COAL ASH DUMPS	2 miles			
HAZARDOUS WASTE INCINERATORS	4 miles			
CEMENT PLANTS	4 miles			
MILITARY BASES AND INSTALLATIONS	4 miles			
MEDICAL WASTE INCINERATORS	4 miles			
HAZARDOUS WASTE DISPOSAL and TRANSPORT	1 for transport 4 for disposal			
FACTORY FARMS	4 minimum up to 8 if multiple			
NUCLEAR REACTORS	10 minimum to 20 miles			
RADIOACTIVE WASTES				

LANDFILLS	4 miles			
AIRPORTS	10 miles			
Comments:				
HAZARD	BUFFER ZONE NEEDED	YES	NO	MAYBE
SECTION TWO HAZARD REVIEW				
CHEMICAL COMPANIES	2-5 miles depending on facility			
INDUSTRIAL OPERATIONS	2-5 miles depending on facility			
FRACKING	2 miles			
MINES	4 miles			
MINE TAILINGS	1-2 miles			
GAS and OIL WELLS	1-2 miles			
HIGHWAYS	1 mile			
AGRICULTURE	1/2 mile			
BROWNFIELDS	1/2 mile			
SEWAGE SLUDGE	1/2 - 1 mile			
SEWAGE TREATMENT PLANTS	2 miles			
GAS STATIONS	1,000 - 2,500 feet			
INJECTION WELLS	1 mile			

DISTANCE FROM HAZARD CHART

SAND and GRAVEL PITS	1/2 - 1 mile			
CONTAMINATED FILL	unsafe if present			
JUNKYARDS	1/2 - 1 mile			
TRASH TRANSFER STATIONS	1/2 mile			
RAILROADS	1/2 mile			
Comments:				
HAZARD	BUFFER ZONE NEEDED	YES	NO	MAYBE
SECTION THREE HAZARD REVIEW				
RADON	unsafe if level high			
PAST USES	unsafe if contamination present			
WELL WATER	unsafe if contamination present			
PUBLIC WATER SUPPLY	unsafe if contamination present			
WATERWAYS	unsafe if contamination present			
DRY CLEANERS	1,000 feet			

GOLF COURSES	be aware of chemicals			
GAS PIPELINES	1/2 mile			
CREMATORIUMS	1/2 mile			
DPW YARDS and HIGHWAY DEPT.	1/2 mile			
TIRE DUMPS	1/2 mile			
CELL TOWERS	out of direct pathway of EMFs			
ELECTRIC POWER LINES	beware of spraying			
CEMETERIES	beware of spraying			
ZONING	watch for changes and new uses			
PROPOSED FACILITIES	watch for new facilities			
Comments:				

Resources For You to Use

There are quite a few places on the internet which have a wealth of information in one place. Some deal with a specific subject such as where a specific hazard is located in areas across the country. Others allow you to check on any company in the country which reports its emissions and view what they submitted to the government.

We have below some of the best sites we have used for our searches. We also have included a list of every state and the main website for the environmental agency. That is always the best place to start any search at.

All these sites were working for us at the time we released the book. The Trump Administration has taken down some sites such as Toxtown which offered comprehensive information of hazardous emissions in your communities. The government apparently did not want the public to have easy access to learn this information.

There are more sites than these if you look around and do Google searches for your specific area and state.

For an updated list of additional sites you can go to our website at www.toxictom.com Good luck with your searching.

http://www.energyjustice.net/map/

This site helps track polluting power plants and helps link you to other people involved in opposing them. There are quite a few great search tools to see where places are located. It is easy to sign up and register but you can use most of the features even if you do not.

https://www.epa.gov/trinationalanalysis

The direct link to the EPA's Toxic Release Inventory. Get information on toxic releases on any of the 21,557 industrial polluters reporting their contribution of toxic chemical emissions. Easy to use and offers a good look into what is in your air from the facility nearby.

https://www.epa.gov/toxics-release-inventory-tri-program

This link takes you to the EPA Toxic Release Inventory. Using the tool you can enter an address and see what facilities are in the area that report their toxic releases. This search tool is not the most user friendly we have seen but with a little playing around in it, you should easily be able to do some basic searches. Keep in mind the only companies required to report to the EPA are those that have to meet minimum thresholds for hazardous waste production. Using this tool you can set a radius to search and easily see the exact location of any facilities reporting and then by searching for the specific company, see what the emissions of each chemical reported are.

http://ft.maps.arcgis.com/apps/MapTools/index.html?appid=e61f65bb8c664b2aa59f47bc175b3c96

This link shows the locations where the wastes from Pennsylvania gas and oil wells are disposed of. It includes fracking wastes as well. It is not just Pennsylvania being used as a disposal area but many other states as well. Click on the map to load it. The loading is a little slow so give it a moment.

https://www.ewg.org/tapwater/#.WXjmntPyvBI

This database shows the results of water analysis for thousands of public water supplies across the country. It is simple to use and navigate. Make certain you have the correct name of the company supplying water to your residence and many companies own other ones and the specific name used may be different than on your bills.

https://www.nukepills.com/nuclear-reactor-maps/

One of the easiest ways to see where the nuclear power plants are located and see how close you would be to them.

https://www.militarytimes.com/2019/07/14/heres-an-updated-map-of-military-sites-where-dod-found-cancer-causing-chemicals-in-the-drinking-water/

This list is just a piece of what is currently known. Almost weekly there are more bases and locations being discovered with contamination in the water. We have no reason to believe this will eventually be every single site in the nation.

http://www.toxicsites.us/index.php

A website which shows over 1,200 of the worst hazardous waste sites in the country. Get information on the locations, what is in the waste site and what is being done with it. This website has a team of great people who put it together including environmental activist Lois Gibbs from Love Canal, New York.

https://www.epa.gov/superfund/search-superfund-sites-where-you-live

This is a direct link to the EPA and has a fairly easy to use located tool. Put in an address and it will search your area bringing up the places in the Superfund list. Keep in mind that the list of Superfund sites frequently changes. There are always additions and deletions according to changes in the existing status of a current site or, a new site contaminated enough to qualify and be added. This is a very good tool meant to be used as part of your search but is certainly not comprehensive and final.

https://www.epa.gov/superfund/search-superfund-sites-where-you-live#map

This is a link to a map that shows what is nearby based on the EPA information. It is a good addition to the link above. This will give you an overall view of the area.

https://echo.epa.gov/detailed-facility-report?fid=110027242320

One of our favorite search tools from the EPA. Just type in an area and up pops all the information on polluters and issues in the area you select. You can also put in a specific facility and it will bring up violations that occurred at the location. A very useful website for digging deeper into an area and specific company.

https://earthjustice.org/features/map-coal-ash-contaminated-sites

A map of the United States showing where hundreds of locations contaminated with coal ash. It also has excellent information on the contamination itself and the extent of it. A truly helpful look at the scope of the coal ash contamination across the country.

https://ecology.wa.gov/Spills-Cleanup/Contamination-cleanup/Cleanup-sites/Locate-contaminated-sites

This is an absolutely fantastic website the Department of Ecology in Washington State developed and has available for the public to use. The department offers up over 13,000 sites across the state which are considered contaminated. They run from underground tanks with gasoline leaks at service stations all the way up to the most toxic sites in the state. What really impresses us is they also offer information on orchards which are a serious threat to people finding themselves in homes unscrupulous developers built on top of old contaminated orchards still poisoned from years ago but never divulged to the buyers. If every state built a website like this it would help millions of people.

https://www.goodjobsfirst.org/violation-tracker

This is one of my favorite websites to review companies on. It list the violations the company has committed and the fines it received. You can see what the violation was, the location if the company has multiple ones and other interesting information.

https://www.npms.phmsa.dot.gov/

A government website mapping the nationwide network of pipelines crossing the country. There is an access level for government only and a public access level. The website is a little finicky at times but try another browser if the one you are using isn't responding well. There is good information here but being a government site they limit your access to some information they consider a classified level.

https://oilandgasthreatmap.com/threat-map/

The website compiled information from a number of other sites showing threat levels of exposure to chemicals from various facilities and compressors. There is more information if you look deeper into any of the specific sites or areas you pull up. A decent tool to use to see the size of the radius around industrial facilities that are impacted.

https://www.fractracker.org/

Fractracker shows not only where fracking is taking place in individual states across the country but also has resources for gas and oil pipeline networks, francs and mines, accident tracking and a few other really good search tools.

http://www.antennasearch.com/default.asp

A super useful tool for finding where almost all the cell towers are located in your area. The site is frequently updated with the latest approved sites to keep it accurate. Between this site and watching your local board meetings for any proposed locations, you should be fairly well covered.

https://www.motherjones.com/environment/2014/04/west-texas-hazardous-chemical-map/

The website lists through a map interface companies which are considered a risk based on the chemicals they handle. There are many more of these industrial facilities which never make the list as they go unreported however, this is a decent starting point for the worst of the group.

https://whatsdown.terradex.com/#9/33.6306/-112.3325/

This site tracks ground and soil contamination at sites across the country. From abandoned munitions to gas station spills and more. Easy to use. Just enter and address and complete information of every potential site in the area you select. Click on a specific site and more detailed information is available.

http://www.ehso.com/

This site is a portal that connects you to a range of resources from federal agencies to state environmental agencies. There are other links to get information on a variety of subjects. We like this site and use it frequently.

https://www.ewg.org/interactive-maps/pfas_contamination/

This site from the Environmental Working Group shows the location of over 2,300 places contaminated with PFAS chemicals. Some are military installation, public water supplies or communities.

https://www.epa.gov/cleanups/cleanups-my-community

This site allows you to search for hazardous waste sites supposedly cleaned up in your community. It has information about the site, the progress of the cleanup and current status.

https://www3.epa.gov/myem/envmap/find.html

My environment by the EPA lets you take a close look into the health of the community and neighborhood you live in. It has information on air quality, pollution, hazardous waste sites and other information. As always, we take the information they give and look at it as a piece of the overall picture.

Health and Environmental Agencies of U.S. States and Territories

Alabama

http://www.adem.state.al.us/default.cnt

Alaska

https://dec.alaska.gov/

Arizona

http://www.azdeq.gov/

Arkansas

https://www.adeq.state.ar.us/

California

https://calepa.ca.gov/

Colorado

https://www.colorado.gov/cdphe

Connecticut

https://portal.ct.gov/deep

Delaware

https://dnrec.alpha.delaware.gov/

District of Columbia

https://doee.dc.gov/

Florida

https://floridadep.gov/

Georgia

https://epd.georgia.gov/

Hawaii

http://dlnr.hawaii.gov/

Idaho

http://dlnr.hawaii.gov/

Illinois

https://www2.illinois.gov/epa/Pages/default.aspx

Indiana

https://www.in.gov/idem/

Iowa

https://www.iowadnr.gov/

Kansas

https://www.kdheks.gov/

Kentucky

http://eqc.ky.gov/Pages/default.aspx

Louisiana

https://www.deq.louisiana.gov/

Maine

https://www.maine.gov/dep/index.html

Maryland

https://mde.maryland.gov/Pages/index.aspx

Massachusetts

https://www.mass.gov/orgs/
massachusetts-department-of-environmental-protection

Michigan

https://www.michigan.gov/egle/0,9429,,00.html

Minnesota

https://www.dnr.state.mn.us/

Mississippi

https://www.mdeq.ms.gov/

Missouri

https://dnr.mo.gov/

Montana

http://deq.mt.gov/

Nebraska

http://www.deq.state.ne.us/

Nevada

https://ndep.nv.gov/

New Hampshire

https://www.des.nh.gov/

New Jersey

https://www.state.nj.us/dep/

New Mexico

https://www.env.nm.gov/

New York

http://www.dec.ny.gov/

North Carolina

https://deq.nc.gov/

North Dakota

https://deq.nd.gov/

Ohio

https://www.epa.state.oh.us/

Oklahoma

https://www.deq.ok.gov/

Oregon

https://www.oregon.gov/DEQ/Pages/index.aspx

Pennsylvania

https://www.dep.pa.gov/Pages/default.aspx

Rhode Island

http://www.dem.ri.gov/

South Carolina

https://www.scdhec.gov/

South Dakota

https://denr.sd.gov/

Tennessee

https://www.tn.gov/environment/

Texas

https://www.tceq.texas.gov/

Utah

https://deq.utah.gov/

Vermont

https://anr.vermont.gov/environmental-conservation

Virginia

https://www.deq.virginia.gov/

Washington

https://ecology.wa.gov/

West Virginia

http://deq.wyoming.gov/

Wisconsin

https://dnr.wi.gov/

Wyoming

http://deq.wyoming.gov/

Conclusion

When I first wrote this book, most of the chapters contained not only information and examples of these environmental risks but also my feelings about how these corporations and industries had such little regard for the way their actions impacted real people, affecting their health and their lives. Dripping with sarcasm, I wrote about how these corporations figuratively owned politicians, using them to write loopholes into regulations to allow the continued wholesale raping and polluting of our environment. I wrote about how industries do their own testing on many highly toxic and hazardous substances and then feed the results to politicians to do their bidding, keeping these poisons in widespread use, such as the golf industry does. Industries such as agriculture, mining, gas and oil, and waste disposal have no problem with allowing the continued poisoning of people so billions of dollars in profits can be raked in by these corporate pirates. Then these monsters turn around and use some of the same money to finance their pet politicians' campaigns to keep them in office.

However, I decided to stick with the facts throughout the body of the book and let the readers get angry on their own, wondering how corporations and the government can get away with this. My trust in corporations and government is minimal to say the least, but I endeavored to stick with the facts and examples of how these hazards affect millions of people every day. Now, hopefully you're as angry as I am.

It appears many industries cared little about the damage they were doing to the environment for decades even though they had full knowledge their methods were faulty all along. Seduced by the billions of dollars these companies were raking in, little thought was given to the damage that was being caused to the environment and the people affected.

Behind it all are the powerful industry groups and their pit bull lobbyists continuing to push for the least amount of government regulation and interference as possible in their business. The industry has been left to police itself like foxes guarding the henhouse. The only people outside these companies who seem to think industries are capable of self-regulation are the politicians who receive huge amounts of cash contributions to keep them in power. Industry 10 points, Joe Citizen, zero.

The right to create pollution is bought and sold like candy in a store. There is a whole system set up whereby polluters are given the right to emit a certain amount of poison into our air, water, and soil. If they reduce the emissions by a certain amount, they are given a credit for the pollution, which they are able to turn around and sell to a polluter who does not want to decrease their emissions and actually may want to emit more than their permits allow. Instead of an overall reduction in the amount of pollution, this system simply shifts it from one place to another.

The buyers of the pollution credits are in many cases companies operating older and outdated plants with antiquated and inadequate pollution control devices. Instead of doing a costly upgrade to add better pollution control, they buy the credits and continue to use the old, inferior equipment. It is much cheaper for the company to just continue running the plant into the ground until it no longer can turn a profit or until they are so out of compliance with regulations that they are forced to shut down or actually upgrade the equipment. The shut down or forced upgrades are almost always the result of relentless pressure by residents and environmental groups on the EPA and state environmental agencies to do their jobs despite politicians trying to ignore the issue.

The big part of the problem is in how the system is tilted in favor of the manufacturers and corporate polluters and not the people being poisoned. Because of industry influencing how the regulations and restrictions are written and enforced by the politicians under their control, the American public is a helpless victim being held hostage by the corporate polluters. Unless a person literally drinks a glass of chemicals containing enough poison to cause an immediate effect, the industry can deny the health effects, even death, arguing these effects were not caused by the chemical. They, therefore, are not accountable for any damage to people or the environment because it cannot be proved they were the source of the poison. In the event they are found to be the source, then all other questions come up. What were the victims predisposed to have? Are there other chemicals present that also had an effect? Were there other sources of possible pollution besides the obvious suspect? These companies have the financial resources to stall, delay, and deny while their victims struggle to pay for medical care. These tactics don't even stop when their victims die.

One popular way to misdirect responsibility is to pass the blame. There is a reason so many industrial facilities are found grouped together in one place. It makes it more difficult to point the finger at any one of the polluters as being responsible. When multiple potential sources are together, the only levels that are usually able to be measured are the group total.

There are quite a few issues inherently wrong with the way industries are allowed to wreak havoc over the environment and poison us on a daily basis. Let's start with the lobbyists and the industry groups they represent. These are some of the most well-financed organizations in the country. These groups represent corporations whose combined earnings total in the trillions for some of the largest business sectors. Their primary purpose is to get politicians to pass laws and regulations that favor their industry and the way they do business. They effectively try to stop any new laws and regulations that would have a negative impact on the industry, and they also lobby for hundreds of billions of dollars to be given to them as subsidies, government contracts, and sweetheart deals where they get rewarded with business worth tens of millions and favorable tax breaks for their industries.

The agriculture industry spent over $35 million dollars on lobbying in 2019. Almost 54% of these lobbyists were ex-politicians who knew the system and how to leverage it to benefit their industry. The average contribution to a member of congress that year was over $82,000. The oil and gas industry spent $125 million in lobbying, and the electric industry, which includes power plants, spent $118 million in 2019. The energy industry as a whole was even more generous to the Senate and House in 2019, spending on average $93,000 per senator. One senator pocketed $717,000 in contributions alone.

This money is given to directly influence the way the senator votes or influences the rules and regulations which will affect the industry the money came from. Sometimes the money greases the levers that control billions in subsidies the government gives select industries, with the largest companies raking in millions of the payout. Like the estimated $1.59 billion allowed to the energy sector to offset the majority of their drilling costs. Or the more than $836 million General Electric has taken in between federal grants and tax credits. Exelon, the company that runs many of the nation's aging and leaking nuclear plants took in over $734 million over the years.

The very agencies that are charged with protecting our health and environment are a part of the problem. It may come as a shock that the agencies at the state and federal level are often influenced by the companies they are supposed to be monitoring and protecting us from. They use tests and studies done by the industry promoting continued use of dangerous chemicals instead of those done independently showing there are risks to people and the environment too great to ignore. They allow the power industry to dump toxic coal ash in giant lagoons, polluting underground aquifers for decades to come. They allow the chemical industry to release millions of gallons of toxic chemicals into our waterways poisoning water supplies. They allow the agriculture industry to cover millions of acres with toxic pesticides and herbicides and then tell us the food is fine to consume.

We're even less protected from accidents. One would expect serious consequences for serious accidents. An accident at a Dupont insecticide plant killed four workers. The government cited Dupont for serious safety violations and lack of training workers. Four people died, and Dupont leaked 20,000 pounds of a highly toxic chemical, only a tiny amount of which kills people. The EPA fined the $85 billion dollar company $99,000 for the violations and stated the workers would still be alive today if Dupont did not fail. By the way, the penalties are most likely a business expense tax write off due to a loophole politicians put in the laws.

On top of lax regulation, the methodology the EPA and industries use for testing exposure limits on people is flawed. The majority of government studies fail to take into account the long-term results of continuous exposure. All the small so-called safe exposure adds up to really large cumulative exposure with serious long-term effects. There is a reason why the X-ray technician always leaves the room. The radiation from one X-ray is negligible, but getting an X-ray every day will kill a person. Real people live near these chemical plants and toxic waste dumps every day. And every day they are again exposed to the toxic poisons that are slowly adding up and causing health effects the government ignores. Since the government doesn't set long-term exposure limits or consider cumulative effects of these chemicals, industries are never held accountable for what the real exposure risks of their poisons are. The government is not only ignoring what is happening, they are in many ways condoning these corporations who are profiting from the continued poisoning of the American public.

Of course, the government allows industries to continue with these unsafe practices; the government does the exact same thing itself on a higher scale. The U.S. government owns the most hazardous waste sites: everything from nuclear waste dumps to lagoons filled with nerve gas and biological weapons waste. These are in addition to all the regular types of toxic waste sites they already have with dioxins, mercury, PCBs, and other deadly chemicals

The terrible fact is the government will not only throw the public under the bus, they will drive the bus. Consider this example which was made public back on May 14, 2018.

The White House and Scott Pruitt (the former head of the EPA) tried to block the publication of a federal health study on the chemicals PFAS and PFOS. The chemicals PFOA and PFOS can cause thyroid issues, pregnancy problems, and some cancers, even at very low levels of exposures. The government allowed the public to be exposed for years to dangerous and unsafe levels of chemicals they refused to study. These chemicals are now being found across the country, contaminating water supplies for millions, especially near military bases, chemical factories, and farm fields where sewage sludge had been used, as well as many other places. The study showed evidence that these chemicals were dangerous to human health at much lower levels than the EPA had previously considered safe but that they never regulated exposure.

These chemicals have been used since the 1970s in items such as fabric coating, non-stick pan coating, and many items used on a daily basis. The 3M Corporation already paid out $1.5 billion in settlements for personal injury claims and contamination. Despite this long history, the chemicals went unstudied by the EPA; tens of millions of people have already been exposed to them.

The chemical industry was in a panic over the soon-to-be-public information showing how the public was being poisoned. Lobby groups for the chemical industry spent tens of millions to keep regulations of these chemicals from going into effect. Industry lobbyists contacted the politicians they had control over to get the administration, who was already in favor of poisoning the environment and the public, to block the report. Unfortunately for the industry and the politicians, public pressure had built up enough to push back against the administration, and eventually the EPA needed to look like they were protecting the public and started to tell the truth. Not by choice, but by public pressure through the media and people being poisoned across the country. Scott Pruitt resigned in disgrace after ethics scandals and other issues showed where his real interests were.

If that single example does not make you angry, here are several more.

Dennis Deziel, the new regional administrator for the New England Region for the EPA, has been barred from participating in many of the issues the agency needs to be involved in there. It has been disclosed that he had been a chemical industry lobbyist and had spent five years working with Dow Chemical on federal government relations issues. The conflict of interest is self-evident. Already, opponents have pointed out he must recuse himself from at least a fifth of the Superfund projects because they involve the chemical interests he previously had dealings with. Dow chemical is involved in 96 Superfund sites across the country; 21 are located in their former employee's region. With a former coal lobbyist, Andrew Wheeler, at the head of the EPA and chemical industry lobbyists in regional positions, the obvious conclusion most people would draw is that the current administration is trying to rapidly dismantle the regulations protecting the public and the environment from industrial rapists.

An investigation in Arizona found that two of the largest utilities, Arizona Public Service and Salt River Project, were using money from their customers' utility bills to lobby against environmental regulations. Membership in two of these lobbying groups cost APS $131,000 per year and SRP $141,000 in 2018. The lobbying groups spent ratepayers' money on a whole host of policies dangerous to their customers' health and the environment's sustainability. They were fighting against stricter regulations for toxic coal ash disposal. They were fighting against rules that would lower levels of mercury, arsenic, and lead from power plants. They were fighting against any limits on effluents released from these plants. Every one of the groups were fighting to weaken all regulations that would protect the people and the environment.

In Murphy Creek, Colorado, outside Aurora, one of the largest homebuilders in the United States is trying to convince the town to allow them to build next to a leaking Superfund site. Lennar Homes wants to build hundreds of homes on 171 acres literally next to the leaking site and over plumes of toxic chemicals underneath the property. Toxic waste at the Superfund site has for decades spread under the properties surrounding the site and continues to migrate further away. Many toxic chemicals are able to enter into buildings on top of the plume as the toxic vapors rise to the surface. In response to the possibility new homeowners will be subjected to the toxic vapors, the developer claims it is going to add a vapor barrier to the homes and take measures to vent the toxic chemicals from the inside of the homes. One of the more toxic chemicals is 1,4 dioxane, a cancer-causing chemical, which has been found over a mile away from the Superfund site. The developer adds they will tell the homeowners about the landfill nearby. They aren't calling it the Superfund site it is; they're calling it a landfill. By downplaying the potential high health risk from the Superfund chemical contamination next to the houses, the developer hopes to convince people how safe it is to live and raise a family there. This is another example of placing people who may not fully grasp the long-term consequences of where they live into really bad situations.

In a neighborhood in New Orleans, Louisiana, a young couple bought a home to raise their children in a quiet, clean, and safe place. Not until after they moved into their home did they find out the property across the street, which appeared to be a nice open lot, was the site of the old city incinerator and that it was contaminated with high levels of lead. Records show that authorities had known about the contamination for over a decade but never warned residents nearby. The property simply sat open and inviting. No restrictive fencing or warnings were even posted. While the city claims the lead will not get into the groundwater or surrounding properties, we have seen many instances where the winds and people carry particles of toxic chemicals contaminated areas hundreds of yards away, and even further, from the site. Why the city chose not to warn nearby residents for over 10 years about this imminent hazard we do not know. Shame on them for putting children and families at risk.

The average person would think that there are rules and regulations in place to prevent these types of situations from happening. They are correct; however, frequently the rules are twisted, bent, ignored, and downright broken by industries and politicians who are more interested in dollars and developments than in protection and people.

In San Francisco, California, contaminated properties such as brownfields are supposed to go through an environmental review process prior to getting approvals for development to be allowed. Despite this, an investigation found a dozen properties that had been fast-tracked and exempted from public environmental reviews. In one case, the property had been a gas station where the soil was contaminated with the chemical benzene, a known carcinogen. The city changed its mind to allow condos on that property after the neighbors hired a lawyer. Environmental reviews and cleanups of toxic, contaminated property to build housing and other uses protects the people who will live on the site and in the neighborhood. Skirting these reviews only helps developers to make lots of money and the city to get tax dollars in their pockets.

A trash incinerator in Brooks, Oregon, is threatening the county and state that it will double the rates it currently charges or close its doors if they are not awarded state tax credits for being considered green renewable energy. Covanta, a nationwide operator of incinerators, will make an extra $5 million from the credits. The company has hired a team of lobbyists who are working the hallways in the state capital trying to get politicians to vote for approval. Calling themselves a generator of renewable energy from burning garbage is a stretch of the imagination. In 2017, the incinerator emitted greenhouse gases totaling 160,843 metric tons into the atmosphere. Now they are trying threats and blackmail to claim they are clean and green renewable energy. Opponents who are fighting against giving the incinerator a renewable energy designation are making public that the company has resorted to threats and extortion to take taxpayer money while they continue to pollute copio

In Buffalo, New York, there was a Superfund site along the Buffalo River that has been designated one of the top most toxic sites in the country. Along came SolarCity, now owned by Tesla, who proposed building a large manufacturing plant on the property. Instead of requiring the massive cleanup of the property and toxic waste site warranted, environmental authorities did an end run around the regulations. They claimed there would be no negative effects caused by the project and went forward with construction. In spite of workers finding radioactive waste and toxic chemical waste during the preparation, nothing was slowed down for environmental agencies to take a closer look. The agency running the project, SUNY Polytechnic Institute (a New York State University), was allowed to again claim the discovery of the waste would not have a negative impact and continued with the construction. So much hazardous waste was discovered, and they were all required to be disposed of at a special landfill out of state. Unsurprisingly, another 50,000 cubic feet of toxic dirt was allowed to simply be buried at the factory site. Much of it was used as contaminated fill for the parking lot.

This Buffalo property had a very toxic past even before it became the Tesla plant. Years earlier, there were quite a few steel manufacturing plants who for decades polluted the property and nearby ones with thousands of tons of toxic material. On those nearby properties, the city of Buffalo allowed developers to build homes on top of those toxic waste dumps in the 1990s, also declaring there would be no environmental impacts caused by not doing a full-scale clean up. Underneath the neighborhood there are chemicals such as lead, benzene, coke waste from industrial furnaces, arsenic, and other toxic chemicals.

The city was forced to finally offer to buy out some residents, but only after many had suffered health effects that will haunt them for life. One resident spoke about pets dying, neighbors who are sick, and children born with birth defects. In addition, there are several types of cancer reported throughout the development. In the meantime, the developers made millions while the residents were never told about the poisons they were living on top of.

In Porter, New York, a hazardous waste landfill that closed a decade ago is seeking to expand its operation and reopen. All the nearby towns, and even the school district that stands to receive $1 million per year from the landfill, are all opposed to the project. The town of Porter is not getting involved because Porter signed a deal with the landfill operators for several million dollars in exchange for not opposing an application from the company. In spite of cancer clusters in the area and and 389 documented spills and leaks of toxic waste at the facility, the town of Porter remains silent.

Fracking produces millions of gallons of a waste product called brine. It is basically contaminated water containing salts, toxic chemicals, and radioactive particles. The waste is supposed to be disposed of according to strict regulations because of the hazardous materials it contains. The fracking industry, however, has a loophole written in the places brine can be disposed of. In at least thirteen states, it is legal to use fracking brine for dust control on unpaved roadways and as a winter de-icer on highways. The Department of Transportation in Ohio has used one half million gallons on its highways. While landfills are prohibited from taking most waste with over 50 picocuries of radiation in it, samples of this de-icer have been measured at levels of up to 2,500 picocuries. The industry is able to contaminate thousands of miles of roadways with radioactive materials because the lobby group for the industry had their pet politicians call this a beneficial use of the waste. Therefore, it is perfectly legal in spite of it being hazardous. A familiar pattern is emerging.

In Baltimore, Maryland, the community must be doing something right with reducing and recycling their trash. The amount of trash in Baltimore County that was sent to the incinerator operated by Wheelabrator in 2018 was down 65,000 tons from the 215,000 expected by the company. The response from the company was to sue Baltimore County for $32 million dollars for having less trash. Wheelabrator already gets millions of dollars in state subsidies for pouring millions of pounds of toxic gases into the atmosphere in the name of creating green renewable energy in the industry. Now, the company wants the county to stop doing such a good job recycling and cutting into the available amount of trash they can burn or pay up.

The current head of the EPA, Andrew Wheeler, for years was a lobbyist representing some very big corporate interests at the firm he worked with. It was revealed that Wheeler's former colleague from the firm had been meeting with the EPA under Wheeler about rules and projects that directly impacted the clients he had represented. Projects being reviewed by the EPA have included issues dealing with several companies Wheeler had lobbied on their behalf and now was the head of the agency they were appearing to with requests on their behalf. The EPA chief is not supposed to be involved in any of the issues a company had lobbied for prior to taking the position. Several high-ranking officials have questioned Wheeler's involvement and whether he is properly following the rules.

Wheeler is known as a revolving door politician. They go from politics to industry and back again, using their contacts to push forward their personal agendas. Someone sends a check directly to a senator for the senator's campaign or other use with the expectation of favors in return. It seems that regular people cannot possibly compete with companies and that type of influence. However, doing so is much easier than we may think. That is what these politicians and corporations hope you do not find out.

We the people actually hold the power, but we need to use it in a way that is effective. While it may appear the slightly improving environmental situation has reversed direction under the Trump administration, that is only temporary. The tide is finally starting to turn against these puppet politicians and the corporations who pull the strings. It is definitely within the control of the people to make the changes needed to the system. In spite of the fact much of the damage already done will take decades and centuries to repair, today we are able to take concrete steps to prevent additional damage from occurring. The war against corporate polluters, lobbyists, and bad politicians is being won daily in victories across the country. Knowing that individuals and groups of people like you can fight big corporations and win is what drives people like yourselves into taking the action needed to force these changes and stop these companies from further destroying our health and environment.

The biggest power we have is the use of factual information. When involved in a battle against a proposed project or trying to get an existing one to comply with something they presently are not, the use of indisputable facts is unmatched as a weapon. This is the biggest fear of any company, whether large or small: that the public will learn the truth and use that information against them. They love when there is a public meeting and all that occurs is people yelling they don't want it here or speaking in general terms about water or air pollution. Once you speak from a position of knowledge, you become a danger to these politicians and corporations. Solid, verifiable, and pertinent facts used by people like you and me have taken down major corporations and stopped projects in their tracks. Creating situations where you are able to publicly put bad players on the defense is a tactic you can become quite good at. It is very effective, and once you are able to have them appear weak and unconvincing with their responses, you begin to weaken them and make their lies more obvious.

Facility tours, public meetings, presentations, and hearings are all places you would expect them to ask if there are any questions; they really mean it. Unless, of course, you ask a tough one that makes them look bad. In that case, you will be ignored, asked to leave, or receive a politically-charged doublespeak answer that never addresses what you asked.

I became quite good at asking those questions and sometimes, not often, got a feeble attempt at an answer in response. Public hearings were always one of my favorites to publicly put politicians and corporate and industry officials on the spot. When there are reporters in the room, they seem to instinctively know when the situation is about to heat up, and once the reporters get to know you and the fact you will liven things up using facts, they love it as it gets their stories run in the papers and on the internet.

Another huge power we have as individuals and groups is the power to create pressure through negative publicity. No politician wants it broadcasted they allowed the oil and gas industry to poison a river because they took $700,000 from them. I have learned over the years to leverage information about politicians that may be perceived as negative as a tool to use. In many cases, the press willingly follows up on that information, pressing the politician further. Continued public pressure does take a toll and will wear down the politician, frequently getting concessions and protections to health and the environment. Remember to always keep it factual and to present the information forcefully and directly. Do your best to not let emotions take over and lose control because once you do, the politicians and industries will point to you as an irrational person who doesn't know anything. You will find politicians, industry representatives, and other opponents will push your buttons trying to get you to lose it in front of people.

People like you have had their fill of this and are beginning to organize and fight back stronger than ever before. The best part of this is the people are beginning to pile up wins. Chemical plants that have poured tons of toxic chemicals into the air of a local community are being closed down because of pressure brought on them by local and area citizens. Proposed landfills and trash incinerators are being stopped in their tracks because of united groups of residents saying you need to find a better solution. Mothers who become experts at the emissions coming from medical waste incinerators and their effects on the nearby community are turning the tide against these polluters. There is no end in sight for what can be accomplished and done by educated and concerned citizens.

In Catskill, New York, opponents of a proposal from Wheelabrator, one of the largest trash incinerator operators in the country, was derailed after an intense two year battle. Residents from all over the area joined forces with environmental groups to protest and defeat the project. Wheelabrator had been trucking the ash to a landfill they owned in neighboring Massachusetts. The new landfill would enable the company to save millions of dollars in transportation costs over its operational life. The proposed location was a former quarry. Thinking they had found the perfect location in a quiet rural area where resistance would be low, Wheelabrator applied to the state for the permit. The plan was to dump 445,000 tons of ash per year for 50 years. Opponents immediately made their position known with huge turnouts at public meetings and many demonstrations and press releases. Residents and environmental groups focused on the fact that the quarry was located in an area where the geology was unsuited for safety and the area underlaid with limestone formations conducive to spreading pollution once the landfill started to leak. The announcement by the company that they were withdrawing the proposal was met with cheers from residents.

Residents in a section of Birmingham, Alabama, had a message for Sherman Industries when they heard the company was looking to locate to their Five Points West community. They were met by the community collectively saying, "No, you cannot place your cement plant next to our homes." The community was able to quickly galvanize enough support to enlist the mayor of Birmingham to join the residents and tell the company it was not welcome in that section of the city. By responding quickly, showing a large amount of resistance, and making public facts about why the location was inappropriate, the residents were able to gain the support of high-level politicians. It swiftly ended the possible protracted battle.

In the Lehigh Valley area of Pennsylvania, people in the Slate Belt section found out a sewage sludge treatment plant was proposing an operation nearby. The plant was going to truck in sewage sludge from treatment plants and process the material by drying it into fertilizer and other products. From the beginning when the proposal became public in 2015, town planning meetings frequently had 100-200 people present. Almost all of the people who spoke at the meetings were opposed to the project. The residents objected based on the impact from the truck traffic, emissions from the processing facility and environmental risks related to drainage from the site and a quarry located on the project property. After four years, the planning commission unanimously voted to have the Board of Supervisors deny the project. Instead, the company withdrew the application just prior to the meeting. Legitimate objections based on facts and continual public presence at the meetings won this battle.

In Baldwin, Pennsylvania, residents heard a toxic waste storage and transfer company wanted to open up a facility in town. The company was applying for a special exemption to operate the business from the Zoning Board. Residents in the town felt that having a business that would continuously be trucking in hazardous waste, storing it, and then trucking it back out again would present a potential danger they were unwilling to take. The risk for an accident was too great. The townspeople mobilized and made their position known to the town hall and the zoning board members. At the public hearing there were well over 200 people packed into the room. It was obvious the board members had done their homework on the potential dangers, especially when the zoning board members asked questions about possible problems, and they did not seem to feel the responses were adequate to assure there would be none. The crowd burst into cheers when the zoning board unanimously denied the special exemption to the hazardous waste company

In Bridgeton, Missouri, north of Saint Louis, residents were overjoyed to hear that after years of demanding it, a nearby landfill holding tons of radioactive waste leftover from the Manhattan Project was finally going to remove the waste. The local landfill had originally accepted this waste transferred from the cleanup of another hazardous landfill elsewhere. Residents fought for years to have the waste removed, citing health issues including nosebleeds, headaches, and rashes. The waste will be dug out down to 20 feet deep, and the remaining contamination covered and sealed. The cost is over $200 million, and the landfill operator is still fighting the project, contending there are no health risks to the residents from the radioactive waste. The cleanup is expected to take three years to complete. Persistence and determination by the residents and citizens groups paid off in getting the toxic waste site cleaned up.

In the state of Montana, the Department of Environmental Quality wanted to raise the level of radioactivity the landfills could accept to 200 picocuries. The proposal would have allowed Montana landfills to become the dumping ground for radioactive fracking waste. Currently, almost all states will only allow radioactive waste up to a level of 50 picocuries to be disposed of. Much of fracking waste contains a level of radioactivity many times that. The fracking industry had lobbied very hard, trying to get the state's landfills opened up as a place to send waste nowhere else would accept. Public pressure on the Department of Environmental Quality of Montana appears to have forced them to back off its proposal. Public outrage and a flood of comments all opposing opening up the state moved the agency to declare the limit of 50 picocuries would stand, and is now actually enforceable if violated. The Montana Petroleum Association tried to have one of their politicians derail the entire project regulating the waste to get the industry the 200 picocuries limit, but in the end, the people won the battle.

In Plainfield Township, Pennsylvania, opponents of a sewage sludge treatment plant fought with the developer for almost three years over the potential plant. Many of the residents used a fact-based opposition approach to the facility, noting the environmental issues already well documented with sewage sludge and land application. After three long years, the company decided it had already spent too much money to lose anymore, and in a face-saving move, pulled their permit request just prior to a meeting where the entire Board of Supervisors appeared to be ready to deny it anyway.

In one of the longest-running battles against a polluter, Stericycle, a so-called medical waste management company, was finally chased out of North Salt Lake City after years of battles with the local residents since the original permit was issued in 1989. Years of emissions, spills, and accidents at the plant plagued the neighborhood. Stericycle was backed into a corner after finally getting caught releasing chemical emissions out the smokestack far beyond their permit and never reporting them to environmental authorities. Not wanting to spend the money to upgrade to proper pollution control devices, the company announced they were closing. Original plans were to move to another county in Utah where perhaps local resistance would not be as strong, but those plans were destroyed when local residents in the new community launched a years-long battle stopping construction plans. Thinking the third try may be their lucky one, Stericycle set their sites on North Las Vegas, Nevada. Resistance there also proved too much for the company. At the time of finishing this book in July, 2020, Stericycle is now making an attempt to convince Reno, Nevada to allow them in. Residents there also are getting ready to mobilize and keep the incinerator out.

The examples of regular people just like you and me taking on industries and politicians are endless. That list goes on further than the list of things the government and industry is getting away with. They are being called out and forced to make changes. Hitting them in the pocketbook is the only way to make them. Change the politician, change the rules, and change the way corporations are allowed to operate. Hold the agencies and people responsible for enforcement accountable. The second the average citizen lets their guard down and stops pressuring industries and politicians to end their polluting ways is the exact second the pollution resumes. Constant vigilance and continuous pressure are vital.

In the meantime, be aware of what hazards are near you. No matter how careful you are and how thoroughly you investigate your surroundings, there will most likely be several hazards. Unless you live in an environment in its own bubble, it is unavoidable. Even in Alaska, there are days the particulate matter is even worse than it is in smog-choked Los Angeles. With so many factors out of our control and until massive, systemic change happens, we must seek to minimize what negative impacts we can on our own. Be aware and protect yourself, or suffer the negative consequences of the ignorance that industries and politicians hope to keep you in a while longer.

Bibliography and Useful References

The following list by chapter is material we either read or used information from for the book. We included all of these as it was all helpful somehow in our project.

Section One:

1. Superfund sites

https://www.petoskeynews.com/charlevoix/black/toxic-land-one-family-s-story/article_0e18578b-66fa-56ff-9f4b-77666799cab1.htmlhttps://www.timesfreepress.com/news/local/story/2019/oct/07/thousands-southside-chattanooga-homes-still-need-testing-toxic-lead/505273/

https://www.recordonline.com/news/20190725/state-to-remediate-superfund-site-in-new-windsor

https://wfpl.org/louisville-hazardous-waste-site-still-leaking-pollution-into-ohio-river/

https://www.theatlantic.com/technology/archive/2019/09/silicon-valley-full-superfund-sites/598531/

https://sanfrancisco.cbslocal.com/2019/09/10/mountain-view-toxic-superfund-site-clean-enough-for-new-housing-construction/

https://time.com/4695109/superfund-sites-toxic-waste-locations/

https://toxmap.nlm.nih.gov/toxmap/faq/2009/08/what-are-the-superfund-site-npl-statuses.html

https://www.forbes.com/2010/04/28/superfunds-environment-pollution-toxic-opinions-contributors-robert-h-weiss.html#10dfae1c7db8

https://www.epa.gov/sites/production/files/2015-09/documents/webpopulationrsuperfundsites9.28.15.pdf

https://www.epa.gov/superfund/superfund-remedial-performance-measures

https://www.newsweek.com/2016/06/24/bunker-hill-superfund-silver-valley-lead-poisoning-469222.html

https://cumulis.epa.gov/supercpad/SiteProfiles/index.cfm?fuseaction=second.cleanup&id=0200663

https://www.coloradotrust.org/content/story/living-near-superfund-sites-may-be-academic-barrier-some-children

https://publicintegrity.org/environment/human-exposure-uncontrolled-at-114-superfund-sites/

https://cumulis.epa.gov/supercpad/SiteProfiles/index.cfm?fuseaction=second.cleanup&id=0200663

https://cronkitenews.azpbs.org/2019/01/03/tucson-water-treatment-plant-contamination/

https://www.theatlantic.com/technology/archive/2019/09/silicon-valley-full-superfund-sites/598531/

https://publicintegrity.org/environment/human-exposure-uncontrolled-at-114-superfund-sites/

https://www.tandfonline.com/doi/full/10.1080/2330443X.2017.1408439

https://www.wbur.org/news/2011/05/22/toxic-sites

https://www.lohud.com/story/tech/science/environment/2016/11/11/dangerous-chemical-again-found-nj-superfund-site/93647320/

https://www.nytimes.com/2007/07/29/business/yourmoney/29spill.html

https://www.atsdr.cdc.gov/HAC/pha/ringwood/PHA%2520Text.pdf

https://www.reuters.com/article/us-usa-at-t-california/att-to-pay-california-52-million-in-hazardous-waste-disposal-settlement-idUSKCN0J50AG20141121

https://qz.com/1017181/silicon-valley-pollution-there-are-more-superfund-sites-in-santa-clara-than-any-other-us-county/

https://apnews.com/9717f8be97db4944a825088dc5eee379/US-prosecutors%253A-Water-company-illegally-dumped-toxic-waste

https://www.usatoday.com/story/money/business/2013/05/28/wal-mart-waste/2366999/

https://news.wttw.com/2019/10/31/epa-adds-abandoned-chicago-dumping-ground-superfund-cleanup-list

https://www.hcn.org/issues/38/1137/

https://www.epa.gov/sites/production/files/2015-09/documents/webpopulationrsuperfundsites9.28.15.pdf

https://www.abc15.com/news/local-news/investigations/frustrations-mount-with-slow-progress-in-phoenix-area-superfund-site

2. Trash incinerators

https://saugus.wickedlocal.com/news/20191018/public-challenges-wheelabrator-saugus-emission-control-plan

https://www.wcbm.com/2019/04/15/mds-new-clean-energy-bill-includes-millions-in-subsidies-for-pollution-spewing-trash-incinerators/

https://www.baltimoresun.com/maryland/baltimore-county/ph-tt-wheelabrator-0417-story.html

https://ilsr.org/wp-content/uploads/2018/12/ILSRIncinerationFInalDraft-6.pdf

https://toxtown.nlm.nih.gov/chemicals-and-contaminants/nitrogen-oxides

https://www.pbs.org/newshour/science/is-burning-trash-a-good-way-to-dispose-of-it-waste-incineration-in-charts

https://saugus.wickedlocal.com/news/20190725/residents-voice-concerns-over-wheelabrator-saugus-noise

https://www.ncbi.nlm.nih.gov/books/NBK233633/

https://www.statesmanjournal.com/story/news/2019/08/15/covanta-garbage-burner-close-contract-marion-county-collapse/2019502001/

https://poststar.com/news/local/as-incinerator-seeks-permit-renewal-air-quality-group-holds-zero/article_1943c1eb-9084-5476-b214-de84dee73339.html

https://www.clf.org/blog/whats-wrong-with-burning-our-trash-anyway/

http://www.energyjustice.net/incineration/worsethancoal

http://www.energyjustice.net/incineration/worsethancoal

https://www.nap.edu/read/5803/chapter/2#2

https://earther.gizmodo.com/trash-burning-power-plants-wind-up-in-our-most-marginal-1834928924

https://www.wusa9.com/article/news/local/lorton-incinerator-fire-causes-regional-concern/65-397053209
https://www.washingtonpost.com/local/md-politics/waste-plant-fires-put-maryland-montgomery-county-and-company-on-hot-seat/2017/01/08/728f58ba-d208-11e6-a783-cd3fa950f2fd_story.html
https://www.climatecentral.org/news/community-activists-rally-against-proposed-trash-incinerator-15605
https://www.ithaca.com/news/activists-prepare-for-battle-against-proposed-flx-trash-incinerator/article_9fe26222-f420-11e7-8f70-df44ca2b3beb.html
https://enviroliteracy.org/environment-society/waste-management/incineration/
https://www.clf.org/blog/whats-wrong-with-burning-our-trash-anyway/
https://www.ithaca.com/news/activists-prepare-for-battle-against-proposed-flx-trash-incinerator/article_9fe26222-f420-11e7-8f70-df44ca2b3beb.html
https://archive.epa.gov/epawaste/nonhaz/municipal/web/html/airem.html
https://www.tishmancenter.org/
https://www.tishmancenter.org/
https://www.wxyz.com/news/local-news/investigations/detroit-incinerator-often-exceeds-pollution-limits-rarely-fined-by-state
https://www.solidwastemag.com/waste-management/canadian-sludge-hauler-killed-in-detroit-area-accident/1000010790/
https://patch.com/new-jersey/newarknj/essex-county-trash-incinerator-s-weird-pink-smoke-ignites-debate
http://energyrecoverycouncil.org/wp-content/uploads/2019/10/ERC-2018-directory.pdf
http://energyrecoverycouncil.org/Downloads/
https://inhabitat.com/burn-the-trash-forget-the-toxins-is-the-usas-new-garbage-policy/
https://apnews.com/a436c2bd7c20420898aec2234bacd21c
https://fox43.com/2019/03/28/pink-smoke-coming-from-waste-to-energy-facility-in-lancaster-co-isnt-dangerous-to-public-or-environment/
https://www.longdom.org/open-access/release-of-dioxins-from-solid-waste-burning-and-its-impacts-on-urbanhuman-population-a-review-2375-4397-1000215.pdf
https://www.tapinto.net/towns/newark/sections/green/articles/a-garbage-incinerator-in-newark-is-releasing-dangerous-emissions-again-activists-say
https://poststar.com/news/local/report-hudson-falls-trash-plant-among-country-s-dirty-dozen/article_233446f9-c4a7-54ca-b371-4ca9c24da9c0.html
https://www.cdc.gov/niosh/topics/lead/health.html
https://patch.com/new-jersey/newarknj/color-purple-newark-residents-fed-incinerator-smoke
https://www.epa.gov/facts-and-figures-about-materials-waste-and-recycling/plastics-material-specific-data
https://archive.epa.gov/epawaste/nonhaz/municipal/web/html/airem.html
https://www.ncbi.nlm.nih.gov/pubmed/17432330
https://www.ncbi.nlm.nih.gov/pubmed/17432330
https://www.washingtonpost.com/local/md-politics/waste-plant-fires-put-maryland-montgomery-county-and-company-on-hot-seat/2017/01/08/728f58ba-d208-11e6-a783-cd3fa950f2fd_story.html
https://ccanactionfund.org/maryland/fighting-toxic-incinerators/
https://www.fairfaxcounty.gov/publicworks/recycling-trash/frequently-asked-questions-about-covanta-fire
https://www.ncbi.nlm.nih.gov/pubmed/27599051
https://www.tandfonline.com/doi/full/10.1080/10962247.2012.669740

3. Hazardous waste sites

https://www.mlive.com/news/2019/11/blue-soils-dug-from-under-polluted-wolverine-tannery.html

https://www.infrastructurereportcard.org/cat-item/hazardous-waste/

https://www.infrastructurereportcard.org/cat-item/hazardous-waste/

https://www.southcoasttoday.com/news/20190809/dep-demands-more-dartmouth-documents-in-bliss-corner-dumping

https://www.azcentral.com/story/news/local/arizona-environment/2019/08/13/toxic-groundwater-beneath-phoenix-west-van-buren-but-cleanup-stalled/3486506002/

https://www.azcentral.com/story/news/local/arizona-environment/2019/08/13/toxic-groundwater-beneath-phoenix-west-van-buren-but-cleanup-stalled/3486506002/

https://timeline.com/this-poor-hispanic-town-in-california-has-been-fighting-a-local-toxic-waste-dump-for-35-years-8f6f20f89f9d

https://www.starcourier.com/news/20191120/trump-rolls-back-epa-oversight-in-midwest-favoring-polluters

http://www.ehso.com/cssepa/tsdflandfills.php

https://www.commondreams.org/news/2007/07/16/depleted-uranium-holds-fast-colonie

https://www.freep.com/story/news/local/michigan/oakland/2019/11/06/madison-heights-chemical-firm-fined-1-4-m-owner-sent-prison/2509632001/

https://www.legalmatch.com/law-library/article/hazardous-waste-accident-statistics.html

https://qz.com/1200214/a-houston-company-dumped-cancer-causing-chemicals-into-a-neighborhood-storm-drain/

https://echo.epa.gov/detailed-facility-report?fid=110027242320

https://www.nrdc.org/experts/caroline-reiser/nuclear-waste-dump-hiding

https://nj1015.com/polluted-new-jersey-the-most-toxic-sites-in-your-neighborhood/

https://health.mo.gov/living/environment/hazsubstancesites/hazwastesites.php

https://www.cityandstateny.com/articles/politics/new-york-city/newsday-state-helped-cover-toxic-plume.html

http://www.toxicsites.us/index.php

https://www.theguardian.com/us-news/2020/jan/10/toxic-chemicals-michigan-wolverine-chemicals-pfas

http://www.poisonwells.com/saddleBrookeRanch/saddleBrookeRanch.htm

https://www.theguardian.com/us-news/2020/jan/10/toxic-chemicals-michigan-wolverine-chemicals-pfas

https://www.propublica.org/article/new-jersey-never-ranked-its-most-contaminated-sites

https://www.state.nj.us/dep/srp/kcsnj/

https://www.nj.com/news/2017/04/the_most_toxic_sites_in_each_new_jersey_county.html

https://www.miaminewtimes.com/news/miami-dade-is-one-of-the-worst-places-for-toxic-soil-and-chemical-contamination-11388553

https://www.miaminewtimes.com/news/miami-dade-found-toxic-chemicals-in-overtown-soil-didnt-bother-telling-residents-8820932

https://www.epa.gov/sites/production/files/2015-09/documents/webpopulationrsuperfundsites9.28.15.pdf

https://www.epa.gov/privatewells

https://news.wttw.com/2019/10/31/epa-adds-abandoned-chicago-dumping-ground-superfund-cleanup-list

https://www.buzzfeednews.com/article/laurageiser/photos-of-toxic-places-in-us

https://www.epa.gov/sites/production/files/2015-05/documents/environments-contaminants-contaminated-lands.pdf

https://www.theadvocate.com/article_34c5c5c6-bea2-11e9-a657-db633e0c92dd.html

https://www.newsli.com/2017/10/20/northrop-grumman-bethpage-underground-toxic-plume-threatens-long-islands-water-supply/

https://www.theadvocate.com/article_34c5c5c6-bea2-11e9-a657-db633e0c92dd.html

https://www.theguardian.com/us-news/2020/jan/10/toxic-chemicals-michigan-wolverine-chemicals-pfas

https://www.cityandstateny.com/articles/politics/new-york-city/newsday-state-helped-cover-toxic-plume.html

https://www.infrastructurereportcard.org/cat-item/hazardous-waste/

https://www.infrastructurereportcard.org/wp-content/uploads/2017/01/Hazardous-Waste-Final.pdf

http://www.ehso.com/cssepa/tsdfincin.php

https://www.epa.gov/hwpermitting/hazardous-waste-management-facilities-and-units

https://www.envcap.org/srl/

https://www.metrowestdailynews.com/news/20200207/state-will-clean-up-part-of-framinghams-general-chemical-site-this-year

http://www.startribune.com/clock-is-ticking-on-andover-landfill-and-chemicals-are-leaking/506400592/

https://www.registerguard.com/article/20140203/NEWS/302039922

https://www.abqjournal.com/1386978/state-fines-abq-hazardous-waste-storage-business-360285-as-part-of-settlement.html

https://www.ttnews.com/articles/wiley-sanders-truck-lines-fined-3-million-illegal-transport-hazmat-materials-california

https://www.webwire.com/ViewPressRel.asp?aId=49217

https://www.newburyportnews.com/news/local_news/salisbury-company-fined-for-hazardous-waste-handling/article_a77794a4-79b7-597d-9b29-87bc34845b94.html

https://www.epa.ohio.gov/ddagw/uic

https://www.tribuneledgernews.com/local_news/investigators-search-for-truck-driver-believed-to-have-dumped-waste/article_cb38fee8-363d-11ea-b2b9-43c2ba371097.html

http://chej.org/wp-content/uploads/LandfillFailures20191.pdf

http://www.dnr.louisiana.gov/assets/OC/im_div/uic_sec/EPAposterofwells.pdf

http://chej.org/wp-content/uploads/Deep%2520Well%2520Injection%2520-%2520PUB%2520056.pdf

https://www.investigativepost.org/2018/04/05/6836/

https://m.metrotimes.com/news-hits/archives/2020/01/31/hazardous-waste-plant-in-detroit-with-history-of-serious-violations-gets-state-approval-to-expand-almost-ninefold

https://www.seattletimes.com/seattle-news/tacoma-hazardous-waste-operation-hit-with-1-9-million-state-fine-for-2018-fire/

https://www.theguardian.com/us-news/2015/oct/25/radioactive-waste-dump-fire-reveals-nevada-troubled-past

https://www.freep.com/story/news/local/michigan/2014/09/07/landfill-taking-radioactive-waste-has-history-of-violations-leaks-fires/15210857/

https://www.cityandstateny.com/articles/politics/new-york-city/newsday-state-helped-cover-toxic-plume.html

https://www.atsdr.cdc.gov/substances/toxsubstance.asp?toxid=155

4. Power plants

https://publicintegrity.org/environment/industrial-waste-pollutes-americas-drinking-water/

https://earthjustice.org/news/press/2019/trump-epa-removes-more-critical-toxic-coal-ash-protections

https://www.scientificamerican.com/article/coal-plants-smother-communities-of-color/

https://www.eia.gov/tools/faqs/faq.php?id=427&t=2

https://insideclimatenews.org/news/04032019/coal-ash-groundwater-contamination-toxic-arsenic-memphis-texas-eip
https://www.eia.gov/energyexplained/coal/coal-and-the-environment.php
https://www.eia.gov/tools/faqs/faq.php?id=77&t=11
https://www.reuters.com/article/us-usa-coal-epa/us-limits-on-coal-plant-mercury-emissions-too-costly-trumps-epa-idUSKCN1OR1BU
https://www.epa.gov/mats/cleaner-power-plants
https://www.ucsusa.org/resources/coal-and-air-pollution
https://www.arb.ca.gov/app/emsinv/facinfo/faccrit. php?sort= NOXHi&showpol=&co_=&ab_=&facid_=&dis_=&city_=&fsic_=4911&fname_=&fzip_=&grp=2&dbyr=2016&all_fac=C&all_tox=&chapis_only=&CERR=&dd=
https://www.reuters.com/article/us-usa-coal-epa/us-limits-on-coal-plant-mercury-emissions-too-costly-trumps-epa-idUSKCN1OR1BU
https://www.eia.gov/electricity/annual/html/epa_04_01.html
https://www.theguardian.com/environment/2019/mar/04/us-coal-plants-contaminating-groundwater-toxins-analysis
https://www.acaa-usa.org/Portals/9/Files/PDFs/Coal-Ash-Production-and-Use-2017.pdf
http://www.seas.columbia.edu/earth/wtert/sofos/Bolate%2520thesis_mercury.pdf
http://www.seas.columbia.edu/earth/wtert/sofos/Bolate%2520thesis_mercury.pdf
https://www.ucsusa.org/resources/coal-and-air-pollution
https://www.omicsonline.org/open-access/health-implications-of-heavy-metal-overload-2329-6879.1000145.php?aid=23332
https://earthjustice.org/features/map-coal-ash-contaminated-sites?gclid=Cj0KCQjwhtT1BRCiARIsAGIY51INvWQJE77tKCgK32IiHmTQ5-jEG9Cs9km2XnB6J9SO1wC0PtrLAJAaAqkbEALw_wcB
https://response.epa.gov/site/site_profile.aspx?site_id=11098
https://www.epa.gov/mats/cleaner-power-plants

5. Coal ash dumps

https://earthjustice.org/news/press/2019/trump-epa-removes-more-critical-toxic-coal-ash-protections
https://www.knoxnews.com/story/news/crime/2019/03/04/tva-coal-ash-landfills-threaten-tennessee-water-report-says/3031263002/
https://www.pilotonline.com/news/crime/article_eec157a3-3b16-567e-bdab-052361dc4e66.html
https://home.howstuffworks.com/home-improvement/construction/materials/fly-ash-concrete1.htm
https://www.pilotonline.com/news/crime/article_eec157a3-3b16-567e-bdab-052361dc4e66.html
https://www.pabulletin.com/secure/data/vol40/40-50/2359.html
https://insideclimatenews.org/news/04032019/coal-ash-groundwater-contamination-toxic-arsenic-memphis-texas-eip
https://www.usatoday.com/story/opinion/voices/2019/09/23/coal-ash-cancer-north-carolina-mother-column/2368665001/
https://insideclimatenews.org/news/28062017/coal-ash-jesup-georgia-landfill-federal-loophole
https://www.epa.gov/coalash/frequent-questions-about-2015-coal-ash-disposal-rule
https://earthjustice.org/sites/default/files/files/National%20Coal%20Ash%20Report%203.4.19.pdf
https://insideclimatenews.org/news/04032019/coal-ash-groundwater-contamination-toxic-arsenic-memphis-texas-eip
https://content.sierraclub.org/coal/disposal-ash-waste
https://massivesci.com/articles/coal-ash-water-health-hazard/

http://www.southeastcoalash.org/about-coal-ash/coal-ash-storage/
https://earthjustice.org/features/map-coal-ash-contaminated-sites
https://www.psr.org/wp-content/uploads/2018/05/coal-ash-hazardous-to-human-health.pdf
https://publicintegrity.org/environment/coal-ash-the-hidden-story/
https://www.environmentalintegrity.org/news/first-comprehensive-national-study-of-coal-ash-pollution-finds-widespread-groundwater-contamination/
https://stateimpact.npr.org/pennsylvania/2019/03/04/report-coal-ash-contamination-widespread-in-u-s-pa/
https://www.nytimes.com/2009/01/07/us/07sludge.html
https://cen.acs.org/articles/94/i7/New-Life-Coal-Ash.html
https://earthjustice.org/news/press/2019/first-comprehensive-national-study-of-coal-ash-pollution-finds-widespread-groundwater-contamination
https://pubs.usgs.gov/fs/fs-040-03/
https://www.mlive.com/news/saginaw-bay-city/2020/04/demise-of-coal-burning-power-means-toxic-ash-landfills-forever-on-great-lakes-state-shores.html
https://www.pewtrusts.org/en/research-and-analysis/blogs/stateline/2019/12/02/southern-states-split-in-response-to-epas-coal-ash-rule-rollbacks

6. Hazardous waste incinerators

https://www.bnd.com/news/politics-government/article237556179.html
https://etc.org/advanced-technologies/high-temperature-incineration/
http://www.ehso.com/cssepa/tsdfincin.php
https://mtstandard.com/news/local/construction-underway-on-million-hazardous-waste-processing-plant-outside-of/article_0aa2b839-5cd5-5cbe-9e72-8b46aa6b87cd.html
https://mtstandard.com/news/local/construction-underway-on-million-hazardous-waste-processing-plant-outside-of/article_0aa2b839-5cd5-5cbe-9e72-8b46aa6b87cd.html
https://www.theintell.com/news/20190730/elcon-continues-push-for-waste-plant-in-falls
https://www.theintell.com/news/20190730/elcon-continues-push-for-waste-plant-in-falls
https://wtov9.com/news/local/after-settlement-east-liverpool-residents-learn-whats-next-for-hazardous-waste-plant
https://www.eldoradonews.com/news/2019/jul/15/fire-clean-harbors-forces-brief-evacuation/
https://www.bnd.com/news/politics-government/article231690738.html
https://www.arkansasonline.com/news/2020/apr/05/military-waste-burned-state-despite-contamination-/
https://www.csb.gov/csb-issues-report-on-2009-explosion-at-veolia-es-technical-solutions-l-l-c-hazardous-waste-facility-waste-industry-urged-to-improve-safety-standa-rds-recommendations-issued-to-ensure-safer-siting-of-buildings/
https://www.dailykos.com/stories/2015/8/18/1396318/-Welcome-to-East-Liverpool-Ohio-home-of-one-of-the-world-s-largest-hazardardous-waste-incinerators
https://www.dispatch.com/article/20150514/NEWS/305149827
https://fox2now.com/2019/06/19/sauget-residents-protest-as-epa-rolls-back-emissions-monitoring-at-local-incineration-plant/
http://peoplestribune.org/pt-news/2019/07/illinois-communities-fight-veolias-poisonous-power-2/
https://archive.epa.gov/epawaste/hazard/tsd/td/web/pdf/increp5.pdf
https://www.eldoradonews.com/news/2020/feb/22/clean-harbors-incineration-facility-cited-lawsuit-/
https://www.timesunion.com/business/article/GE-fined-2-25M-fine-for-hazardous-waste-burning-6671768.php

https://www.epa.gov/sites/production/files/2019-12/documents/final_2019_capacity_assessment_report_20191217v1.pdf

https://www.timesunion.com/business/article/GE-fined-2-25M-fine-for-hazardous-waste-burning-6671768.php

https://www.politico.com/states/new-york/newsletters/politico-new-york-new-jersey-energy/2020/04/28/pfas-contamination-found-near-incinerator-334312

https://www.benningtonbanner.com/stories/bennington-college-teamfinds-elevated-levelsof-pfas-around-ny-plant,603396

https://books.google.com/ books?id=oPBbjGEKsmQC&pg=PA333&lpg=PA333&dq=where+is+oldest+hazardous+ waste+incinerator+in+the+us&source=bl&ots=fRrlnB9_aY&sig =ACfU3U29DV3TGKJqlDmjtwBCOPJg Er1KUQ&hl=en&sa=X&ved=2ahUKEwjekvXw96zpAhXN7Z4KHZ9QAWkQ6AEwD3oECAoQAQ#v=onepage&q=where%2520is%2520oldest%2520hazardous%2520waste%2520incinerator%2520in%2520the%2520us&f=false

https://epa.ohio.gov/News/Online-News-Room/News-Releases/ohio-epa-heritage-thermal-services-settle-violations-from-ash-release

http://edocpub.epa.ohio.gov/publicportal/ViewDocument.aspx?docid=979798

https://cen.acs.org/articles/95/i14/New-hazardous-waste-incinerator-comes.html

https://www.epa.gov/hwcorrectiveaction/hazardous-waste-cleanup-dupont-experimental-station-wilmington-de

https://theintercept.com/2019/01/27/toxic-firefighting-foam-pfas-pfoa/

https://www.waste360.com/medical-waste/city-says-curtis-bay-will-rebuild-medical-waste-facility-after-boiler-blast-crippled

https://www.opednews.com/articles/2/With-194-violations-for-em-by-Samuel-Vargo-ATTITUDE_Cancer_Cancer_Health-150604-269.html

https://epa.ohio.gov/News/Online-News-Room/News-Releases/ohio-epa-heritage-thermal-services-settle-violations-from-ash-release

https://www.dailykos.com/stories/2015/8/18/1396318/-Welcome-to-East-Liverpool-Ohio-home-of-one-of-the-world-s-largest-hazardardous-waste-incinerators

https://cen.acs.org/articles/95/i14/New-hazardous-waste-incinerator-comes.html

https://www.epa.gov/hwpermitting/assessment-national-capacity-hazardous-waste-management

https://www.sltrib.com/news/2019/02/22/medical-waste-company/

https://www.timesunion.com/business/article/GE-fined-2-25M-fine-for-hazardous-waste-burning-6671768.php

7. Cement plants

https://www.huffpost.com/entry/vallejo-cement-plant_n_5a99b95ae4b0479c0252730f

https://earthjustice.org/sites/default/files/library/reports/ej_eip_kilns_web.pdf

https://www.timesunion.com/local/article/Health-risk-under-study-553246.php

https://blog.timesunion.com/green/as-ravena-study-finds-mercury-in-cement-plant-neighbors-gop-moves-to-block-cement-plant-mercury-limits/2742/

https://www.timesunion.com/local/article/Health-risk-under-study-553246.php

https://blog.timesunion.com/green/as-ravena-study-finds-mercury-in-cement-plant-neighbors-gop-moves-to-block-cement-plant-mercury-limits/2742/

https://www.cement.org/docs/default-source/fc_concrete_technology/is548-optimizing-the-use-of-fly-ash-concrete.pdf

https://www.epa.gov/enforcement/cemex-fairborn-plant-clean-air-act-settlement

http://bayareacleanenvironment.net/articles/

https://www.cement.org/docs/default-source/market-economics-pdfs/more-reports/2013-plant-info-summary-sample.pdf?sfvrsn=cd760abf_2

http://www.downwindersatrisk.org/2012/10/midlothian-cement-plants-linked-to-higher-child-asthma-rates/

https://www.waste360.com/mag/waste_technology_new_technology

https://www.timesunion.com/business/article/Cement-plant-eyes-tire-burning-plan-in-Coeyemans-13505586.php

https://www.climate-policy-watcher.org/scrap-tires/tire-and-tdf-use-in-portland-cement-kilns.html

https://gcisolutions.com/HWF&CKS.htm

https://www.courier-journal.com/story/tech/science/environment/2017/08/01/clark-county-indiana-cement-plant-blocked-burning-hazardous-waste-fuel/525993001/

https://www.courier-journal.com/story/money/2013/11/29/cemex-wants-to-burn-recycling-leftovers-at-louisville-cement-plant/3782653/

https://www.washingtonpost.com/archive/business/1990/09/17/lafarge-plants-fuel-debate-over-hazardous-waste-disposal/25239d2a-17c3-4ede-b36b-8617d2448e80/?noredirect=on

https://www.energyjustice.net/files/cementkilns/cementkiln.ppt

https://www.inquirer.com/business/trash-electricity-energy-coal-west-virginia-20190429.html

https://www.fhwa.dot.gov/pavement/recycling/fach03.cfm#MixDesignandSpecificationRequirements

http://www.environmentalintegrity.org/pdf/publications/Cementing_A_Toix_Legacy2.pdf

https://precast.org/2010/08/fly-ash-a-hazardous-material/

https://slate.com/technology/2011/11/cement-plant-pollution-leads-to-a-fight-in-chanute-kansas.html

https://iaspub.epa.gov/triexplorer/industry.html?pYear=2017&pLoc=327310&pParent=TRI&pDataSet=TRIQ1

https://www.cemexusa.com/documents/27329108/45560536/cement-kiln-dust-2017.pdf/91694407-1944-6f83-74a5-cdbef1e29b43

https://www.heraldmailmedia.com/news/tri_state/west_virginia/martinsburg-city-councll-denies-zoning-change-requested-for-apartment-development/article_f553d425-cae3-53a0-80a3-c4c2a84e6d0b.html

https://www.timesunion.com/business/article/Plant-is-hit-with-fine-5961662.php

https://www.timesunion.com/local/article/Town-sees-mercury-spike-942371.php

https://www.npr.org/2011/11/10/142183546/epa-regulations-give-kilns-permission-to-pollute

https://www.brighthub.com/environment/green-living/articles/43655.aspx

https://retrohousewifegoesgreen.com/cement-plants-and-cancer/

https://www.hcn.org/issues/42.1/mountains-of-mercury

http://www.environmentalintegrity.org/pdf/publications/Cementing_Toxic_Legacy.pdf

https://www.npr.org/2011/11/10/142183546/epa-regulations-give-kilns-permission-to-pollute

https://www3.epa.gov/ttn/chief/ap42/ch11/final/c11s06.pdf Link to above and below

https://www3.epa.gov/ttn/chief/ap42/ch11/final/c11s06.pdf

https://www.justice.gov/opa/pr/cement-manufacturer-cemex-reduce-harmful-air-pollution-five-plants-under-settlement-epa-and

https://docquery.fec.gov/cgi-bin/fecimg/

http://www.capitolaggregates.com/s/CA_SDS_CEMENT_KD.pdf?v=3

https://dailygazette.com/article/2008/03/07/0307_mercury

http://www.energyjustice.net/files/tires/files/greenpeaceletter.html

https://www.downwindersatrisk.org/2012/10/midlothian-cement-plants-linked-to-higher-child-asthma-rates/

https://earthjustice.org/sites/default/files/library/reports/ej_eip_kilns_web.pdf

https://www.sciencedirect.com/science/article/pii/S0045653518321957?via%3Dihub

https://www.ncbi.nlm.nih.gov/pmc/articles/PMC4306904

https://poststar.com/news/local/lehigh-cement-gets-state-ok-to-burn-paper-and-plastic-mixture/article_a6b97aeb-3a4e-58c4-a789-6188b82e13ae.html

https://www.timesunion.com/business/article/Cement-plant-waste-burning-letter-uncovered-13700109.php

8. Military bases and government installations

https://www.militarytimes.com/news/your-military/2019/07/12/why-dod-is-still-using-burn-pits-even-while-now-acknowledging-their-danger/

https://www.recordonline.com/news/20190802/state-ag-joins-push-to-take-more-action-on-pfas-toxic-chemicals

https://www.newsday.com/long-island/environment/u-s-navy-bethpage-plume-1.36270702

https://kvoa.com/news/2019/09/21/tucson-water-finds-more-water-contamination/

https://www.militarytimes.com/2019/07/14/heres-an-updated-map-of-military-sites-where-dod-found-cancer-causing-chemicals-in-the-drinking-water/

https://www.militarytimes.com/news/your-military/2018/09/16/toxic-waste-sites-military-bases-in-florences-path-under-close-watch/

https://www.gao.gov/new.items/d04147.pdf

https://grist.org/article/military-families-can-face-their-biggest-threats-at-home/

https://www.hillandponton.com/military-base-exposures-united-states/

https://truthout.org/articles/military-hazardous-waste-sickens-land-and-people/

https://www.pbs.org/newshour/show/hazardous-military-waste-making-americans-sick

https://www.militarytimes.com/news/your-military/2018/04/26/dod-126-bases-report-water-contaminants-harmful-to-infant-development-tied-to-cancers/

https://grist.org/article/the-water-crisis-the-trump-administration-didnt-want-you-to-know-about/

https://ieer.org/resource/factsheets/incineration-radioactive-mixed/

http://dissidentvoice.org/Mar05/Marshall0329.htm

https://www.propublica.org/article/military-pollution-open-burns-radford-virginia

https://toxtown.nlm.nih.gov/chemicals-and-contaminants/perchlorate

https://www.spokesman.com/stories/2018/jun/26/shawn-vestal-airway-heights-water-worse-than-feare/

https://www.courthousenews.com/newburgh-hauled-court-toxic-water/

https://qz.com/1655268/us-military-is-a-bigger-polluter-than-140-countries-combined/

https://www.newsweek.com/2014/07/25/us-department-defence-one-worlds-biggest-polluters-259456.html

https://www.atsdr.cdc.gov/PHS/PHS.asp?id=771&tid=150

https://www.reuters.com/investigates/special-report/usa-military-legacy/

https://www.cnic.navy.mil/content/dam/cnic/cnrma/pdfs/Oceana/NAS%2520Oceana%2520Fuel%2520Spill%2520FAQ%252023%2520JUNE%2520ogc.pdf

https://www.reuters.com/investigates/special-report/usa-military-legacy/

https://apps.dtic.mil/dtic/tr/fulltext/u2/a219640.pdf

https://www.change.org/p/paducah-gaseous-diffusion-plant-usec-justice-for-uranium-enrichment-plant-victims-in-paducah-kentucky

https://www.kctv5.com/news/investigations/soldiers-push-for-accountability-leads-to-va-investigation-into-old-army-base/article_a6df74ca-8992-11ea-ba44-db9cfe77aaa0.html

https://www.inquirer.com/news/pfas-water-contamination-bucks-county-pennsylvania-20190628.html

https://www.militarytimes.com/news/your-military/2019/11/20/the-list-of-military-sites-with-suspected-forever-chemicals-contamination-has-grown/

https://www.washingtonpost.com/archive/politics/2004/01/04/cancer-cluster-haunts-nevada-town/159e7684-77d4-4f6e-89dd-fe025314eef8/

https://www.cnn.com/2020/05/14/politics/environmental-protection-agency-water-regulation-perchlorate/index.html

https://www.waaytv.com/content/news/Experts-work-to-remove-toxic-chemicals-dumped-on-Redstone-Arsenal-after-World-War-II-564175711.html

https://kutv.com/news/local/groups-object-to-open-air-burning-of-hazardous-waste-near-salt-lake-city

https://projects.propublica.org/bombs/state/CA#b=30.12224970425994,-138.203091375,44.068581864065,-100.410122625&c=shrink

https://www.atsdr.cdc.gov/phs/phs.asp?id=410&tid=72

https://eec.ky.gov/Environmental-Protection/Waste/hazardous-waste/Pages/paducah-gaseous-diffusion-plant.aspx

https://southerlymag.org/2019/12/04/nuclear-waste-in-paducah-kentucky-poses-extra-threat-to-region-facing-historic-flooding/

9. Medical waste incinerators

https://www.epa.gov/stationary-sources-air-pollution/hospital-medical-and-infectious-waste-incinerators-hmiwi-new-source

https://truthout.org/articles/hypocritical-smoke-the-scandal-of-medical-waste-incineration/

https://www.wastedive.com/news/oregon-covanta-facility-to-begin-accepting-new-medical-waste-from-stericycl/424481/

https://www.huffpost.com/entry/utah-medical-waste-incinerator-stericycle-pollution_n_3909083

http://www.nhlink.net/enviro/scp/medical.html

https://www.wsoctv.com/news/local/matthews-biomedical-waste-incinerator-closing-door/334741693

https://www.wfae.org/post/inside-meckcos-only-medical-waste-incinerator

https://www.govinfo.gov/content/pkg/FR-2013-05-13/pdf/2013-09427.pdf

https://lasvegassun.com/news/2019/mar/13/controversial-medical-waste-incinerator-las-vegas/

https://www.wastedive.com/news/oregon-covanta-facility-to-begin-accepting-new-medical-waste-from-stericycl/424481/

https://p2infohouse.org/ref/15/14349.pdf

http://www.seas.columbia.edu/earth/wtert/sofos/llewellyn_thesis.pdf

https://thenevadaindependent.com/article/proposal-to-build-medical-waste-incinerator-at-industrial-park-faces-opposition

https://www.waste360.com/medical-waste/how-healthcare-industry-addressing-fast-growing-medical-waste-problem

https://lasvegassun.com/news/2019/mar/13/controversial-medical-waste-incinerator-las-vegas/

http://www.bredl.org/pdf2/Stericycle_Overview_of_Stericycle.pdf

https://www.plasticstoday.com/medical/path-recycling-medical-plastic-waste-down-under/143541910360208

https://www.who.int/news-room/fact-sheets/detail/dioxins-and-their-effects-on-human-health

https://www.kuer.org/post/north-salt-lake-residents-upset-about-unfiltered-emissions-medical-waste-incinerator

https://www.statesmanjournal.com/story/news/2016/06/24/medical-waste-could-mean-cash-marion-county/86181770/

https://www.foxnews.com/us/tissue-from-aborted-fetuses-used-to-power-oregon-homes

https://www.fredericknewspost.com/news/environment/fort-detrick-considering-new-incinerator/article_dbdca424-0be0-575c-a029-825c6aba80cd.html

10. Hazardous waste disposal and transport
https://www.sbsun.com/2019/10/10/rialto-company-that-recycles-hazardous-waste-gets-unacceptable-grade-from-state-regulators/
https://www.freep.com/story/news/local/michigan/2014/09/07/landfill-taking-radioactive-waste-has-history-of-violations-leaks-fires/15210857/
https://www.freep.com/story/news/local/michigan/2014/09/07/landfill-taking-radioactive-waste-has-history-of-violations-leaks-fires/15210857/
https://www.theguardian.com/us-news/2015/oct/25/radioactive-waste-dump-fire-reveals-nevada-troubled-past
https://www.seattletimes.com/seattle-news/tacoma-hazardous-waste-operation-hit-with-1-9-million-state-fine-for-2018-fire/
https://m.metrotimes.com/news-hits/archives/2020/01/31/hazardous-waste-plant-in-detroit-with-history-of-serious-violations-gets-state-approval-to-expand-almost-ninefold
https://www.investigativepost.org/2018/04/05/6836/
http://chej.org/wp-content/uploads/Deep%20Well%20Injection%20-%20PUB%20056.pdf
https://www.epa.ohio.gov/ddagw/uic
http://www.dnr.louisiana.gov/assets/OC/im_div/uic_sec/EPAposterofwells.pdf
https://laist.com/2019/05/02/exide_cleanup_contaminants_lawsuit.php
https://www.newburyportnews.com/news/local_news/salisbury-company-fined-for-hazardous-waste-handling/article_a77794a4-79b7-597d-9b29-87bc34845b94.html
https://www.abqjournal.com/1386978/state-fines-abq-hazardous-waste-storage-business-360285-as-part-of-settlement.htm
lhttps://www.webwire.com/ViewPressRel.asp?aId=49217
https://www.ttnews.com/articles/wiley-sanders-truck-lines-fined-3-million-illegal-transport-hazmat-materials-california
https://www.registerguard.com/article/20140203/NEWS/302039922
http://www.startribune.com/clock-is-ticking-on-andover-landfill-and-chemicals-are-leaking/506400592/
https://www.metrowestdailynews.com/news/20200207/state-will-clean-up-part-of-framinghams-general-chemical-site-this-year
https://www.tribuneledgernews.com/local_news/investigators-search-for-truck-driver-believed-to-have-dumped-waste/article_cb38fee8-363d-11ea-b2b9-43c2ba371097.html
https://www.fedcenter.gov/Bookmarks/index.cfm?id=8706&pge_prg_id=29624&pge_id=1868
https://www.envcap.org/srl/
https://www.epa.gov/hwpermitting/hazardous-waste-management-facilities-and-units
http://www.ehso.com/cssepa/tsdfincin.php

11. Factory farms
https://publicintegrity.org/environment/farming-activity-contaminates-water-despite-best-practices/
https://www.delawareonline.com/story/news/local/2018/03/28/allen-harim-faces-241-000-fine-previous-water-pollution/467802002/
https://www.delmarvanow.com/story/news/local/2018/03/20/sussex-residents-protest-chicken-company-plans-millsboro-area-plant/439852002/
https://www3.epa.gov/npdes/pubs/sector_table.pdf
https://libraryguides.law.pace.edu/factoryfarming
https://en.wikipedia.org/wiki/Concentrated_animal_feeding_operation
https://www.theguardian.com/books/2015/sep/25/industrial-farming-one-worst-crimes-history-ethical-question
https://www.thebodypro.com/article/mycobacterium-avium-complex-mac#cause
https://www.cdc.gov/nceh/ehs/docs/understanding_cafos_nalboh.pdf
https://www.cdc.gov/nceh/ehs/docs/understanding_cafos_nalboh.pdf

https://www.farmsanctuary.org/learn/factory-farming/chickens/
https://www.foodandwaterwatch.org/insight/factory-farm-nation-2015-edition
https://thinkprogress.org/5-big-meat-companies-produce-a-combined-162-million-tons-of-manure-each-year-c3acced8f51e/
https://www.sentienceinstitute.org/us-factory-farming-estimates
https://www.newyorker.com/news/dispatch/after-florence-manure-lagoons-breach-and-residents-brace-for-the-rising-filth
https://eos.org/editors-vox/is-living-near-a-farm-bad-for-your-health
https://madison.com/wsj/news/local/govt-and-politics/dna-evidence-traces-drinking-water-hazards-back-to-farms-and/article_daf21447-0b24-5b0e-a1a6-ab3f5f17094c.html
https://eos.org/editors-vox/is-living-near-a-farm-bad-for-your-health
https://www.michfb.com/mi/news/policy_and_politics/legislation_introduced_to_ban_wintertime_spreading/
https://www.ecowatch.com/factory-farms-antibiotic-resistance-2641508323.html?rebelltitem=3#rebelltitem3
https://publicintegrity.org/environment/farming-activity-contaminates-water-despite-best-practices/
https://www.jhsph.edu/news/news-releases/2012/feingold-livestock-associated-mrsa.html
https://americanpromise.net/blog/2019/06/18/big-money-the-farm-bill-and-family-farms-vs-industrial-agriculture/
https://health.howstuffworks.com/human-body/systems/digestive/smell-poop-inhaling-poop-molecules.htm
https://www.motherjones.com/food/2015/01/dust-factory-farms-antibiotic-resistant-genes/
https://eos.org/editors-vox/is-living-near-a-farm-bad-for-your-health
https://www.foodandwaterwatch.org/sites/default/files/factory-farm-nation-report-may-2015.pdf
https://www.scientificamerican.com/article/how-drug-resistant-bacteria-travel-from-the-farm-to-your-table/
https://www.mlive.com/news/2020/04/new-michigan-factory-farm-permit-restricts-winter-manure-application.html
https://www.nytimes.com/1983/10/15/us/dow-halts-fight-to-sell-herbicide.html
https://qz.com/433750/the-world-eats-cheap-bacon-at-the-expense-of-north-carolinas-rural-poor/
https://foodprint.org/issues/how-industrial-agriculture-affects-our-air/
https://www.foodandwaterwatch.org/insight/factory-farm-nation-2015-edition
https://www.ocj.com/2015/02/a-cautionary-tale-of-manure-insurance-and-the-neighbors-well/
https://foodprint.org/issues/how-industrial-agriculture-affects-our-air/
https://qz.com/433750/the-world-eats-cheap-bacon-at-the-expense-of-north-carolinas-rural-poor/
https://foodandwaterwatch.org/problems/factory-farming-food-safety
https://e360.yale.edu/features/as_dairy_farms_grow_bigger_new_concerns_about_pollution
https://www.foodandwaterwatch.org/sites/default/files/factory-farm-nation-report-may-2015.pdf
https://www.cdc.gov/nceh/ehs/docs/understanding_cafos_nalboh.pdf
https://www.npr.org/sections/thesalt/2018/08/29/642646707/investigators-track-contaminated-lettuce-outbreak-to-a-cattle-feedlot
https://www.fda.gov/food/outbreaks-foodborne-illness/environmental-assessment-factors-potentially-contributing-contamination-romaine-lettuce-implicated
https://www.centerforfoodsafety.org/press-releases/5718/epa-stands-behind-conclusion-that-industrial-dairies-contaminate-lower-yakima-valley-drinking-water

https://lacrossetribune.com/news/local/millon-gallon-manure-spill-fouls-root-river-in-minnesota/article_1d32bf50-a715-11e2-b4ac-001a4bcf887a.html
https://www.centerforfoodsafety.org/press-releases/5718/epa-stands-behind-conclusion-that-industrial-dairies-contaminate-lower-yakima-valley-drinking-water
12. Nuclear reactors
https://www.eia.gov/tools/faqs/faq.php?id=207&t=3
https://www.energy.gov/ne/articles/nuclear-101-how-does-nuclear-reactor-work
https://www.statista.com/statistics/267158/number-of-nuclear-reactors-in-operation-by-country/
https://www.bbc.com/news/world-13047267
https://www.cnbc.com/2011/03/16/11-Nuclear-Meltdowns-and-Disasters.html
http://www.lutins.org/nukes.html
https://psmag.com/environment/50-years-after-nuclear-meltdown-3510
http://www.beyondnuclear.org/routine-releases/2013/5/7/entergys-palisades-leaks-79-gallons-of-radioactive-water-int.html
https://www.motherearthnews.com/nature-and-environment/living-near-nuclear-power-plants-part-2-zmaz78jazgoe
https://projects.bettergov.org/power-struggle/leaks.html
http://www.beyondnuclear.org/routine-releases/
https://www.nrc.gov/about-nrc/regulatory/research/soar.html
https://www.citizen.org/news/how-close-is-your-home-to-a-nuclear-power-plant/
http://www.environmentandsociety.org/tools/keywords/brookhaven-report-wash-740
https://study.com/academy/lesson/risks-of-nuclear-power-plants-and-radioactive-waste-safety-and-health-concerns.html#transcriptHeader
https://study.com/academy/lesson/risks-of-nuclear-power-plants-and-radioactive-waste-safety-and-health-concerns.html#transcriptHeader
https://www.theworldincrisis.com/possible-threat-to-us-nuclear-plants-from-boone-dam-failure/
https://noozsaurus.com/is-another-fukushima-nuclear-disaster-brewing-in-eastern-tennessee/
https://www.wbir.com/article/tech/science/environment/threats-found-in-major-dams-in-tennessee/408534583
https://sfbayview.com/2013/04/sacramento-cancer-rate-dropped-after-shutdown-of-rancho-seco-reactor-4319-cancer-cases-prevented/
http://www.ca.allgov.com/news/controversies/less-cancer-found-after-rancho-seco-nuclear-plant-closes-130402?news=849611
https://www.hcn.org/articles/is-yucca-mountain-back-from-the-dead
https://www.pressofatlanticcity.com/news/top_three/oyster-creek-s-solution-to-tritium-leak-water-it-down/article_6b45b29a-dfeb-11df-9777-001cc4c03286.html
https://www.pressofatlanticcity.com/news/top_three/oyster-creek-s-solution-to-tritium-leak-water-it-down/article_6b45b29a-dfeb-11df-9777-001cc4c03286.html
http://nukefreetexas.org/downloads/health_risks_nuclear_power.pdf
https://www.nukepills.com/nuclear-reactor-maps/
https://www.timesonline.com/article/20130328/News/303289882
https://cdn.allthingsnuclear.org/wp-content/uploads/2014/10/20140818-Routine-Releases-Rev-1.pdf
https://www.ucsusa.org/resources/regulatory-roulette
https://www.nrc.gov/reactors/operating/ops-experience/tritium/plant-info.html
https://cdn.allthingsnuclear.org/wp-content/uploads/2014/10/20140818-Routine-Releases-Rev-1.pdf
https://www.ucsusa.org/resources/regulatory-roulette
https://www.pressofatlanticcity.com/news/top_three/oyster-creek-s-solution-to-tritium-leak-water-it-down/article_6b45b29a-dfeb-11df-9777-001cc4c03286.html

https://www.stltoday.com/news/local/illinois/investigation-radioactive-leaks-at-illinois-nuclear-plants/article_5afd12ac-e54b-5b20-be98-b72233c1075c.html

13. Radioactive waste

https://www.rollingstone.com/politics/politics-features/oil-gas-fracking-radioactive-investigation-937389/

https://www.thestate.com/news/local/environment/article242667201.html

https://fortress.wa.gov/ecy/publications/documents/1505013.pdf

https://books.google.com/books?id =LCBoDwAAQBAJ&pg=PT865&lpg=PT865&dq=accident+at+waste+control+specialists+and rews+texas&source=bl&ots=QhW6btq1qC&sig=ACfU3U2V0uM_pmtQyd5WZFzc_DNoFrY3IA &hl=en&sa=X&ved=2ahUKEwiknvXn8bDpAhUKr54KHZqJDPEQ6AEwBXoECAoQAQ#v=onepage&q=accident%2520at%2520waste%2520control%2520specialists%2520andrews%2520texas&f=false

https://www.sltrib.com/news/environment/2018/01/11/utah-fines-radioactive-waste-handler-50k-for-failure-to-control-dust-but-its-not-the-dust-that-worries-environmentalists/

https://www.thestate.com/news/politics-government/article228487274.html

https://www.npr.org/blogs/thetwo-way/2014/05/23/315279895/organic-kitty-litter-chief-suspect-in-nuclear-waste-accident

https://www.reviewjournal.com/local/local-nevada/a-year-after-fiery-accident-at-radioactive-waste-dump-in-nevada-the-meter-is-running-on-a-fix/

https://www.theguardian.com/us-news/2015/oct/25/radioactive-waste-dump-fire-reveals-nevada-troubled-past

https://www.spokesman.com/stories/2019/mar/11/state-is-taking-a-new-look-at-this-richland-radioa/

https://lasvegassun.com/news/2011/feb/13/yucca-or-not-nuclear-waste-resides-here/

https://aeon.co/essays/downwinders-the-noxious-legacy-of-the-hanford-nuclear-site

https://www.thespectrum.com/story/news/2019/12/18/nevada-us-government-runs-secret-plutonium-smuggling-near-las-vegas/2689768001/

https://www.thestate.com/news/local/environment/article236794528.html

https://www.nrc.gov/reading-rm/doc-collections/fact-sheets/blending-llrw.html

https://www.courthousenews.com/feds-give-thumbs-up-to-west-texas-nuke-waste-plan/

https://www.nrc.gov/reading-rm/doc-collections/fact-sheets/blending-llrw.html

https://www.latimes.com/nation/la-na-hanford-plutonium-exposure-20180330-story.html

https://www.keizertimes.com/posts/1238/radioactive-waste-in-oregon

https://www.newhavenindependent.org/index.php/archives/entry/nuclear_corp/

https://www.latimes.com/nation/la-na-hanford-plutonium-exposure-20180330-story.html

https://violationtracker.goodjobsfirst.org/parent/waste-management

https://www.thestate.com/news/local/environment/article240309966.html

https://publicherald.org/government-failed-you-pittsburgh-state-rep-drafts-bill-to-stop-radioactive-fracking-waste-tenorm-from-entering-public-waters/

https://www.currentargus.com/story/news/local/2018/06/13/holtec-allay-fears-nuclear-site-near-carlsbad/695442002/

https://www.currentargus.com/story/news/local/2018/03/26/new-energy-fuel-waste-nuclear-storage-texas-new-mexico-facility/454677002/

https://www.nirs.org/wp-content/uploads/factsheets/llwnolowrisk.pdf

https://www.nrc.gov/reading-rm/doc-collections/fact-sheets/radwaste.html

https://www.dm.usda.gov/ohsec/rsd/radwaste.htm

https://www.nirs.org/radioactive-waste/llw/

https://www.conserve-energy-future.com/types-of-radioactive-waste.php

https://www.nrc.gov/reading-rm/doc-collections/maps/radioactive-waste-sites.html

https://www.nrc.gov/waste/llw-disposal/licensing/locations.html
https://www.nrc.gov/reading-rm/doc-collections/maps/radioactive-waste-sites.html
https://www.kxlh.com/news/missoula-county/radioactive-fracking-waste-could-end-up-in-missoula-landfill
https://www.smithsonianmag.com/science-nature/radioactive-wastewater-from-fracking-is-found-in-a-pennsylvania-stream-351641/
https://stateimpact.npr.org/pennsylvania/2019/09/11/how-did-fracking-contaminants-end-up-in-the-monongahela-river-a-loophole-in-the-law-might-be-to-blame/
https://bismarcktribune.com/news/state-and-regional/alexander-landfill-operator-found-with-radioactive-waste/article_9839ebd9-ae50-5e23-af1b-ed53d20da3b6.html
https://www.rgj.com/story/news/politics/2019/07/10/sisolak-slams-illegal-nuclear-waste-delivery-doe-rick-perry-yucca-las-vegas-nevada/1697983001/
https://news.stlpublicradio.org/post/confused-about-bridgeton-west-lake-landfills-heres-what-you-should-know
https://www.nrdc.org/onearth/small-towns-battle-against-radioactive-fracking-waste
https://sf.curbed.com/2019/9/20/20875795/treasure-island-radioactive-toxic-contaminated-dirt-home-navy
https://www.grandforksherald.com/news/radioactive-waste-north-dakota-monitor
https://www.businessinsider.com/radioactive-object-found-in-8-billion-san-francisco-neighborhood-2018-9
http://www.earthisland.org/journal/index.php/articles/entry/kentucky-towns-illegally-dumped-radioactive-waste/
https://abc6onyourside.com/news/local/traces-of-carcinogen-found-at-school-leads-to-concerns-about-nuclear-waste-facility
http://www.washingtonpost.com/wp-srv/national/daily/aug99/paducah29.htm
https://www.nytimes.com/1979/02/16/archives/radioactive-uranium-dump-discovered-under-denver-radium-was-for.html
https://sfbayview.com/2016/09/treasure-island-residents-choking-on-toxic-dust-released-into-high-winds-as-navy-excavations-fast-track-island-redevelopment/
https://earther.gizmodo.com/this-town-didnt-want-to-be-a-radioactive-waste-dump-th-1834789264
https://www.freep.com/story/news/local/michigan/2018/11/23/fracking-waste-radioactive-michigan/2072103002/
https://www.thestate.com/latest-news/article209093444.html
http://uptownmessenger.com/2019/07/hazardous-waste-containers-removed-in-gert-town-city-and-state-report/
https://www.thestate.com/latest-news/article209093444.html
https://www.thedailybeast.com/energy-department-plans-to-reclassify-high-level-toxic-waste-as-low-level-angering-environmental-advocates

14. Landfills

https://www.danielshealth.com/knowledge-center/disease-from-improper-disposal
https://newstalk870.am/state-plans-another-big-cleanup-at-toxic-pasco-landfill/
https://apps.ecology.wa.gov/gsp/Sitepage.aspx?csid=1910
https://www.tennessean.com/story/news/2019/04/15/decatur-county-landfill-lawsuit-toxic-leachate/3425243002/
https://www.sfchronicle.com/science/article/Toxic-soil-went-from-SF-s-Hunters-Point-to-12854269.php#photo-15426875
http://chej.org/wp-content/uploads/LandfillFailures20191.pdf
https://www.heraldmailmedia.com/news/usa_today/a-landfill-operator-tried-to-abandon-an-environmental-disaster-now-a-small-town-is-fighting/article_8d60cf64-000b-52b0-9015-c4f76169b4aa.html
https://www.who.int/water_sanitation_health/medicalwaste/002to019.pdf
https://www.mwatoday.com/news/garbage/weather-affects-the-landfill.aspx
https://www.tri-cityherald.com/news/local/article218736290.html

https://www.scdhec.gov/environment/your-land/landfills-overview/how-landfills-work

http://www.fox9.com/news/6-600-drums-of-toxic-waste-to-be-removed-from-former-wde-landfill-in-andover

https://www.courant.com/news/connecticut/hc-news-pfas-landfill-worries-20190815-ozvn6kfidfhu5apowususpmusy-story.html

https://www.courant.com/news/connecticut/hc-news-pfas-landfill-worries-20190815-ozvn6kfidfhu5apowususpmusy-story.html

https://newstalk870.am/state-plans-another-big-cleanup-at-toxic-pasco-landfill/

https://krocnews.com/olmsted-county-preparing-to-expand-kalmar-ash-landfill/

https://www.sltrib.com/news/politics/2019/08/19/landfill-tip-promontory/

https://www.krdo.com/news/locals-worried-about-an-old-landfill-contaminating-mesa-creek/1107307553

http://indepthnh.org/2019/08/16/new-law-compels-state-action-at-coakley-landfill/

https://www.gazettenet.com/ENVLeverettWater-25206303

https://signalscv.com/2019/08/companies-plead-no-contest-in-waste-cover-up/

https://www.sciencedaily.com/releases/2016/05/160524211817.htm

https://www.courant.com/news/connecticut/hc-news-pfas-ellington-well-20190913-i36bkl7vzfhnpdds4j2gkgrxd4-story.html

https://www.tennessean.com/story/news/2019/04/15/decatur-county-landfill-lawsuit-toxic-leachate/3425243002/

https://www.readingeagle.com/news/article/in-boyertown-bad-vibrations-over-potential-uptick-in-truck-traffic

https://agruamerica.com/hdpe-liner-history/

https://www.wgbh.org/news/2016/04/15/news/contaminated-wells-force-central-mass-residents-live-bottled-water-months

https://billingsgazette.com/news/state-and-regional/montana/western-n-d-landfill-operator-found-with-radioactive-waste/article_2bff9dd0-a96e-5980-85ad-4d105cd8241b.html

https://www.stltoday.com/news/local/metro/missouri-identifies-new-groundwater-contamination-at-and-surrounding-bridgeton-landfill/article_f0446341-8449-541e-add0-bd35ad0d5bcf.html

http://chej.org/wp-content/uploads/LandfillFailures20191.pdf

https://www.theguardian.com/environment/2020/feb/01/like-movie-students-uncovered-toxic-waste-scandal

https://www.gwinnettdailypost.com/archive/old-dump-blamed-for-explosion/article_1209a30f-626f-5ea8-8111-02209cb2a274.html

https://lbpost.com/news/mobile-home-sinking-closed-north-long-beach-landfill-lawsuit/

https://www.wivb.com/news/plaintiffs-in-nash-road-landfill-lawsuit-get-favorable-ruling/

https://www.miamiherald.com/news/local/community/miami-dade/article239154158.html

https://www.mlive.com/news/grand-rapids/2016/08/landfill_gas_could_cause_explo.html

https://news.stlpublicradio.org/post/bridgeton-landfill-pay-16-million-settle-state-lawsuit-over-underground-smoldering-fire

https://www.vcstar.com/story/news/2017/09/29/hospitals-sent-untreated-medical-waste-landfill-records-show/584227001/

https://keyt.com/news/ventura-county/2020/01/29/businessman-pled-guilty-for-illegal-disposal-of-solid-waste-in-ventura-county/

https://www.post-gazette.com/business/healthcare-business/2016/06/21/UPMC-fined-451-000-for-illegal-disposal-of-medical-waste/stories/201606210141

https://www.nps.gov/subjects/air/humanhealth-sulfur.htm

http://chej.org/wp-content/uploads/LandfillFailures20191.pdf

https://weather.com/science/environment/news/2020-02-20-oregon-landfill-illegally-accepted-2-million-pounds-of

https://www.danielshealth.com/knowledge-center/disease-from-improper-disposal

https://www.health.ny.gov/environmental/outdoors/air/landfill_gas.htm

https://www.sciencedaily.com/releases/2016/05/160524211817.htm

https://weather.com/science/environment/news/2020-02-20-oregon-landfill-illegally-accepted-2-million-pounds-of

http://www.gfredlee.com/Landfills/Comblandfillsupdate.pdf

https://www.google.com/search?sxsrf=ALeKk02X1XPY8G3DpkaXO4GuJS7vjfusTw%253A1588884223642&ei=_3K0XobjJtfh-gSDxlj4BA&q=health+effects+of+endotoxins&oq=health+effects+of+endotoxins&gs_lcp=CgZwc3ktYWIQAzIGCAAQFhAeMgYIABAWEB46BQgAEJECOgUIABCDAToCCABQxokCWPa9AmCjvwJoAHAAeAGAAe8CiAGcM5IBCDIuMC4yMy40mAEAoAEBqgEHZ3dzLXdperABAA&sclient=psy-ab&ved=0ahUKEwjG3vT9zqLpAhXXsJ4KHQMiAk8Q4dUDCAs&uact=5

https://www.nps.gov/subjects/air/humanhealth-sulfur.htm

https://learn.kaiterra.com/en/air-academy/sulfur-oxides-sox

http://www.gfredlee.com/Landfills/Comblandfillsupdate.pdf

https://www.google.com/search?sxsrf=ALeKk02X1XPY8G3DpkaXO4GuJS7vjfusTw%253A1588884223642&ei=_3K0XobjJtfh-gSDxlj4BA&q=health+effects+of+endotoxins&oq=health+effects+of+endotoxins&gs_lcp=CgZwc3ktYWIQAzIGCAAQFhAeMgYIABAWEB46BQgAEJECOgUIABCDAToCCABQxokCWPa9AmCjvwJoAHAAeAGAAe8CiAGcM5IBCDIuMC4yMy40mAEAoAEBqgEHZ3dzLXdperABAA&sclient=psy-ab&ved=0ahUKEwjG3vT9zqLpAhXXsJ4KHQMiAk8Q4dUDCAs&uact=5

https://www.nps.gov/subjects/air/humanhealth-sulfur.htm

https://www.northjersey.com/story/news/hudson/2019/09/13/agency-says-reduced-fumes-kearny-nj-landfill-mayor-not-convinced/2310584001/

https://www.danielshealth.com/knowledge-center/disease-from-improper-disposal

https://www.health.ny.gov/environmental/outdoors/air/landfill_gas.htm

https://www.sciencedaily.com/releases/2016/05/160524211817.htm

15. Airports

https://www.indoordoctor.com/health-issues-caused-by-living-near-an-airport/

https://catalog.camfil.us/living-near-an-airport-could-be-a-matter-of-life-and-death.html

https://www.nhs.uk/news/lifestyle-and-exercise/living-near-an-airport-may-be-bad-for-your-health/

https://www.lung.org/our-initiatives/healthy-air/outdoor/air-pollution/highways.html

https://www.businessinsider.com/high-blood-pressure-if-living-near-airports-2017-6

https://www.faa.gov/airports/environmental/airport_noise/noise_exposure_maps/

https://www.minnpost.com/second-opinion/2014/06/airport-pollution-may-have-been-seriously-underestimated-study-suggests/

https://www.indoordoctor.com/health-issues-caused-by-living-near-an-airport/

https://www.sun-sentinel.com/news/fl-xpm-1987-08-30-8703100429-story.html#

https://www.ajc.com/business/atlanta-airport-plans-prompt-new-push-noise/3KADEqIJY7u43zkESo9f6M/

https://yourvalley.net/yourvalley/government/sky-harbor-send-planes-glendale/

https://www.nhs.uk/news/lifestyle-and-exercise/living-near-an-airport-may-be-bad-for-your-health/

https://www.nbclosangeles.com/news/local/FAA-Changes-to-Flight-Pattern-Over-Mid-City-Central-LA-511758182.html

https://www.skyharbor.com/about/Information/AirportFacts

https://thequietcoalition.org/airplane-noise-health-hazard/

https://www.baltimoresun.com/opinion/op-ed/bs-ed-op-0812-air-noise-20190812-d5sf3qsbh5c3jk7q3jp4lr7sgu-story.html

https://www.newhavenindependent.org/index.php/archives/entry/tweed_CAA/

https://www.baltimoresun.com/opinion/op-ed/bs-ed-op-0812-air-noise-20190812-d5sf3qsbh5c3jk7q3jp4lr7sgu-story.html

https://www.minnpost.com/second-opinion/2014/06/airport-pollution-may-have-been-seriously-underestimated-study-suggests/

https://pubs.acs.org/doi/full/10.1021/es5001566

https://www.forbes.com/sites/larryhusten/2013/10/08/people-who-live-near-airports-at-increased-risk-for-cardiovascular-disease/#8efd1a14141b

https://www.ncbi.nlm.nih.gov/pmc/articles/PMC5437751/

https://www.thoughtco.com/health-effects-of-airport-noise-and-pollution-1204091

http://www.us-caw.org/

https://inhabitat.com/infographic-the-toxic-dangers-of-living-near-an-airport/

https://pubs.acs.org/doi/pdf/10.1021/es5001566

http://www.restud.com/wp-content/uploads/2015/09/MS17397manuscript.pdf

http://www.restud.com/wp-content/uploads/2015/09/MS17397manuscript.pdf

https://www.dailymail.co.uk/health/article-6355039/Living-near-noisy-road-airport-increases-risk-heart-attack-stroke-THREE-TIMES.html

https://airwaysmag.com/airports/san-luis-obispo-regional-airport-opens-new-terminal/

https://www.sanluisobispo.com/news/local/article232906067.html

https://www.wsoctv.com/news/local/neighbors-to-deal-with-more-noise-as-airport-construction-shifts-flight-paths/935921996/

https://www.biologicaldiversity.org/news/press_releases/2019/sonoma-county-lead-pollution-04-30-2019.php

https://www.mtexpress.com/news/bellevue/bellevue-moves-to-protect-water-sources-from-airport-contamination/article_d024397c-582f-11ea-ab18-3b221bef8ddc.html

https://www.newhavenindependent.org/index.php/archives/entry/tweed_appeal/

https://www.seattletimes.com/business/boeing-aerospace/sea-tac-airport-expansion-to-gather-speed-as-surrounding-communities-feel-the-impact/

https://www.motherjones.com/politics/2013/01/private-planes-still-use-leaded-gasoline/

https://catalog.camfil.us/living-near-an-airport-could-be-a-matter-of-life-and-death.html

https://www.acs.org/content/acs/en/pressroom/presspacs/2009/acs-presspac-november-18-2009/hidden-threat-elevated-pollution-levels-near-regional-airports.html

https://www.acs.org/content/acs/en/pressroom/presspacs/2009/acs-presspac-november-18-2009/hidden-threat-elevated-pollution-levels-near-regional-airports.html

Section Two:

16. Chemical companies

https://news.google.com/articles/CBMifWh0dHBzOi8vd3d3Lm5iY25ld3MuY29tL25ld3MvdXMtbmV3cy82Ni1wZW9wbGUtcmVjZWl2aW5nLW1lZGljYWwtYXR0ZW50aW9uLWFmdGVyLWV4eG9ubW9iaWwtcmVmaW5l cnktZXhwbG9zaW9uLXRleGFzLW4xMDM3NTgx0gEsaHR0cHM6Ly93d3cubmJjbmV3cy5jb20vbmV3cy9hbXAvbmNuYTEwMzc1ODE?hl=en-US&gl=US&ceid=US%3Aen

https://www.chemicalsafetyfacts.org/chemistry-context/debunking-myth-chemicals-testing-safety/

https://www.cbsnews.com/news/cancer-alley-reserve-louisiana-denka-plant-health-risk-higher-national-average-2019-07-24/

https://www.humanrightsnetwork.org/waiting-to-die

http://www.earthisland.org/journal/index.php/articles/entry/do-you-live-in-a-chemical-disaster-danger-zone/

https://www.epa.gov/enforcement/exxon-mobil-corporationexxonmobil-oil-corporation-clean-air-act-settlement

https://www.statista.com/topics/1526/chemical-industry-in-the-us/
https://www.thecanadianencyclopedia.ca/en/article/petrochemical-industry
https://theintercept.com/2019/02/01/chemours-genx-north-carolina-netherlands/
https://www.cnn.com/2019/11/27/us/port-neches-plant-explosion/index.html
https://www.foxnews.com/us/fire-rages-at-texas-chemical-plant-50000-people-under-evacuation-orders
https://www.foxnews.com/world/chemical-plant-explosion-spain-triggers-emergency
https://www.gosanangelo.com/story/news/2020/02/24/texas-sues-company-whose-port-neches-chemical-plant-exploded/4858871002/
https://www.wwltv.com/article/news/health/formosa-air-quality-appeal/289-58fb4838-76a3-4c08-8b5f-eb1bdf600886
https://www.nola.com/news/environment/article_39b8e060-58c4-11ea-8597-433bbb66a486.html
https://dfw.cbslocal.com/2020/02/27/houston-chemical-plant-toxic-pollutants-harvey/
https://www.npr.org/sections/health-shots/2018/03/06/583973428/after-decades-of-air-pollution-a-louisiana-town-rebels-against-a-chemical-giant
https://theintercept.com/2019/02/24/epa-response-air-pollution-crisis-toxic-racial-divide/
https://wfpl.org/hexion-faces-fine-for-repeated-toxic-releases-as-neighbors-lose-hope/
https://nypost.com/2020/02/29/south-carolina-factory-to-produce-chemical-banned-by-172-countries/
https://www.thestate.com/news/local/environment/article240600806.html
https://nypost.com/2020/02/29/south-carolina-factory-to-produce-chemical-banned-by-172-countries/
https://www.wnyc.org/story/cleaning-more-100-years-spilled-oil-new-jersey/
https://www.northjersey.com/story/news/watchdog/2018/02/14/dupont-chemical-contaminate-pompton-lakes-homes/947802001/
https://d3n8a8pro7vhmx.cloudfront.net/unitedchurchofchrist/pages/24840/attachments/original/1582721312/FINAL_BreathToThePeople_2.26.2020.pdf?1582721312
https://www.ucc.org/breathtothepeople
https://www.texastribune.org/2018/03/30/arkema-disaster-harvey-regulations-texas-crosby/
http://www.motherjones.com/environment/2014/04/west-texas-hazardous-chemical-map/
https://www.npr.org/sections/health-shots/2017/11/02/558313744/in-the-wake-of-chemical-fires-texans-worry-about-toxic-effects
https://www.arnolditkin.com/crosby-kmco-chemical-plant-explosion-attorneys/
https://www.colorado.edu/law/sites/default/files/attached-files/mulhern_web_edition_pdf.pdf
https://www.nbcboston.com/news/local/explosions-reported-at-manufacturing-plant-in-newburyport-mass/2076248/
https://www.news5cleveland.com/news/local-news/akron-canton-news/all-clear-issued-for-those-who-evacuated-after-akron-chemical-explosion

17. Industrial operations

https://www.northjersey.com/story/news/environment/2020/02/11/polluters-nj-released-6-million-pounds-toxic-chemicals-2018/4723661002/
https://www.thestate.com/news/local/environment/article241068271.html
https://www.sltrib.com/news/politics/2019/07/08/ogden-bought-an-old/
https://news.usc.edu/156523/lead-in-baby-teeth-exide-battery-plant/
https://www.northjersey.com/story/news/environment/2019/03/27/dupont-nj-chemours-over-pompton-lakes-nj-pollution-chambers-works-phil-murphy-nrd/3279022002/

https://www.thestate.com/news/local/environment/article241068271.html
https://wwjnewsradio.radio.com/articles/news/michigan-man-sentenced-illegally-storing-hazardous-waste
https://www.sfgate.com/news/bayarea/article/State-Fines-Steel-Plant-Uss-Posco-In-Lead-Dust-14993638.php
https://www.sgvtribune.com/2020/05/08/quemetco-battery-recycling-plant-pays-600000-for-emissions-violations/
https://www3.epa.gov/ttn/chief/ap42/ch02/final/c02s03.pdf
https://wfpl.org/hexion-faces-fine-for-repeated-toxic-releases-as-neighbors-lose-hope/
https://www.beaumontenterprise.com/news/article/Imelda-cited-in-release-of-almost-100-000-pounds-14465542.php
https://www.epa.gov/hwgenerators/biennial-hazardous-waste-report
https://www.ualrpublicradio.org/post/next-steps-whirlpools-contamination-fort-smith
https://www.news10.com/news/local-news/working-toward-a-clean-water-solution-in-hoosick-falls/
https://www.timesunion.com/local/article/Hoosick-Falls-plant-declared-federal-Superfund-11720248.php
https://gis.cdc.gov/Cancer/USCS/DataViz.html
https://records.tceq.texas.gov/cs/idcplg?IdcService=TCEQ_SEARCH
https://www.epa.gov/national-air-toxics-assessment/2014-nata-assessment-results
https://www.chicagotribune.com/news/breaking/ct-met-sterigenics-epa-ethylene-oxide-testing-20190307-story.html
https://www.chicagotribune.com/news/breaking/ct-met-sterigenics-epa-ethylene-oxide-testing-20190307-story.html
https://www.ehn.org/air-pollution-cancer-michigan-ontario-2645647668.html?rebelltitem=2#rebelltitem2
https://www.kokomotribune.com/news/a-super-polluter-haynes-international-named-one-of-nation-s/article_4a045724-5fe5-11ea-8190-53879ed3dbec.html
https://wicomicocountypollution.weebly.com/industries.html
https://wgntv.com/news/sterigenics-to-close-willowbrook-facility/
https://www.webmd.com/special-reports/ethylene-oxide/20190719/residents-unaware-of-cancer-causing-toxin-in-air
https://www.webmd.com/cancer/news/20191002/county-delays-georgia-sterigenics-reopening
https://www.ajc.com/news/local-govt--politics/fda-urges-state-reopen-sterigenics-residents-skeptical/AhhKxtxbuUHaZHJ3JeGJ7N/
https://www.mdjonline.com/news/chemical-group-testing-seven-georgia-locations-for-ethylene-oxide/article_e6b74102-3d8b-11ea-940d-af49683677b9.html
https://www.mdjonline.com/news/chemical-group-testing-seven-georgia-locations-for-ethylene-oxide/article_e6b74102-3d8b-11ea-940d-af49683677b9.html
https://www.epa.gov/ground-water-and-drinking-water/drinking-water-health-advisories-pfoa-and-pfos
https://www.newsobserver.com/news/business/article235963052.html
18. Fracking
https://www.workers.org/2020/02/46184/
https://www.desmogblog.com/2020/01/23/oil-gas-well-casing-failure-fracking-xto-ohio-blowout
https://stateimpact.npr.org/pennsylvania/2020/06/12/range-resources-pleads-no-contest-to-environmental-crimes-at-southwest-pa-well-sites/
https://www.technologynetworks.com/analysis/news/toxic-chemicals-identified-in-fracking-wastewater-335291
https://beyondpesticides.org/dailynewsblog/2020/06/the-pesticide-atrazine-and-200-other-toxic-chemicals-found-in-fracking-wastewater-contamination-goes-unregulated/

https://www.greenpeace.org/usa/global-warming/issues/fracking/environmental-impacts-water/
https://www.centerforfoodsafety.org/press-releases/5718/epa-stands-behind-conclusion-that-industrial-dairies-contaminate-lower-yakima-valley-drinking-water
https://oilprice.com/Energy/Crude-Oil/The-Real-History-Of-Fracking.html
https://www.nrdc.org/media/2014/141216
https://www.pri.org/stories/2019-07-31/fracking-causes-environmental-damage-and-birth-defects-new-study-shows
https://newsinteractive.post-gazette.com/blog/fracking-and-health-2/
https://www.greenpeace.org/usa/global-warming/issues/fracking/environmental-impacts-water/
https://pittsburgh.cbslocal.com/2019/11/22/pennsylvania-research-funding-fracking-health-impacts-cancer/
https://today.duke.edu/2015/09/frackfoot
https://www.fractracker.org/resources/oil-and-gas-101/
https://www.gao.gov/products/GAO-14-555
https://www.forbes.com/sites/rrapier/2017/04/21/how-the-shale-boom-turned-the-world-upside-down/#199f76f677d2
https://www.forbes.com/sites/rrapier/2017/04/21/how-the-shale-boom-turned-the-world-upside-down/#778e04a477d2
https://psmag.com/environment/epa-failing-properly-oversee-hundreds-thousands-toxic-injection-wells-87407
https://www.watercalculator.org/water-news-events/fracking-wastewater-dispose/
https://www.forbes.com/sites/rrapier/2017/04/21/how-the-shale-boom-turned-the-world-upside-down/#2bd94dde77d2
https://www.greenpeace.org/usa/colorado-fracking-companies-admit-to-major-air-pollution-problem-emissions-rules-proposed/
https://www.forbes.com/sites/rrapier/2017/04/21/how-the-shale-boom-turned-the-world-upside-down/#2bd94dde77d2
https://www.climatechangenews.com/2014/10/30/fracking-may-release-cancer-causing-air-pollution-study/
https://www.propublica.org/article/federal-report-appears-to-undercut-epa-assurances-water-safety-pennsylvania
https://www.livescience.com/34464-what-is-fracking.html
https://cfpub.epa.gov/ncea/hfstudy/recordisplay.cfm?deid=332990
https://toxtown.nlm.nih.gov/chemicals-and-contaminants/volatile-organic-compounds-vocs
https://cfpub.epa.gov/ncea/hfstudy/recordisplay.cfm?deid=332990
https://www.psr.org/wp-content/uploads/2018/05/fracking-and-air-pollution.pdf
https://archive.sltrib.com/article.php?id=56692751&itype=CMSID
https://www.rollingstone.com/politics/politics-features/oil-gas-fracking-radioactive-investigation-937389/
https://extension.psu.edu/water-withdrawals-for-development-of-marcellus-shale-gas-in-pennsylvania
https://www.indybay.org/newsitems/2020/04/04/18832107.php
https://www.fractracker.org/
https://www.businessinsider.com/chemicals-used-in-fracking-2015-5#unknown-chemicals-10
http://www.allgov.com/news/controversies/us-companies-have-dumped-30-trillion-gallons-of-toxic-liquids-into-the-earthare-they-coming-back-to-haunt-us?news=844678
https://www.eastoregonian.com/opinion/columnists/kullberg-radioactive-fracking-waste-a-threat-to-public-health/article_83a69fea-5447-11ea-bc5e-b72d6ce4c76b.html
https://www.eastoregonian.com/opinion/columnists/kullberg-radioactive-fracking-waste-a-threat-to-public-health/article_83a69fea-5447-11ea-bc5e-b72d6ce4c76b.html

https://www.workers.org/2020/02/46184/
https://papost.org/2020/02/11/natural-gas-driller-cited-by-dep-for-fracking-wastewater-spill-in-lycoming-county-now-faces-criminal-charges/
https://www.desmogblog.com/2020/01/23/oil-gas-well-casing-failure-fracking-xto-ohio-blowout
19. Mines
https://www.mining-technology.com/projects/morenci/
https://www.statista.com/statistics/238405/us-gold-production-since-2005/
https://apnews.com/8158167fd9ab4cd8966e47a6dd6cbe96
http://www.theupa.org/uranium_technology/conventional_mining/
https://www.chicagotribune.com/nation-world/ct-us-mining-wastewater-pollution-20190220-story.html
https://www.usgs.gov/special-topic/water-science-school/science/mining-and-water-quality?qt-science_center_objects=0#qt-science_center_objects
https://www.dcourier.com/news/2019/nov/13/epa-officials-update-d-h-residents-about-mine-smel/
https://www.mtexpress.com/news/environment/deq-says-triumph-mine-pollution-is-being-curtailed/article_3396d300-f509-11e9-b89f-1f52af74c41a.html
https://www.washingtonpost.com/national-security/a-tiny-towns-long-struggle-to-rid-itself-of-toxic-waste-reaches-the-supreme-court/2019/11/30/24d0e41e-1084-11ea-9cd7-a1becbc82f5e_story.html
https://missoulian.com/opinion/columnists/montana-conservation-elders-oppose-proposed-smith-river-mine/article_6115f399-7c3b-5d6e-b51d-24c68353f9ca.html
https://www.environment.co.za/mining/effects-of-mining.html
https://www.sciencedaily.com/releases/2008/03/080326201751.htm
https://www.safewater.org/fact-sheets-1/2017/1/23/miningandwaterpollution
https://www.dailyinterlake.com/article/20191201/AP/312019947
https://www.asbestos.com/jobsites/libby/
https://www.safewater.org/fact-sheets-1/2017/1/23/miningandwaterpollution
https://www.chicagotribune.com/nation-world/ct-us-mining-wastewater-pollution-20190220-story.html
http://theconversation.com/before-the-us-approves-new-uranium-mining-consider-its-toxic-legacy-91204
https://helenair.com/news/state-and-regional/u-s-mining-sites-dump-million-gallons-of-fouled-wastewater/article_d127ce8e-71aa-5bff-8f6c-006da54f7275.html
https://www.rgj.com/story/news/2020/04/14/nevada-mine-expansion-native-tribes-protest-groundwater-pumping-plan/2993906001/
20. Mine tailings
https://en.wikipedia.org/wiki/Church_Rock_uranium_mill_spill
https://www.pbs.org/wgbh/frontline/article/tailings-dams-where-mining-waste-is-stored-forever/
https://superfund.arizona.edu/hazards-emphasized
https://www.nbcnewyork.com/news/local/epa-proposes-cleanup-for-ny-town-contaminated-with-arsenic-due-to-mine-from-1800s/1483560/
https://www.kunc.org/post/colorado-rivers-banks-toxic-pile-continues-shrink
https://superfund.arizona.edu/hazards-emphasized
https://www.kunc.org/post/colorado-rivers-banks-toxic-pile-continues-shrink
https://geoinfo.nmt.edu/resources/uranium/mining.html
http://www.tailings.info/basics/tailings.htm
https://www.reuters.com/article/us-mining-tailings-insurance/pollution-insurance-costs-to-jump-for-u-s-tailings-dams-after-vale-disaster-idUSKBN1ZK214
https://tucson.com/news/arizona_news/pima-county-issues-freeport-another-violation-for-dust-from-mine/article_9d922b29-5991-5bd7-8003-c68d88c398f3.html

https://mtstandard.com/news/local/timber-butte-residents-say-not-in-my-backyard-to-mine/article_d6a1853f-dfc4-5460-8afb-9033de3cc854.html

https://www.epa.gov/hw/legislative-and-regulatory-timeline-mining-waste

http://www.bountyheadbebop.com/forum/index.php?topic=24.0

https://www.nbcnewyork.com/news/local/epa-proposes-cleanup-for-ny-town-contaminated-with-arsenic-due-to-mine-from-1800s/1483560/

https://www.mining-technology.com/projects/morenci/

https://www.eia.gov/uranium/production/annual/

https://www.eia.gov/todayinenergy/detail.php?id=39352

https://www.eacourier.com/copper_era/news/freeport-buys-k-acres-of-land-for-mine-waste-storage/article_a1274688-460e-11ea-a913-cf7580bc906c.html

https://www.google.com/search?sxsrf=ALeKk02mZKsEt-vAeubjpl0OnlFpXPuzyw%253A1588855460137&ei=pAK0XvPxB8rY-wTNtao4&q=health+issues+of+heavy+metals+in+blood&oq=health+issues+of+heavy+metals+in+blood&gs_lcp=CgZ wc3ktYWIQAzIFCCEQqwl6DggAEOoCELQCEJoB EOUCOgQIABBDOgUIABCRAjoCCAA6BQgAEIMBOg gIIRAWEB0QHjoG CAAQFhAeOgUIIRCgAToHCCEQChCgAVDU3B5YkKwfYLewH2gCcAB4AI AB7gGIAelikgEHMTEuMjUuMpgBAKABAaoBB2d3cy13aXqwAQY&sclient=psy-ab&ved=0ahUKEwjz07Pq46HpAhVK7J4KHc2aCgcQ4dUDCAs&uact=5

https://www.who.int/news-room/fact-sheets/detail/arsenic

https://www.osha.gov/SLTC/cadmium/healtheffects.html

https://www.denverpost.com/2019/02/20/colorado-mining-industry-water-pollution/

https://news.stlpublicradio.org/post/jefferson-county-epa-test-more-residential-yards-lead-contamination#stream/0

https://nmpoliticalreport.com/2019/10/11/another-mining-spill-on-the-animas-river/

21. Gas and oil wells

https://www.vice.com/en_us/article/pgxbzn/study-links-cancer-to-living-within-a-mile-of-oil-wells

https://oilandgasthreatmap.com/threat-map/

https://www.sciencedaily.com/releases/2018/04/180409103920.htm

https://www.washingtonpost.com/graphics/national/united-states-of-oil/

https://www.cardiovascularbusiness.com/topics/healthcare-economics/living-near-oil-gas-wells-means-worse-cv-health

https://www.desmogblog.com/2020/02/13/los-angeles-vista-hermosa-cap-orphan-oil-wells-leaking-doggr

https://www.sciencedaily.com/releases/2018/04/180409103920.htm

https://theconversation.com/living-near-active-oil-and-gas-wells-in-california-tied-to-low-birth-weight-and-smaller-babies-140034

https://earthworks.org/cms/assets/uploads/archive/files/publications/Oilandgaspollution.pdf

https://www.oilandgaslawyerblog.com/end-road-dish/

http://www.gaslandthemovie.com/whats-fracking/faq/water-used

https://ehp.niehs.nih.gov/doi/full/10.1289/EHP2553

https://www.cpr.org/2019/10/18/everything-you-need-to-know-about-colorados-new-health-study-on-oil-and-gas/

https://thinkprogress.org/study-finds-chemicals-in-residents-living-near-gas-wells-20b8d2134b02/

https://www.epa.gov/sites/production/files/2018-04/documents/ghgemissions_abandoned_wells.pdf

https://thinkprogress.org/study-finds-chemicals-in-residents-living-near-gas-wells-20b8d2134b02/

https://thinkprogress.org/study-finds-chemicals-in-residents-living-near-gas-wells-20b8d2134b02/

http://www.oleantimesherald.com/news/pa-allegheny-national-forest-face-new-wave-of-abandoned-oil-gas-wells/article_ad2d15cc-2e38-506d-997d-d4b28467b36c.html

https://publicintegrity.org/environment/wells-run-dry/deserted-oil-wells-haunt-los-angeles-with-toxic-fumes-and-enormous-cleanup-costs/

https://earthworks.org/cms/assets/uploads/archive/files/publications/FLOWBACK-TXOGAP-HealthReport-lowres.pdf

https://www.oilandgaslawyerblog.com/end-road-dish/

https://www.scientificamerican.com/article/health-effects-of-oil-and-gas-emissions-investigated-in-texas/

https://www.americangeosciences.org/geoscience-currents/spills-oil-and-natural-gas-fields

https://preservethebeartoothfront.com/2014/08/25/gas-flaring-is-dangerous-to-our-health-and-environment-theres-one-way-to-stop-it-locally/

https://earthworks.org/issues/pit_rule/

https://environmentnewmexicocenter.org/reports/nmc/fracking-numbers

22. Highways

https://www.scienceabc.com/humans/why-are-vehicles-exhaust-fumes-harmful-to-humans.html

https://www.fhwa.dot.gov/policyinformation/pubs/hf/pl11028/chapter1.cfm

https://ephtracking.cdc.gov/showProximityToHighways.action

https://www.bostonglobe.com/metro/2016/04/13/new-evidence-dangers-living-near-highways/hVyqTnY4iyn9YRoNSwWtGI/story.html

https://www.statista.com/statistics/859950/vehicles-in-operation-by-quarter-united-states/

https://www.transportation.gov/mission/health/proximity-major-roadways

https://sandiego.urbdezine.com/2015/05/28/what-is-a-safe-distance-to-live-or-work-near-high-auto-emission-roads/

https://www.lung.org/our-initiatives/healthy-air/outdoor/air-pollution/highways.html

https://www.epa.gov/air-research/research-near-roadway-and-other-near-source-air-pollution

https://www.ncbi.nlm.nih.gov/pmc/articles/PMC4243514/

https://www.webmd.com/cancer/lymphoma/news/20041202/low-level-benzene-exposure-may-be-harmful#1

https://www.ncbi.nlm.nih.gov/pmc/articles/PMC6121301/

https://www.uvm.edu/~empact/air/CO_traffic.php3

https://t2conline.com/how-to-improve-the-air-quality-in-your-home/

https://www.latimes.com/local/california/la-me-freeway-pollution-what-you-can-do-20171230-htmlstory.html

https://sandiego.urbdezine.com/2015/05/28/what-is-a-safe-distance-to-live-or-work-near-high-auto-emission-roads/

https://vault.sierraclub.org/sprawl/report04_highwayhealth/

https://dailyddose.wordpress.com/2012/09/25/car-exhaust-fumes-air-pollution-are-contributing-factors-to-increase-in-autism/

https://www.uvm.edu/~empact/air/CO_traffic.php3

(https://www.ncbi.nlm.nih.gov/pmc/articles/PMC6121301/)

https://www.latimes.com/local/california/la-me-freeway-pollution-what-you-can-do-20171230-htmlstory.html

https://www.theaa.com/driving-advice/fuels-environment/emissions

https://airqualitynews.com/2020/04/16/air-pollution-from-brake-dust-may-be-as-harmful-as-diesel-exhaust-on-immune-cells/

23. Agriculture

https://www.sciencedirect.com/topics/agricultural-and-biological-sciences/organochlorine-pesticides

https://www.desertsun.com/story/news/education/2016/05/25/grower-fined-after-pesticides-make-valley-students-sick/84912140/
https://www.ewg.org/enviroblog/2017/08/atrazine-harmful-weedkiller-taints-tap-water-millions-us
https://www.nrdc.org/sites/default/files/atrazine10.pdf
https://www.usgs.gov/special-topic/water-science-school/science/pesticides-groundwater?qt-science_center_objects=0#qt-science_center_objects
https://grist.org/food/you-get-my-drift-pesticides-cause-big-problems-when-they-go-where-theyre-not-wanted/
https://www.harvestpublicmedia.org/post/wind-blown-pesticides-issue-courtrooms-communities-across-us
https://www.yakimaherald.com/news/local/report-farms-orchards-should-warn-neighbors-before-they-spray-chemicals/article_6d0301b8-ff11-11e6-919f-dbecb4a5aac0.html
https://www.wpr.org/residents-concerned-about-pfas-contamination-farm-fields-marinette-peshtigo-area
https://www.sej.org/headlines/sewage-sludge-spreading-leads-farm-groundwater-pfas-contamination
https://www.wpr.org/residents-concerned-about-pfas-contamination-farm-fields-marinette-peshtigo-area
https://americanpromise.net/blog/2019/06/18/big-money-the-farm-bill-and-family-farms-vs-industrial-agriculture/
https://www.farmprogress.com/good-and-bad-applying-sewage-sludge-farmland-1
https://publicintegrity.org/environment/unintended-consequences-farming-fertilizer-climate-health-water-nitrogen/
https://www.pri.org/stories/2020-01-21/farming-s-growing-problem
https://www.politico.com/newsletters/morning-agriculture/2019/10/25/epa-rolls-out-pesticide-zone-rule-781524
http://www.pesticide.org/shelter_in_place_rule_change
http://www.panna.org/our-campaigns/stop-drift
https://thehill.com/policy/energy-environment/467357-trump-administration-rule-to-shrink-exclusion-boundaries-near
https://www.opb.org/news/article/contaminated-soil-lingers-where-apples-once-grew/
https://biologicaldiversity.org/w/news/press-releases/united-states-uses-85-pesticides-outlawed-in-other-countries-2019-06-06/
https://www.ehn.org/op-ed-yes-food-is-grown-in-sewage-waste-thats-a-problem-2641639016.html?rebelltitem=1#rebelltitem1
https://www.opb.org/news/article/washington-orchards-northwest-pesticides-chlorpyrifos/
https://www.hcn.org/issues/42.21/farmings-toxic-legacy
https://www.latimes.com/business/la-fi-pesticide-schools-20171110-story.html
https://thefern.org/2014/03/warning-signs-how-pesticides-harm-the-young-brain/
https://www.latimes.com/archives/la-xpm-2000-jan-02-mn-49877-story.html
https://www.pbs.org/newshour/science/arsenic-laced-soil-lingers-where-children-play-in-washington
https://content.sierraclub.org/grassrootsnetwork/team-news/2018/08/another-reason-biosolids-should-never-be-used-fertilizer-pfas
https://www.sciencemag.org/news/2017/01/why-tomatoes-got-bland-and-how-make-them-sweet-again
http://www.ottawacitizen.com/technology/Rise+mutant+tasteless+tomato/7245790/story.html
https://www.ecowatch.com/dairy-contamination-pfas-sludge-2640973890.html?rebelltitem=1#rebelltitem1

http://www.vancouversun.com/health/
Metro+sewage+fertilizer+Nicola+Valley+arms/11285525/story.html
https://www.livescience.com/57647-why-store-tomatoes-are-tasteless.html
http://www.scielo.br/scielo.php?script=sci_arttext&pid=S0103-90162003000400029
24. Brownfields
https://www.eli.org/sites/default/files/eli-pubs/d9.11.pdf
https://ny.curbed.com/2019/3/28/18282766/
south-street-seaport-howard-hughes-water-street-brownfield-dec
https://www.buildingsaltlake.com/contamination-of-former-dry-cleaning-site-complicates-an-ambitious-development-in-salt-lake-citys-central-9th/
https://www.ncbi.nlm.nih.gov/pmc/articles/PMC1447146/
http://nycosh.org/wp-content/uploads/2015/03/FS-01-Brownfield2.pdf
https://stateimpact.npr.org/pennsylvania/2019/10/16/
how-philadelphia-built-a-neighborhood-on-toxic-soil-of-former-clearview-landfill/
https://buffalonews.com/2020/02/18/
at-tonawanda-coke-there-are-four-cleanups-not-one/
https://thedcline.org/2020/04/24/
decades-of-setbacks-stall-the-development-of-a-landfill-turned-park-in-ward-7/
https://www.tceq.texas.gov/remediation/vcp/vcp.html
https://www.eastbaytimes.com/2019/09/28/pouring-concrete-over-toxic-soil-where-4000-apartments-may-rise-is-not-ok-activists-tell-richmond-council/
http://indiana.gov/ifa/brownfields/files/Brownfields_Basics_rev_12-4-09.pdf
https://observer-reporter.com/news/localnews/contamination-monitoring-continues-near-washington-brownfield-site/article_586b1a12-485e-11ea-a852-032852c44f72.html
http://indiana.gov/ifa/brownfields/files/Brownfields_Basics_rev_12-4-09.pdf
http://southernbrownfield.blogspot.com/2010/08/austin-tx-crestview-transit-station.html
25. Sewage sludge
https://wtop.com/loudoun-county/2018/10/
stink-raised-over-use-of-biosolids-in-loudoun-co/
https://earthjustice.org/our_work/cases/2013/
cleaning-up-sewage-sludge-incinerators
https://www.epa.gov/stationary-sources-air-pollution/
sewage-sludge-incineration-units-ssi-new-source-performance
http://www.sludgefacts.org/Ref98.pdf
https://www.epa.gov/stationary-sources-air-pollution/sewage-sludge-incineration-units-ssi-new-source-performanhttps://www.federalregister.gov/documents/2016/04/29/2016-09292/federal-plan-requirements-for-sewage-sludge-incineration-units-constructed-on-or-before-october-14ce
https://www.federalregister.gov/documents/2016/04/29/2016-09292/federal-plan-requirements-for-sewage-sludge-incineration-units-constructed-on-or-before-october-14
https://www.ncbi.nlm.nih.gov/pmc/articles/PMC4232481/
https://www.epa.gov/newsreleases/manchester-new-hampshire-municipal-waste-incinerator-reduce-mercury-emissions-under
https://extension.psu.edu/what-is-sewage-sludge-and-what-can-be-done-with-it
https://epa.ohio.gov/portals/51/pdf/Sludge%2520advisory%2520final.pdf
https://www.epa.gov/biosolids/frequent-questions-about-biosolids
http://css.umich.edu/factsheets/us-wastewater-treatment-factsheet
https://extension.psu.edu/what-is-sewage-sludge-and-what-can-be-done-with-it
https://www.hindawi.com/journals/jeph/2016/8467023/
https://healthfully.com/dangers-near-wastewater-treatment-facility-7648140.html
https://www.dovertownship.org/faqs/what-is-sewage-and-where-does-it-come-from/
https://www.ncbi.nlm.nih.gov/pmc/articles/Phttps://www.lacitysan.org/san/faces/home/portal/s-lsh-sp/s-lsh-sp-tire;jsessionid=HfZAEIdQOWd2tb3aMI9k9jB6digYPXmaM

EmpK8CBbmTORdBEDLbn!957961601!-786936155?_afrLoop=1502553446395623&_afrWindowMode=0&_afrWindowId=null&_adf.ctrl-state=2uwnlgm8x_1#!%2540%2540%253F_afrWindowId%253Dnull%2526_afrLoop%253D1502553446395623%2526_afrWindowMode%253D0%2526_adf.ctrl-state%253D2uwnlgm8x_5MC4232481/
https://www.sfgate.com/nation/article/Concerns-grow-over-tainted-sewage-sludge-spread-14435391.php
https://apnews.com/28f5b0563b1441b6ac5d29f7dfd7f5f7/%2527Poop-Train%2527-full-of-NYC-sewage-raises-stink-in-Alabama-town
https://vtdigger.org/2019/05/06/landfill-owner-blames-odor-complaint-spike-tactics-expansion-opponents/
https://www.sfgate.com/nation/article/Concerns-grow-over-tainted-sewage-sludge-spread-14435391.php
https://www.sej.org/headlines/sewage-sludge-spreading-leads-farm-groundwater-pfas-contamination
https://vtdigger.org/2012/10/01/out-of-state-sludge-shipped-to-moretown-landfill-raises-questions-about-expansion-proposal/
https://abc3340.com/news/abc-3340-news-iteam/sewer-sludge-complaints-growing-as-the-tanker-dumping-goes-into-overdrive
https://www.fosters.com/article/20100108/gjopinion02/701089997
https://www.who.int/news-room/fact-sheets/detail/mercury-and-health
https://www.mauinews.com/news/local-news/2020/04/maui-county-loses-supreme-court-case-over-sewage-plant-injection-wells/
https://abc3340.com/news/abc-3340-news-iteam/adem-moves-to-regulate-the-dumping-of-sewer-sludge-on-alabama-farmland
https://www.wxpr.org/post/rhinelander-spread-tons-sludge-near-site-contaminated-wells-wxpr-investigation-finds
https://content.sierraclub.org/grassrootsnetwork/sites/content.sierraclub.org.activistnetwork/files/teams/documents/PFASinBiosolidsdocx-formatted.pdf

26. Gas stations

https://www.wnep.com/article/news/local/carbon-county/leak-from-gas-station-may-have-contaminated-water/523-090dcb28-e3f7-4ac3-ad3c-df535323076b
https://phys.org/news/2018-10-gas-stations-vent-toxic-fumes.html
https://ww3.arb.ca.gov/ab2588/rrap-iwra/gasiwra.pdf
https://www.medicalnewstoday.com/articles/323426.php
https://www.atsdr.cdc.gov/PHS/PHS.asp?id=466&tid=83
https://ag.ny.gov/environmental/oil-spill/what-are-health-effects-exposure-petroleum-products
https://www.sciencedaily.com/releases/2018/10/181004110021.htm
https://www.conservationmagazine.org/2014/10/all-those-drops-add-up-small-spills-at-the-gas-station/
https://link.springer.com/article/10.1007/s40572-015-0074-8
https://www.nbcchicago.com/investigations/Underground-Storage-Tanks-Creating-Toxic-Soil-231846081.html
https://www.des.nh.gov/organization/commissioner/pip/factsheets/dwgb/documents/dwgb-22-20.pdf
https://education.seattlepi.com/pollution-gas-stations-3472.html
https://theporchpress.com/2011/03/the-gas-station-—-an-inconvenient-neighbor/
https://theporchpress.com/2011/03/the-gas-station-—-an-inconvenient-neighbor/
https://www.treehugger.com/cars/prevent-gasoline-contaminating-soil-groundwater.html
https://hub.jhu.edu/2014/10/07/gas-station-spills/
https://ag.ny.gov/environmental/oil-spill/what-are-health-effects-exposure-petroleum-products

https://ag.ny.gov/environmental/oil-spill/what-are-health-effects-exposure-petroleum-products
https://www.conservationmagazine.org/2014/10/all-those-drops-add-up-small-spills-at-the-gas-station/
https://ww3.arb.ca.gov/ab2588/rrap-iwra/gasiwra.pdf
https://www.sciencedaily.com/releases/2011/02/110204130315.htm
https://www.scientificamerican.com/article/is-it-safe-to-live-near-gas-station/
https://www.statista.com/statistics/859950/vehicles-in-operation-by-quarter-united-states/
https://www.fueleconomy.gov/feg/quizzes/answerQuiz16.shtml
https://phys.org/news/2018-10-gas-stations-vent-toxic-fumes.html
http://bcn.boulder.co.us/basin/waterworks/lust.html
https://www.baltimoresun.com/maryland/carroll/bal-ca.mtbe30oct30-story.html
https://www.sciencedaily.com/releases/2018/10/181004110021.htm
https://www.wnep.com/article/news/local/carbon-county/leak-from-gas-station-may-have-contaminated-water/523-090dcb28-e3f7-4ac3-ad3c-df535323076b
https://www.healthline.com/health/gasoline
https://www.scientificamerican.com/article/is-it-safe-to-live-near-gas-station/
https://link.springer.com/article/10.1007/s40572-015-0074-8
https://www.indianaenvironmentalreporter.org/posts/gas-stations-emit-10-times-more-benzene-than-previously-recorded
https://ww3.arb.ca.gov/ab2588/rrap-iwra/gasiwra.pdf
https://www.cancer.org/cancer/cancer-causes/benzene.html
https://tdn.com/news/local/shell-station-to-clean-up-contamination-from-leaking-storage-tanks/article_1e5abcf8-d3d6-5e72-bfb3-625c0f521d22.html
https://www.publichealth.columbia.edu/public-health-now/news/gas-stations-vent-far-more-toxic-fumes-previously-thought
https://www.publichealth.columbia.edu/public-health-now/news/gas-stations-vent-far-more-toxic-fumes-previously-thought
27. Sewage treatment plants
https://www.hindawi.com/journals/jeph/2016/8467023/
https://www.theeagle.com/news/local/flies-invade-homes-after-problem-at-wastewater-treatment-plant/article_a1aac2c4-c3b5-5adc-ae91-e7c9972162c4.html
https://uajsa.com/combined-sewer-overflow/
https://www.kxan.com/news/local/williamson-county/liberty-hill-facing-violations-for-chlorine-and-sewage-found-in-the-san-gabriel-river/
https://www.atsdr.cdc.gov/mmg/mmg.asp?id=385&tid=67#bookmark01
https://www.iwapublishing.com/news/industrial-wastewater-treatment
http://web.deu.edu.tr/atiksu/ana52/atiksu.html
https://www.pressconnects.com/story/news/local/watchdog/2019/01/31/sewage-overflows-new-york-rivers-creeks/2678607002/
https://healthfully.com/dangers-near-wastewater-treatment-facility-7648140.html
https://wateruseitwisely.com/journey-pipes-happens-wastewater/
https://www.kxan.com/top-stories/liberty-hill-neighbors-at-a-loss-over-overwhelming-smell-in-their-neighborhood/
https://wateruseitwisely.com/journey-pipes-happens-wastewater/
https://www.kxan.com/top-stories/liberty-hill-neighbors-at-a-loss-over-overwhelming-smell-in-their-neighborhood/
https://www.kxan.com/top-stories/liberty-hill-neighbors-at-a-loss-over-overwhelming-smell-in-their-neighborhood/
https://www.omaha.com/news/nebraska/omaha-s-raw-sewage-flows-into-missouri-river-after-flood/article_2039b7c8-0ba2-5bdf-ac20-a2af01f928a8.html
https://fox17.com/news/local/group-to-sue-city-claiming-it-let-47m-gallons-of-untreated-sewage-run-into-cumberland

https://publicherald.org/pennsylvania-is-discharging-radioactive-fracking-waste-into-rivers-as-landfill-leachate-impacting-the-chesapeake-bay-ohio-river-watersheds/

https://myfox47.com/2019/07/23/horrible-smell-in-pine-island-leads-to-disgusting-discovery/

https://www.statesmanjournal.com/story/news/local/silverton/2019/07/23/dream-house-paraplegic-woman-uninhabitable-after-sewer-backup/1710445001/

https://www.craigdailypress.com/news/cruddy-cleanup-craigs-wastewater-plant-backed-up-after-system-fried-by-lightning/

https://www.jsonline.com/story/news/local/milwaukee/2018/03/12/relief-sight-walkers-point-residmmsd-block-odors-jones-island-sludge-tanks-after-complaints-walkers/404243002/

https://www.marinij.com/2018/09/23/mill-valley-sewage-plant-stench-agitates-students-parents/

https://wlos.com/news/ask-13/ask-13-odor-from-wastewater-treatment

https://www.theeagle.com/news/local/flies-invade-homes-after-problem-at-wastewater-treatment-plant/article_a1aac2c4-c3b5-5adc-ae91-e7c9972162c4.html

https://www.pressconnects.com/story/news/local/watchdog/2019/01/31/sewage-overflows-new-york-rivers-creeks/2678607002/

https://myfox47.com/2019/07/23/horrible-smell-in-pine-island-leads-to-disgusting-discovery/

https://www.rep-am.com/local/news-local/2019/07/25/waterbury-will-spend-100000-identifying-fixes-to-sewer-plant4/

https://www.pressconnects.com/story/news/local/watchdog/2019/01/31/sewage-overflows-new-york-rivers-creeks/2678607002/

https://www.pressconnects.com/story/news/local/watchdog/2019/01/31/sewage-overflows-new-york-rivers-creeks/2678607002/

https://www.usgs.gov/special-topic/water-science-school/science/a-visit-a-wastewater-treatment-plant?qt-science_center_objects=0#qt-science_center_objects

28. Injection wells

https://www.miamiherald.com/news/local/community/miami-dade/article70629122.html

https://www.publicnewsservice.org/2019-08-12/water/live-near-a-proposed-injection-well-look-online-to-find-out/a67427-1

http://catskillcitizens.org/learnmore/injectionwells.pdf

https://stateimpact.npr.org/pennsylvania/tag/deep-injection-well/

https://www.propublica.org/article/injection-wells-the-poison-beneath-us

https://oilprice.com/Energy/Energy-General/The-Facts-about-Fracking-Fluid-and-its-Disposal.html

https://www.fractracker.org/2019/03/wicked-witch-of-the-fracking-wastewater/

https://www.miamiherald.com/news/local/community/miami-dade/article70629122.html

https://www.epa.gov/uic/class-ii-oil-and-gas-related-injection-wells

https://www.propublica.org/article/injection-wells-the-poison-beneath-us

https://www.dallasnews.com/news/2019/07/25/wastewater-injection-can-make-faults-twice-as-likely-to-fail-quake-study-says/

https://www.epa.gov/uic

http://chej.org/wp-content/uploads/Deep%2520Well%2520Injection%2520-%2520PUB%2520056.pdf

https://www.michiganradio.org/post/hazardous-waste-deep-injection-wells-might-reopen

https://www.michiganradio.org/post/hazardous-waste-deep-injection-wells-might-reopen

https://wreg.com/news/carrier-withdraws-request-to-dump-wastewater-into-aquifer/

https://vcreporter.com/2020/01/oxnard-oil-field-waste-site-closed-anterra-disposal-wells-shut-after-violations/

https://www.wkbn.com/news/local-news/wastewater-well-overflows-in-coitsville-twp/

https://www.webcenter11.com/content/news/City-2-businesses-reach-clean-water-settlements-with-EPA-567364511.html

https://www.southernenvironment.org/news-and-press/news-feed/tennessee-county-board-rebuffs-proposal-to-put-superfund-waste-in-drinking-water-source

https://www.wxpr.org/post/former-director-city-injected-wastewater-sludge-near-site-contaminated-wells

http://chej.org/wp-content/uploads/Deep%2520Well%2520Injection%2520-%2520PUB%2520056.pdf

https://pubs.usgs.gov/wsp/2281/report.pdf

http://catskillcitizens.org/learnmore/InjectionWells.pdf

29. Sand and gravel pits

http://www.hugoneighborhood.org/ELUT%25207A1b%2529%25281%2529%2520Silica%2520Dust%2520-%2520Gravel%2520Pit%2520Flyer.pdf

https://www.nytimes.com/1991/06/30/nyregion/new-regulations-restrict-gravel-pits.html

https://www.ewg.org/research/sandstorm/health-concerns-silica-outdoor-air

https://nypost.com/2019/09/20/father-and-son-plunge-to-their-deaths-at-abandoned-quarry/

https://www.woodtv.com/news/target-8/the-pit-next-door-law-stops-gravel-mine-opposition/

https://www.statesman.com/opinion/20190927/editorial-dont-let-quarry-owners-pollute-and-walk-away

https://www.gazettenet.com/Gravel-company-reaches-settlement-with-EPA-over-emissions-violations-18467341

https://www.torontoenvironment.org/gravel/impacts

https://prd-wret.s3-us-west-2.amazonaws.com/assets/palladium/production/atoms/files/mcs-2019-sandc.pdf

https://geology.com/articles/abandoned-mines.shtml

https://www.columbian.com/news/2011/sep/22/neighbors-fault-yacolt-quarry-for-well-woes-despit/

https://www.mcall.com/news/mc-xpm-2002-03-29-3388751-story.html

https://www.osha.gov/dsg/topics/silicacrystalline/background_info_silica.html

https://www.cbs8.com/article/news/local/mission-gorge-quarry-has-neighbors-living-in-the-blast-zone/509-4118f042-af4b-4520-813b-2777703732fd

https://www.mdislander.com/maine-news/quarry-neighbors-want-noise-ordinance

https://www.wsmv.com/news/neighbors-near-quarry-raise-concerns-about-damage-to-homes/article_062129fb-5239-527c-ad84-47287699f361.html

https://www.ecori.org/public-safety/2017/4/15/quarry-dust-coats-westerly-in-concern

https://vtdigger.org/2017/06/27/neighbors-fault-state-lax-quarry-regulation/

https://www.wbez.org/stories/_/4871cfbc-422d-4b6d-a95c-394ac4b5ddc9

https://www.latimes.com/nation/nationnow/la-na-nn-fracking-sand-mining-20140925-story.html

https://www.nhs.uk/conditions/carbon-monoxide-poisoning/

https://www.wpr.org/frac-sand-mining-causing-metal-contamination-groundwater

https://earthworks.org/issues/frac_sand_health_and_environmental_impacts/

30. Contaminated fill

https://www.insidernj.com/press-release/dirty-dirt-bill-goes-govs-desk/

https://www.njspotlight.com/stories/19/02/18/measure-to-keep-dirt-brokers-out-of-recycling-industry-moves-ahead-but-slowly/

https://www.usatoday.com/story/opinion/voices/2019/09/23/coal-ash-cancer-north-carolina-mother-column/2368665001/

https://www.ctpost.com/local/article/Former-Fairfield-official-pleads-not-guilty-in-14541846.php

https://abc3340.com/news/abc-3340-news-iteam/how-did-contaminated-soil-get-into-three-north-birmingham-neighborhoods

https://www.indianaenvironmentalreporter.org/posts/tainted-dreams

https://www.flamboroughreview.com/news-story/9730295-hamilton-eyes-new-bylaw-crackdown-on-development-fill-dumping/

https://www.ajc.com/news/local/danger-the-ground-lead-contaminates-westside-atlanta-neighborhood/r1wiLsN2tyUIlbuTI7yxEO/

https://www.charlotteobserver.com/news/local/article239985418.html

https://www.charlotteobserver.com/news/local/article239985418.html

https://stateimpact.npr.org/pennsylvania/2019/10/16/how-philadelphia-built-a-neighborhood-on-toxic-soil-of-former-clearview-landfill/

https://mcgowanprograms.com/blog/clean-fill-contractors-clean-actually-clean/

https://healthybuilding.net/blog/460-filled-with-uncertainty-toxic-dirt-in-building-construction

https://www.google.com/search?sxsrf=ALeKk017IeN5Z9TOfgx8NkIK5rpYYPyAVA%253A1587156563118&source=hp&ei=UxaaXrTrBIO6-gSYzZrACA&q=pennsylvania+2007+beneficial+american+ash +recycling+corp.+of+pennsylvania&oq=pennsylvania+2007+beneficial+american+ash+recyclin&gs_lcp=CgZwc3ktYWIQARgBMgUIIRCrAjIFCCEQqwIyBQghEKsCOgUIABCDAToCCAA6BQgAEM0COgYIABAWEB46BQghEKABOgcIIRAKEKABOggIIRAWEB0QHjoECCEQCko1CBcSMTBnMzMwZzEzOGcxNDRnMTM0ZzE2N2cxNTRnMTQ0ZzEzOWcxMjFnMTY0ZzEzNWcxNDhKHggYEhowZzFnMWcxZzFnMWcxZzFnMWcxZzVnNWczMlDjJVj75wFg3_cBaAFwAHgAgAGpAogB2DiSAQYwLjQ5LjKYAQCgAQGqAQdnd3Mtd2l6&sclient=psy-ab

https://www.miamiherald.com/news/local/environment/article234228687.html

https://www.ctpost.com/local/article/Toxic-secrets-5553424.php

https://www.miamiherald.com/news/local/environment/article234228687.html

https://patch.com/new-jersey/longbranch/mob-tied-dumping-toxic-waste-new-jersey-state-0

https://fox13now.com/2014/04/23/residents-upset-after-discovering-homes-were-built-above-citys-old-landfill/

https://www.southbendtribune.com/news/local/cleanup-of-coal-ash-gets-underway-in-the-pines/article_ca3c92e0-14e2-53c0-b29b-cd3abf67ce71.html

https://abcnews.go.com/US/house-built-north-carolina-landfill-sinking-feeling/story?id=19274453

31. Junkyards

https://www.thereporter.com/2020/01/03/solano-appliance-recycling-firm-fined-for-mishandling-burning-hazardous-waste/

https://www.capecodtimes.com/article/20020303/news01/303039978

https://www.northcountrypublicradio.org/news/story/38752/20190528/cleanup-work-to-begin-at-former-potsdam-junkyard

https://www.kstk.org/2018/08/08/resident-suspects-medical-issues-linked-to-living-near-byford-junkyard/

https://myautostore.com/hidden-dangers-you-pull-it-junkyards/

https://www.hotcars.com/photos-of-the-biggest-car-junkyards-around-the-world/

https://toxicsaction.org/issues/waste/illegal-junkyards/

https://www.capecodtimes.com/article/20020303/news01/303039978

https://www.epa.gov/sites/production/files/2014-03/documents/toxicological_and_human_health_effects_following_exposure_3v.pdf

http://planphilly.com/articles/2017/01/06/kensington-community-living-with-scrap-yards-wants-a-future-without-these-operations-nextdoor
http://www7.nau.edu/itep/main/HazSubMap/twrap_HzSubMap_Salvage.asp
https://www.amazonfrontlines.org/chronicles/health-oil/
https://cupola.gettysburg.edu/cgi/viewcontent.cgi?article=1304&context=student_scholarship
https://poststar.com/news/local/kingsbury-junkyard-owner-settles-pollution-case-with-state/article_c576b3e0-bd3f-11e0-a9d3-001cc4c002e0.html
https://www.delawareonline.com/story/news/local/2014/06/01/pollution-risks-claymont-plant-studied/9846027/
https://gizmodo.com/the-worlds-biggest-auto-shredder-eats-450-cars-an-hour-5893440
http://www.mennoworld.org/archived/2009/11/2/love-neighbor-spurred-opposition-car-crusher/?print=1
https://www.eia.gov/todayinenergy/detail.php?id=36614
https://www.houstonchronicle.com/news/houston-texas/houston/article/Danger-in-air-near-metal-recyclers-4154951.php
https://www.itemlive.com/2018/07/17/lynn-files-contempt-order-against-western-avenue-junkyard/
https://www.itemlive.com/2019/07/28/lynn-junkyard-shut-down-last-year-for-illegal-dumping-now-under-purchase-agreement/
https://www.medicalnewstoday.com/articles/322165#what-is-refrigerant-poisoning
https://americanrecycler.com/8568759/
https://lamprecycling.veoliaes.com/newsletter/April2014/4
https://www.bts.gov/content/motor-vehicles-scrapped
https://www.nj.com/business/2013/04/behold_the_mega_shredder_jerse.html
https://lamprecycling.veoliaes.com/newsletter/April2014/4
http://www7.nau.edu/itep/main/HazSubMap/twrap_HzSubMap_Salvage.asp
https://www.postbulletin.com/news/local/flooding-brings-up-questions-about-oronoco-salvage-yard-pollution/article_9be19fd4-9dbc-11e9-925a-df1d5be0984b.html
https://www.mesothelioma.com/blog/asbestos-in-drinking-water-the-danger-of-old-asbestos-pipes-and-natural-disasters/
https://www.chicagotribune.com/news/environment/ct-general-iron-pollution-rahm-emanuel-20190702-w7sla352cjgknbs7jnlcmrsxvi-w7sla352cjgknbs7jnlcmrsxvi-story.html
https://www.inquirer.com/science/superfund-site-epa-matteo-west-deptford-gloucester-county-20191112.html
https://www.chicagotribune.com/news/breaking/ct-met-clybourn-corridor-scrap-yard-20171228-story.html
https://www.capecodtimes.com/article/20020303/news01/303039974
https://www.amazonfrontlines.org/chronicles/health-oil/
https://www.mesothelioma.com/blog/asbestos-in-drinking-water-the-danger-of-old-asbestos-pipes-and-natural-disasters/
https://www.statista.com/statistics/274931/number-of-vehicles-scrapped-in-the-us/
https://www.thestarpress.com/story/news/local/2019/07/19/recycling-facilitys-mercury-emissions-scrutinized/1758254001/

32. Trash transfer stations

https://www.epa.gov/sites/production/files/2016-03/documents/r02002.pdf
https://archive.epa.gov/epawaste/nonhaz/municipal/web/pdf/r02002.pdf
https://ceds.org/landfills-transfer-stations/
https://www.scdhec.gov/environment/your-land/landfills-overview/how-landfills-work
https://www.wastedive.com/news/eminent-domain-waste-transfer-station-DC/543107/
http://www.washingtonpost.com/wp-dyn/content/article/2007/07/24/AR2007072402241.html?noredirect=on

https://www.wastedive.com/news/eminent-domain-waste-transfer-station-DC/543107/

https://www.epa.gov/sites/production/files/2016-03/documents/wtsguide.pdf

https://www.theguardian.com/cities/2016/oct/27/new-york-rubbish-all-that-trash-city-waste-in-numbers

https://www.sun-sentinel.com/news/fl-xpm-1985-03-29-8501120562-story.html

https://nypost.com/2013/03/11/people-fleeing-their-ues-neighborhood-because-of-incoming-trash-transfer-station/

https://queenseagle.com/all/2019/9/27/opinion-private-sanitation-is-disfiguring-black-and-brown-neighborhoods-in-queens

https://www.waste360.com/blog/how-green-was-my-garbage-truck

https://evanstonnow.com/story/news/bill-smith/2010-10-28/neighbors-raise-a-stink-about-transfer-station

33. Railroads

http://www.bluejersey.com/2012/11/toxic-waste-causes-lock-down-in-three-nj-towns/

https://www.cnn.com/2012/11/30/us/new-jersey-train-derail/index.html

http://www.vancouversun.com/health/Homes+near+rail+lines+face+exposure+harmful+emissions+study/9652314/story.html

https://pocketsense.com/dangers-of-living-near-railroad-tracks-12591644.html

https://www.sciencedaily.com/releases/2010/05/100506141634.htm

https://www.thelancet.com/journals/lancet/article/PIIS0140-6736(13)61613-X/fulltext

https://www.thelancet.com/journals/lancet/article/PIIS0140-6736(13)61613-X/fulltext

https://www.kendallcountynow.com/2015/10/12/idling-trains-in-oswego-leave-neighbors-awake/avkx6c7/

https://www.gcnews.com/articles/lirr-neighbors-object-to-idling-trains-night-noise/

https://www.dailyemerald.com/archives/railroad-s-herbicide-use-elicits-complaints/article_7f23d35e-0644-5505-b1f7-1d18cce9165a.html

https://beyondpesticides.org/dailynewsblog/2007/07/children-sprayed-at-day-care-with-railroad-herbicides/

https://thinkprogress.org/lamberts-point-coal-dust-pollution-6dbec60d1e9e/

https://www.eagletribune.com/news/haverhill/trains-still-idle-in-their-bradford-neighborhood-so-they-re/article_9a5dd116-e5b5-5601-8726-50013f018f16.html

https://www.sfchronicle.com/bayarea/article/Some-Mission-Bay-neighbors-fuming-over-12383764.php

https://safetydata.fra.dot.gov/OfficeofSafety/publicsite/Query/AccidentByRegionStateCounty.aspx

http://www.vancouversun.com/health/Homes+near+rail+lines+face+exposure+harmful+emissions+study/96 52314/story.html

https://dot111.info/2019/09/11/smoke-billows-into-the-air-after-train-derailment-near-dupo-illinois/

https://www.ble-t.org/pr/news/headline.asp?id=12729

https://www.ecori.org/pollution-contamination/2019/4/1/railroad-company-cited-for-herbicide-use-in-coastal-zone

https://www.iarc.fr/news-events/iarc-diesel-engine-exhaust-carcinogenic/

https://www.gao.gov/products/GAO-19-443

https://plastics.americanchemistry.com/Automotive/

https://www.headlice.org/comb/what-are-head-lice-and-nits/chemical-concerns/signs-and-symptoms-of-pesticide-poisoning/#rssp

https://www.cdc.gov/nceh/hearing_loss/what_noises_cause_hearing_loss.html

https://www.hearingaidknow.com/how-loud-is-too-loud-decibel-levels-of-common-sounds

https://www.eagletribune.com/news/bradford-neighbors-to-get-relief-after-decades-fighting-idling-trains/article_b48fbcc2-92a5-11e9-a104-f7abb5e272b1.html

https://www1.health.gov.au/internet/publications/publishing.nsf/Content/ohp-enhealth-manual-atsi-cnt-l~ohp-enhealth-manual-atsi-cnt-l-ch5~ohp-enhealth-manual-atsi-cnt-l-ch5.16
https://www.webmd.com/cancer/herbicide-glyphosate-cancer#1
Section Three:
34. Radon
https://www.health.harvard.edu/newsletter_article/Radon_revisited
https://www.wesa.fm/post/pa-has-especially-high-levels-radon-new-bill-would-mandate-testing-it-schools
https://health.usnews.com/health-care/patient-advice/articles/2017-09-14/does-radon-exposure-cause-lung-cancer
https://activerain.com/blogsview/1100259/have-you-heard-the-story-of-stanley-watras-
https://archive.epa.gov/water/archive/web/html/basicinformation-2.html
https://www.ncbi.nlm.nih.gov/pmc/articles/PMC4721325/
https://pubs.usgs.gov/gip/7000018/report.pdf
https://www.atsdr.cdc.gov/csem/csem.asp?csem=8&po=7
https://archive.epa.gov/water/archive/web/html/basicinformation-2.html
https://archive.epa.gov/water/archive/web/html/basicinformation-2.html
35. Past uses
https://www.berkshireeagle.com/stories/great-barrington-brownfield-finally-gets-a-cleanup,588613
https://dailyreporter.com/2019/10/28/uw-scientists-map-areas-of-historic-lead-zinc-contamination/
https://www.nj.com/news/2009/08/nj_lawsuit_puts_farmland_devel.html
https://www.miaminewtimes.com/news/aventura-isles-residents-sue-over-arsenic-pesticide-contamination-in-miami-dade-development-11375713
https://news.bloomberglaw.com/environment-and-energy/new-orleans-sued-over-inhumane-living-conditions-atop-landfill
36. Well water
https://www.nhpr.org/post/mid-century-dartmouth-hazardous-waste-site-contaminates-drinking-water#stream/0
https://www.epa.gov/privatewells/potential-well-water-contaminants-and-their-impacts
https://www.cdc.gov/healthywater/drinking/public/drinking-water-faq.html
https://www.wqa.org/resources/news-releases/id/187/private-wells-need-annual-water-quality-testing
https://www.groundwater.org/get-informed/basics/testing.html
https://www.health.state.mn.us/communities/environment/water/wells/disclosures/index.html
https://lancasteronline.com/opinion/editorials/for-our-health-and-safety-some-oversight-is-needed-for/article_879c6528-e168-11e9-ad34-d7396b3639ab.html
https://pubs.usgs.gov/gip/gw_ruralhomeowner/
https://www.cdc.gov/healthywater/observances/gaw.html
https://www.gazettenet.com/ENVLeverettWater-25206303
https://pubs.usgs.gov/gip/gw_ruralhomeowner/
https://www.democratandchronicle.com/story/news/2014/05/26/batavia-contamination-lamb-farms/9599231/
https://www.nytimes.com/2009/09/18/us/18dairy.html?mtrref=www.google.com&gwh=815F9689F0D865BF79D1B5620 DB94240&gwt=pay&assetType=REGIWALL
https://www.nytimes.com/2009/09/18/us/18dairy.html?mtrref=www.google.com&gwh=815F9689F0D865BF79D1B 5620DB94240&gwt=pay&assetType=REGIWALL
https://www.usgs.gov/news/study-estimates-about-21-million-people-using-wells-high-arsenic

https://pediatrics.aappublications.org/content/123/6/1599

https://semspub.epa.gov/work/05/918981.pdf

https://www.baltimoresun.com/news/environment/bs-gr-salisbury-well-contamination-20121116-story.html

https://mde.maryland.gov/programs/LAND/MarylandBrownfieldVCP/Documents/Morris%2520Mill.pdf

https://www.pottsmerc.com/news/local/toxins-linger-on-berks-farm-from-long-ago-environmental-sins/article_b09e8a6c-e30e-11e9-8f66-378eb98d0602.html

https://www.startribune.com/baby-s-death-sparks-water-safety-fight-with-the-ag-industry/502477031/

https://www.wandtv.com/news/assumption-water-could-be-deadly-to-children-under-months/article_fa1e2a7c-d647-11e9-9dec-833fcaf175db.html

https://www.sfreporter.com/news/coverstories/2005/01/05/oh-crap/

https://www.democratandchronicle.com/story/news/2014/05/26/batavia-contamination-lamb-farms/9599231/

https://19january2017snapshot.epa.gov/il/wedron-groundwater-site-wedron-illinois_.html

https://www.mywebtimes.com/articles/tn/2013/01/28/07b97b28fbdf57fabbbc88f9c95fcf5f/index.xml

https://www.gazettenet.com/ENVLeverettWater-25206303

https://marengo-uniontimes.com/news/1553-well-water-contamination

https://www.nytimes.com/2009/09/18/us/18dairy.html?auth=login-email&login=email

https://www.wmdt.com/2019/12/dnrec-mountaire-sign-agreement-to-resolve-violations-at-poultry-processing-facilities/

https://www.delawareonline.com/story/news/local/2019/12/17/delaware-says-new-deal-address-pollution-resolve-issues-mountaire/2670234001/

https://marengo-uniontimes.com/news/1553-well-water-contamination

https://www.wwdmag.com/groundwater/illinois-epa-takes-over-groundwater-concern-wedron-ill

https://odh.ohio.gov/wps/wcm/connect/gov/3c6c97da-58be-44fe-81e5-7a62a3e77afe/btex.pdf?MOD=AJPERES&CONVERT_TO=url&CACHEID=ROOTWORKSPACE.Z18_M1HGGIK0N0JO00QO9DDDDM3000-3c6c97da-58be-44fe-81e5-7a62a3e77afe-mjHzkJR

37. Public water supplies

https://www.theloquitur.com/155525-2/

https://www.usatoday.com/story/news/2016/12/13/broken-system-means-millions-of-rural-americans-exposed-to-poisoned-or-untested-water/94071732/

https://news.google.com/articles/CBM
ibWh0dHBzOi8vYWJjbmV3cy5nby5jb20vUG9saXRp
Y3MvbmV3YXJrLWhhbmRpbmctYm90dGxlZC13YXRlc
i1maWx0ZXJzLWZhaWwtcHJvdGVjdC1yZXNpZGVudHMv
c3Rvcnk_aWQ9NjQ5MjUyMzfSAXFodHRwczovL2FiY2
5ld3MuZ28uY29tL2FtcC9Qb2xpdGljcy9uZXdhcmst aGFuZGluZy1ib3R0bGVkLXdhdGVy
LWZpbHRlcnMtZmFpbC1wcm90ZWN0LXJlc2lkZW50cy9zdG9yeT9pZD02NDkyNTIzNw
?hl=en-US&gl=US&ceid=US%253Aen

https://news.google.com/articles/CAIiEAhxFVBQ_nBkSUhFJNcM0-MqFwgEKg8IAC
oHCAowjuuKAzCWrzwwloIY?hl=en-US&gl=US&ceid=US%253Aen

https://www.delawareonline.com/story/news/local/2017/11/30/poisoned-wells-found-around-mountaire-farms/911645001/

https://www.epa.gov/sdwa/how-epa-regulates-drinking-water-contaminants

https://ecode360.com/10584713

https://www.miaminewtimes.com/news/poisoned-well-6363391

https://www.spokesman.com/stories/2018/jun/26/shawn-vestal-airway-heights-water-worse-than-feare/
https://www.wamc.org/post/hoosick-falls-study-finds-more-illnesses-linked-pfoa-exposure
https://www.12news.com/article/news/local/arizona/research-into-arizona-towns-uranium-contaminated-water-supply-sparks-change/124812713
https://www.timesunion.com/news/article/EPA-lays-out-plan-to-regulate-PFOA-in-drinking-13616439.php
https://www.epa.gov/sdwa/how-epa-regulates-drinking-water-contaminants
https://qz.com/1039535/in-the-us-most-people-are-drinking-water-that-is-legally-safe-but-isnt-actually-risk-free/
https://www.npr.org/sections/thesalt/2014/01/24/265520673/drinking-water-not-tested-for-tens-of-thousands-of-chemicals
https://www.theguardian.com/environment/2016/mar/06/water-utilities-lead-testing-analysis-us-cities
https://amp.freep.com/amp/2618868001?__twitter_impression=true
https://www.epa.gov/dwreginfo/drinking-water-regulations
https://patch.com/new-jersey/bloomfield/bloomfield-continues-struggle-water-contamination
https://www.pbs.org/newshour/health/even-if-your-drinking-water-gets-a-passing-grade-it-may-not-be-safe
http://www.ncpolicywatch.com/2019/11/18/selc-plans-to-sue-burlington-over-pfas-14-dioxane-pollution-in-drinking-water-sludge/
https://www.motherjones.com/environment/2016/06/rural-water-contamination-uranium-navajo-sanders/
https://www.usatoday.com/story/news/2016/12/13/broken-system-means-millions-of-rural-americans-exposed-to-poisoned-or-untested-water/94071732/
https://www.usatoday.com/story/news/2016/12/13/broken-system-means-millions-of-rural-americans-exposed-to-poisoned-or-untested-water/94071732/
https://www.cnn.com/2016/06/28/us/epa-lead-in-u-s-water-systems/index.html
https://www.nbcnews.com/news/us-news/cancer-causing-chemical-tcp-plagues-california-drinking-water-n783726
http://www.msnbc.com/all/west-virginia-chemical-spill-water
https://www.ewg.org/tapwater/?zip5=10001&searchtype=zip#.WXZc1_8rKPQ
https://www.epa.gov/sdwa/how-epa-regulates-drinking-water-contaminants
https://patch.com/new-york/portwashington/cancer-causing-pollutants-found-albertson-water-district-drinking-water
https://www.nwherald.com/2013/06/20/marengo-residents-unfazed-by-water-pollution-allegations/aqwx7cw/
https://qz.com/1039535/in-the-us-most-people-are-drinking-water-that-is-legally-safe-but-isnt-actually-risk-free/
https://www.webmd.com/digestive-disorders/news/20120914/tainted-tap-water-sickens-1.1-million-each-year#1
https://www.ewg.org/tapwater/#.WXjmntPyvBI
https://www.webmd.com/digestive-disorders/news/20120914/tainted-tap-water-sickens-1.1-million-each-year#1
https://www.webmd.com/digestive-disorders/news/20120914/tainted-tap-water-sickens-1.1-million-each-year#1
https://publicintegrity.org/environment/industrial-waste-pollutes-americas-drinking-water/
https://www.northcarolinahealthnews.org/2019/08/15/pittsboro-agrees-to-notify-residents-of-drinking-water-contamination/
https://www.ewg.org/tapwater/state-of-american-drinking-water.php
https://www.cdc.gov/healthywater/drinking/public/index.html

https://www.theguardian.com/us-news/2016/jan/27/michigan-water-testing-rules-pre-flushing-taps-flint-lead
https://www.epa.gov/sites/production/files/2018-03/documents/dwtable2018.pdf
https://www.epa.gov/ground-water-and-drinking-water/national-primary-drinking-water-regulations
38. Waterways
https://www.southernmarylandchronicle.com/2019/11/27/potomac-riverkeeper-network-files-intent-to-sue-owners-of-defunct-luke-paper-mill/
https://lancasteronline.com/news/local/brunner-island-power-plant-to-pay-million-fine-for-ash/article_ab692f58-b2cf-11e9-a7f5-83eecc557a83.html
https://news.google.com/articles/CAIiEPhYddy2IGMKDuE58sG6AylqFwgEKg8IACoHCAowjuuKAzCWrzwwt4QY?hl=en-US&gl=US&ceid=US%253Aen
http://www.southernmarylandchronicle.com/2019/11/27/potomac-riverkeeper-network-files-intent-to-sue-owners-of-defunct-luke-paper-mill/
https://publicintegrity.org/environment/industrial-waste-pollutes-americas-drinking-water/
https://www.app.com/story/news/health/2018/06/12/nj-beach-water-clean-bacteria-safe-swim/693807002/
https://www.nbcconnecticut.com/news/local/the-dangers-of-pfas-why-a-chemical-spill-into-farmington-river-is-still-a-concern-months-later/171330/
https://www.riverkeeper.org/campaigns/stop-polluters/pcbs/
https://www.news5cleveland.com/news/local-news/cleveland-metro/142-600-gallons-of-sewage-and-stormwater-discharged-into-lake-erie-after-tuesday-nights-severe-weather
https://www.northcountrypublicradio.org/news/story/34300/20170714/playing-whack-a-mole-with-pollution-at-lake-george-s-million-dollar-beach
https://www.citizen-times.com/story/news/local/2018/09/24/french-broad-river-sewage-spill-people-advised-stay-out-river/1409745002/
https://atlantic2.sierraclub.org/content/pcb-cleanup-hudson-river-far-complete
https://www.northcountrypublicradio.org/news/story/28460/20150525/remembering-the-historic-fight-over-pcb-dredging-on-the-hudson-river
https://poststar.com/news/local/officials-good-swim-season-on-million-dollar-beach-so-far/article_f2063498-d15f-5981-86d9-87d70fcd56b9.html
https://dhhr.wv.gov/News/chemical-spill/Documents/ElkRiverMedicalRecordSummary.pdf
https://www.startribune.com/minneapolis-lake-nokomis-beaches-close-indefinitely-after-e-coli-sickens-3-kids/540614722/
https://www.epa.gov/npdes/report-congress-combined-sewer-overflows-great-lakes-basin
https://bringmethenews.com/minnesota-news/69-now-sickened-in-e-coli-outbreak-at-lake-nokomis
https://www.csmonitor.com/Environment/2014/0210/Coal-ash-spill-How-certain-that-Dan-River-is-safe-for-drinking-wildlife
https://patch.com/new-jersey/pointpleasant/47-nj-beaches-unsafe-bacteria-brink-closing
https://www.cdc.gov/mmwr/volumes/67/wr/mm6725a1.htm
https://www.chicagotribune.com/news/environment/ct-indiana-chemical-spill-beaches-closed-20190816-znazwueorrcvfipo4hfy5ze7vi-story.html
https://www.bridgemi.com/michigan-environment-watch/two-five-large-michigan-industrial-plants-exceeded-water-pollution-limits
https://www.fox2detroit.com/news/piles-of-sludge-have-washed-up-on-st-clair-shores-coastline
https://www.chicagotribune.com/news/environment/ct-indiana-chemical-spill-beaches-closed-20190816-znazwueorrcvfipo4hfy5ze7vi-story.html

https://www.cleanwateraction.org/features/dangers-hexavalent-chromium-chromium-6-california-drinking-water

https://www.oregon.gov/oha/PH/HealthyEnvironments/DrinkingWater/Monitoring/Documents/health/ammonia.pdf

https://www.oregon.gov/oha/PH/HealthyEnvironments/DrinkingWater/Monitoring/Documents/health/ammonia.pdf

https://cbs6albany.com/news/local/grant-awarded-for-lake-george-sewage-treatment-plant

39. Dry cleaners

https://nypost.com/2016/05/09/the-hidden-danger-of-living-near-your-neighborhood-dry-cleaners/

https://www.hindawi.com/journals/jeph/2009/183920/#abstract

https://www.kunm.org/post/no-cleanup-plan-abq-dry-cleaning-spill

https://boltoncenter.wordpress.com/2011/10/21/comparing-dpw-barn-contamination-to-smith-contamination/

https://www.ccim.com/cire-magazine/articles/cleaning-after-dry-cleaners/?gmSsoPc=1

https://www-1.kansas.com/news/politics-government/article216625720.html

https://www.chicagotribune.com/news/chi-perc-pollution-07262009-story.html

https://billingsgazette.com/news/local/largely-forgotten-superfund-site-in-heart-of-billings-could-earn/article_537d169a-166f-5b3c-a326-70e4ab20e3d0.html

https://www.kunm.org/post/no-cleanup-plan-abq-dry-cleaning-spill

https://www.southbendtribune.com/news/local/south-bend-sues-former-dry-cleaners-for-significant-contamination-of/article_d82a3388-f3dd-5d85-9c37-f06e8493d6f6.html

https://www.chicagotribune.com/news/ct-xpm-2009-07-26-0907250254-story.html

https://www.popsci.com/dry-cleaning-chemicals/

https://www.mlive.com/news/ann-arbor/2016/11/epa_testing_for_toxic_vapors_i.html

https://www.chicagotribune.com/news/chi-perc-pollution-07262009-story.html

40. Golf courses

https://www.miamiherald.com/news/local/environment/article234228687.html

https://medium.com/@wgarris/could-the-golf-course-green-be-poisoning-you-and-your-child-5646511965c7

https://www.beyondpesticides.org/programs/golf-and-the-environment/organic-alternatives

https://www.oregonlive.com/health/2019/06/son-nagged-by-golfing-dads-death-from-a-rare-cancer-were-pesticides-at-fault.html

https://www.orlandosentinel.com/news/seminole-county/os-seminole-county-rolling-hills-cleanup-20180511-story.html

https://www.orlandosentinel.com/news/os-xpm-2011-06-21-os-tainted-wells-insecticide-20110615-story.html

https://www.sun-sentinel.com/local/broward/tamarac/fl-sb-tamarac-woodmont-dirt-debate-20180110-story.html

https://www.mda.state.mn.us/sites/default/files/inline-files/golfcoursecontamination_1.pdf

https://www.chicagotribune.com/suburbs/northbrook/ct-nbs-mission-arsenic-tl-0707-20160707-story.html

https://www.sun-sentinel.com/health/fl-reg-golf-courses-decontamination-20170619-story.html

https://tamaractalk.com/report-shows-dangerous-contamination-levels-lie-beneath-golf-course-13346

https://www.mda.state.mn.us/sites/default/files/inline-files/golfcoursecontamination_1.pdf

http://greenfertility.blogspot.com/2007/07/toxic-golf-courses.html

https://www.golfdigest.com/story/environment_feldman

https://www.beyondpesticides.org/assets/media/documents/documents/toxic-fairways-1995.pdf

https://www.thengfq.com/2018/02/golfs-9-billion-turf-course-supplies-industry/

https://www.gcsaa.org/media/news-release/2019-news-releases/2019/02/27/amvac-chemical-corporation-pledges-to-EIFG

http://psep.cce.cornell.edu/Tutorials/core-tutorial/module04/index.aspx

https://extension.psu.edu/potential-health-effects-of-pesticides

https://www.opensecrets.org/industries/indus.php?ind=N13++

https://medium.com/@wgarris/could-the-golf-course-green-be-poisoning-you-and-your-child-5646511965c7

https://www.lohud.com/story/news/investigations/2016/07/27/pesticides-threaten-golf-courses-homes-businesses-new-york/86194386/

https://cumulis.epa.gov/supercpad/SiteProfiles/index.cfm?fuseaction=second.Cleanup&id=0510389#bkground

https://www.beyondpesticides.org/programs/golf-and-the-environment/organic-alternatives

https://clubandresortbusiness.com/golf-course-on-landfill-to-be-permanently-closed/

https://www.discovermagazine.com/environment/at-play-on-a-field-of-trash

https://thebradentontimes.com/beruff-suing-activist-over-claims-on-landfill-adjacent-development-p19749-137.htm

https://www.nrdc.org/onearth/coal-ash-hazardous-coal-ash-waste-according-epa-coal-ash-not-hazardous-waste#:~:text=Coal%20ash%20is%20incredibly%20dangerous,and%20a%20variety%20of%20cancers.

https://www.lohud.com/story/news/investigations/2016/07/27/pesticides-threaten-golf-courses-homes-businesses-new-york/86194386/

41. Gas pipelines

https://theweek.com/articles/882813/america-needs-stop-natural-gas-pipeline-mania

https://earthworks.org/media-releases/pollution-found-at-nexus-gas-transmission-wadsworth-compressor-station/

https://news.google.com/articles/ CBMibGh0dHBzOi8vd3d3Lm5iY25ld3MuY29tL25ld3MvdXMtbmV3cy9vbmUtZGVhZC1rZW50dWNreS1h ZnRlci1nYXMtbGluZS1leHBsb3Npb24tc2hvb3RzLWZpcmViYWxsLXNreS1uMTAzODA5NtlBLGh0dHBzOi8vd3d3Lm5iY25ld3MuY29tL25ld3MvYW1wL25jbmExMDM4MDk2?hl=en-US&gl=US&ceid=US%253Aen

https://www.ecowatch.com/kentucky-pipeline-explosion-2639622771.html?rebelltitem=1#rebelltitem1

https://www.ingaa.org/File.aspx?id=10753

https://www.chicagotribune.com/news/ct-xpm-1988-12-08-8802230052-story.html

https://www.bts.gov/content/us-oil-and-gas-pipeline-mileage

https://www.latimes.com/local/la-xpm-2011-aug-30-la-me-0831-san-bruno-20110831-story.html

https://hingham.wickedlocal.com/news/20170203/commentary-compressor-station-threatens-hingham

https://www.npr.org/2018/12/14/677003961/pg-e-falsified-gas-pipeline-safety-records-regulators-say

https://www.propublica.org/article/oil-leak-is-latest-mishap-for-troubled-alaska-pipeline-system

https://www.rigzone.com/news/oil_gas/a/31383/two_more_corrosion_leaks_in_bps_alaska_pipeline/

http://blogs.edf.org/energyexchange/files/2016/07/PHMSA-Blowdown-Analysis-FINAL.pdf

http://preservethenrv.com/articles/natural_gas_compressor_stations.pdf

https://www.nytimes.com/1987/02/26/us/pcb-dumping-by-4-pipelines-reported.html

https://www.latimes.com/business/la-fi-aliso-canyon-gas-pipeline-explosion-20190711-story.html
https://www.npms.phmsa.dot.gov/
https://www.justice.gov/archive/opa/pr/2002/February/02_enrd_053.htm
https://www.mothersoutfront.org/gas_leaks_in_my_new_neighborhood
http://blogs.edf.org/energyexchange/files/2016/07/PHMSA-Blowdown-Analysis-FINAL.pdf
https://www.nrdc.org/experts/amy-mall/va-pipeline-compressor-station-threatens-nearby-community
https://www.nytimes.com/1987/02/26/us/pcb-dumping-by-4-pipelines-reported.html
https://journals.plos.org/plosone/article?id=10.1371/journal.pone.0174310
https://journals.plos.org/plosone/article?id=10.1371/journal.pone.0174310
https://www.farmvilleherald.com/services/about-us/
https://journals.plos.org/plosone/article?id=10.1371/journal.pone.0174310
https://www.delawareriverkeeper.org/sites/default/files/DRN%2520Compressor%2520station%2520fact%2520sheet.pdf
https://gallery.mailchimp.com/288ec92ff904f6e9cb80bdb20/files/cd1e56fb-508a-4edd-9e8f-36ee43142c14/Health_Effects_Associated_with_Stack_Chemical_Emissions_from_NYS_Natural_Gas_Compressor_Stations_2008_2014.pdf
https://www.justice.gov/archive/opa/pr/2002/February/02_enrd_053.htm
https://www.washingtonpost.com/archive/politics/1987/02/26/epa-widens-probe-of-pcb-dumping-by-pipeline-firms/fa5a9058-09ed-4a7e-b1ad-cec503d1c5b4/
https://www.chicagotribune.com/news/ct-xpm-1988-12-08-8802230052-story.html
https://theweek.com/articles/882813/america-needs-stop-natural-gas-pipeline-mania
https://www.albany.edu/about/assets/Complete_report.pdf

42. Crematoriums

https://www.smithsonianmag.com/smart-news/radioactive-material-transferred-cancer-patients-body-contaminated-arizona-crematorium-180971580/
https://www.epa.sa.gov.au/files/477146_info_cremation.pdf
https://www.ejnet.org/crematoria/
https://www.timescall.com/2013/08/31/controversy-surrounding-crematoriums-near-neighborhoods-is-nothing-new/
http://www.nbcnews.com/id/16656749/ns/us_news-environment/t/cremation-pollution-neighbors-nervous/
https://www.telegraph.co.uk/science/2019/02/26/cremating-cancer-patients-have-had-radiotherapy-may-send-radioactive/?WT.mc_id=tmg_share_em
https://www.telegraph.co.uk/science/2019/02/26/cremating-cancer-patients-have-had-radiotherapy-may-send-radioactive/
http://www.swcleanair.org/docs/permits/Final/17-3241TSD.PDF
https://www.sciencedaily.com/releases/2016/09/160927143042.htm
https://www.ncbi.nlm.nih.gov/pmc/articles/PMC6555253/
https://www.smithsonianmag.com/science-nature/could-funeral-future-help-heal-environment-180957953/
https://www.weatherforddemocrat.com/opinion/letters_to_the_editor/the-crematorium-next-to-my-home/article_71e6f13d-ddaf-5f7c-a7fc-e057b98a1efc.html
https://www.nfda.org/news/media-center/nfda-news-releases/id/2511/nfda-cremation-and-burial-report-shows-rate-of-cremation-at-all-time-high
https://www.cremationassociation.org/page/IndustryStatistics
https://www.latimes.com/archives/la-xpm-2007-dec-26-na-cremate26-story.html
http://www.woodstockga.gov/DocumentCenter/View/1781/DPH---Crematories?bidId=
https://www.baaqmd.gov/~/media/files/engineering/public-notices/2019/29716/a2315_nsr_29716_eval_091219-pdf.pdf?la=fil-ph

https://medium.com/s/story/are-crematory-emissions-safe-simply-because-theyre-unregulated-fed494ea9f9c
https://www.latimes.com/archives/la-xpm-2007-dec-26-na-cremate26-story.html
https://www.ncbi.nlm.nih.gov/pmc/articles/PMC1279940/
43. DPW yards
https://www.fairfieldcitizenonline.com/news/article/Fairfield-seeks-3-million-in-suit-over-11139296.php
https://www.fairfieldcitizenonline.com/news/article/Fill-pile-costs-climb-to-almost-800k-13227117.php
https://www.tapinto.net/towns/new-brunswick/sections/government/articles/city-investigates-pollution-at-public-works-yard
https://www.fairfieldcitizenonline.com/news/article/Cost-of-DPW-tank-removal-project-fueled-by-4219630.php
https://www.ctpost.com/local/article/Court-docs-Fairfield-CFO-stole-fill-pile-files-15026928.php
https://www.northjersey.com/story/news/bergen/teaneck/2019/12/12/teaneck-nj-study-contamination-dpw-site-considered-housing/4409208002/
https://boltoncenter.wordpress.com/2011/10/21/comparing-dpw-barn-contamination-to-smith-contamination/
https://www.courant.com/breaking-news/hc-br-fairfield-public-works-illegal-dumping-kickback-arrests-20190807-6mzflp6y4zaw3ggpzh4745xevq-story.html
https://www.fairfieldcitizenonline.com/news/article/Cost-of-DPW-tank-removal-project-fueled-by-4219630.php
https://www.timesunion.com/news/article/Coxsackie-confronts-fuel-spill-at-DPW-garage-13415879.php
https://dailygazette.com/article/2009/01/08/0108_dec
44. Tire dumps
https://qz.com/961991/a-week-long-toxic-tire-fire-overwhelmed-firefighters-in-texas-so-they-called-in-the-epa-to-help-put-it-out/
https://www.momtastic.com/webecoist/2013/05/28/burn-rubber-the-worlds-9-worst-tire-fires/2/
https://www.sacurrent.com/the-daily/archives/2016/07/15/state-rep-says-south-side-tire-dump-would-burn-for-nine-months-if-it-caught-fire
https://upnorthlive.com/news/local/hundreds-of-tires-burn-in-massive-blaze-at-cheboygan-county-junkyard
https://www.thestarpress.com/story/news/local/2019/10/18/waste-tire-dump-behind-muncie-fresh-thyme-being-cleaned-up-idem/4008357002/
https://walterborolive.com/2019/11/state-funds-sought-for-tire-removal/
https://www.newfalconherald.com/DisplayPrintArticle.php?ArticleID=6194
https://www.energyjustice.net/tires/burners
https://www.popularmechanics.com/cars/car-technology/a22553570/waste-tires/
https://www.ustires.org/system/files/USTMA_scraptire_summ_2017_072018.pdf
https://www.energyjustice.net/tires/burners
https://www.timesunion.com/news/article/Comptroller-says-tire-cleanup-has-worked-but-some-14414056.php
https://blog.epa.gov/2015/08/26/improperly-stored-tires-lead-to-big-mosquito-problems/
https://www.11alive.com/article/news/crime/serial-tire-dumper-caught-in-clayton-county-after-allegedly-beating-girlfriend-with-handgun/85-6e8d9114-5f80-44c1-8b6e-182a52e1cbae
https://www.ksdk.com/article/news/local/neighbors-tired-illegal-tire-dumper-florissant-shopping-center/63-c8c5d61b-acf8-4d77-8924-8b2310bc1dfc
https://www.texarkanagazette.com/news/texarkana-region/story/2019/dec/16/illegal-dump-site-now-estimated-50000-tires-officials-say-cleanup-could-cost-150000/808471/

https://www.theintell.com/news/20180708/dep-eyes-1986-fire-as-potential-cause-of-rockhill-water-contamination

https://www.phoenix.gov/firesite/Documents/20218a.pdf

https://www.postandcourier.com/news/lax-rules-left-mountains-of-mosquito-infested-flammable-tires-in-south-carolina/article_98fccbfe-4417-11e8-9ea4-c765b5a2c64c.html

https://www.tires-easy.com/blog/the-wheel-truth-about-tire-recycling-fees/

https://www.amazon.com/gp/help/customer/display.html?nodeId=200989080

https://archive.epa.gov/epawaste/conserve/materials/tires/web/html/laws.html

https://www.tires-easy.com/blog/the-wheel-truth-about-tire-recycling-fees/

https://www.record-eagle.com/news/local_news/pfas-found-in-blair-residential-wells/article_6666e93b-7450-543b-b6de-454f4e9ddab9.html

https://www.record-eagle.com/news/local_news/pfas-found-in-blair-residential-wells/article_6666e93b-7450-543b-b6de-454f4e9ddab9.html

https://cumulis.epa.gov/supercpad/SiteProfiles/index.cfm?fuseaction=second.Cleanup&id=0302772#bkground

http://freerepublic.com/focus/f-news/644212/posts

https://cen.acs.org/articles/91/i43/Tire-Inferno.html

45. Cell towers

https://www.reporternewspapers.net/2020/05/26/sandy-springs-resident-sues-verizon-over-5g-poles-alleging-property-devaluation-and-health-risks/

https://www.celltowerleaseexperts.com/cell-tower-lease-news/cell-tower-industry-facts-figures-2016/

https://www.fcc.gov/consumers/guides/human-exposure-radio-frequency-fields-guidelines-cellular-and-pcs-sites

https://sacramento.cbslocal.com/2019/03/12/school-cell-tower-causing-cancer/

http://www.avaate.org/spip.php?article2827

https://www.vice.com/en_ca/article/7x4dnd/the-people-who-claim-theyre-allergic-to-wifi

https://www.rt.com/news/450775-massive-health-experiment-5g-cancer/

https://www.vice.com/en_us/article/pa8bpk/5g-wireless-rekindles-fight-over-cellular-health-risks

https://emfscientist.org/

https://ecfsapi.fcc.gov/file/10711815002508/FCCREPLYCommentsEnvironmentalHealthTrust%20PDF.pdf

http://www.antennasearch.com/sitestart.asp

https://www.ehstoday.com/archive/article/21905443/emf-exposure-linked-to-increased-risk-of-miscarriage?fbclid=IwAR0gC4e6SGcA4FiDg1RSBTB7ecK3sLbgu3LWJwTPtoySPCnRkwrog5Et6Y4

https://www.nature.com/articles/s41598-017-16623-8#Sec9

https://www.abc10.com/article/news/cell-phone-tower-to-be-removed-from-ripon-school/103-7fdfc7f4-01fa-4c53-8e20-bdf0769c3449

https://www.epa.gov/radtown/electric-and-magnetic-fields-power-lines#about-electric-magnetic-fields-from-power-lines

https://www.sciencedaily.com/releases/2017/12/171213095534.htm

https://www.who.int/peh-emf/publications/facts/fs296/en/

https://www.ncbi.nlm.nih.gov/pubmed/26372109

https://www.fatherly.com/health-science/flame-retardants-pbde-dangerous-for-kids/

https://www.cbsnews.com/news/chemical-banned-in-70s-from-kids-pajamas-resurfaces-in-couches/

https://www.nytimes.com/1977/04/08/archives/usbans-a-flame-retardant-used-in-childrens-sleepwear.html

https://www.fastcompany.com/90212342/3m-knew-your-non-stick-pan-was-poisoning-you-in-the-70s

https://www.npr.org/2011/03/02/134196209/
study-most-plastics-leach-hormone-like-chemicals
https://www.who.int/peh-emf/meetings/archive/en/keynote3ng.pdf
46. Electric power lines
https://ceds.org/tl/
http://appvoices.org/2018/06/07/hazardous-spray/
https://www.nytimes.com/1983/10/15/us/dow-halts-fight-to-sell-herbicide.html
https://www.recorder.com/my-turn-connell-EversourceHerbicideSpraying-26571267
http://appvoices.org/2018/06/07/hazardous-spray/
https://www.ncbi.nlm.nih.gov/pubmed/21946043
https://www.ncbi.nlm.nih.gov/pubmed/21946043
https://appvoices.org/2018/06/07/hazardous-spray/
47. Cemeteries
https://eponline.com/articles/2006/09/01/arsenic-and-old-graves.aspx
https://archive.org/stream/tilDeathDoWePollute
AndBeyondThePotentialPollutionOfCemeteriesAnd/TillDeathD
http://www.ncbi.nlm.nih.gov/pmc/articles/PMC3315260/
http://www.doeni.gov.uk/niea/cemeteryguidance.pdf
http://villagememorial.blogspot.ca/2015/05/pollution-from-embalming-and-
cemeteries.html
http://www.wspgroup.com/en/WSP-UK/Who-we-are/Newsroom/features/
Groundwater-near-cemeteries/
http://www.mass.gov/eohhs/gov/departments/dph/programs/environmental-health/
comm-sanitation/burial-and-
https://www.utne.com/environment/arsenic-contamination-ze0z1306zpit
https://www.utne.com/environment/arsenic-contamination-ze0z1306zpit
https://www.disabled-world.com/health/cemetery.php
http://www.nymoon.com/pubs/undertone/dead/
https://science.howstuffworks.com/science-vs-myth/afterlife/embalming4.htm
https://www.mercurynews.com/2019/05/22/
mosque-cemetery-proposal-stirs-controversy-in-san-martin/
https://www.fredericksburg.com/news/local/stafford/stafford-
leaders-vote-to-maintain-cemetery-ordinance-amid-doj-investigation/
article_69a0c8c8-0576-59e8-987a-99e9391fbe88.html
https://eponline.com/articles/2006/09/01/arsenic-and-old-graves.aspx?m=2
https://www.planningni.gov.uk/index/advice/northern_ireland_environment_agency_
guidance/practice_guide_-_cemeteries_burials_and_the_water_environment.pdf
https://modernfarmer.com/2014/12/graveyard-soil-science/
https://www.motherjones.com/politics/1999/01/dead-water/
https://www.smithsonianmag.com/science-nature/
arsenic-and-old-graves-civil-war-era-cemeteries-may-be-leaking-toxins-180957115/
https://pubs.usgs.gov/sir/2018/5120/sir20185120.pdf
https://www.disabled-world.com/health/cemetery.php
https://www.atlasobscura.com/articles/cemetery-soil-human-remains
https://www.researchgate.net/
publication/297275185_Potential_groundwater_pollutants_from_cemeteries
https://pubs.usgs.gov/sir/2018/5120/sir20185120.pdf
https://link.springer.com/article/10.1023/A:1005186919370
48. Zoning
https://people.uwec.edu/ivogeler/w270/what_is_zoning.htm
https://www.bozemandailychronicle.com/news/city/residents-sue-bozeman-over-
zoning-change/article_022e458a-d2bd-5038-9b2f-681f97bfac68.html
https://www.indystar.com/story/news/2018/06/06/
nora-residents-sue-city-over-controversial-zoning-decision/676556002/

https://www.wnky.com/stone-crest-residents-prepare-to-fight-zoning-change/
https://www.heraldmailmedia.com/news/tri_state/west_virginia/martinsburg-city-councll-denies-zoning-change-requested-for-apartment-development/article_f553d425-cae3-53a0-80a3-c4c2a84e6d0b.html
https://people.uwec.edu/ivogeler/w270/what_is_zoning.htm
https://www.bozemandailychronicle.com/news/city/residents-sue-bozeman-over-zoning-change/article_022e458a-d2bd-5038-9b2f-681f97bfac68.html
https://www.indystar.com/story/news/2018/06/06/nora-residents-sue-city-over-controversial-zoning-decision/676556002/
https://www.wnky.com/stone-crest-residents-prepare-to-fight-zoning-change/
49. Proposed facilities
https://www.anchoragepress.com/news/worst-air-pollution-in-america/article_e9035fd0-49f0-11e8-b4c4-f3f6bd518135.html
https://radio.wosu.org/post/coal-declines-rural-town-fights-against-proposed-strip-mine
https://missoulian.com/opinion/columnists/montana-conservation-elders-oppose-proposed-smith-river-mine/article_6115f399-7c3b-5d6e-b51d-24c68353f9ca.html
https://www.wastedive.com/news/arizona-supreme-court-clears-way-for-controversial-trash-burning-plant/413831/
https://www.nbc12.com/story/38844020/residents-fighting-back-against-proposed-landfill-on-cumberland-powhatan-line/
https://www.inlander.com/spokane/a-proposed-silicon-smelter-in-northeast-washington-hasneighbors-worried-about-pollution-and-potential-health-impacts/Content?oid=9504983
https://mtstandard.com/news/local/timber-butte-residents-say-not-in-my-backyard-to-mine/article_d6a1853f-dfc4-5460-8afb-9033de3cc854.html
https://mtstandard.com/news/local/timber-butte-residents-say-not-in-my-backyard-to-mine/article_d6a1853f-dfc4-5460-8afb-9033de3cc854.html
https://grist.org/beacon/how-one-town-said-no-to-fracking-waste/
https://princetoninfo.com/proposed-plant-poses-a-toxic-threat/
https://www.sandiegouniontribune.com/business/energy-green/story/2019-06-10/new-mexicos-governor-comes-out-against-nuclear-waste-site
http://www.theprogressnews.com/progress_news/no-dump-county-citizens-oppose-landfill-despite-positive-economic-impact/article_a4e139b4-5092-54d8-a9c4-b98eee057a1f.html
https://www.climatecentral.org/news/community-activists-rally-against-proposed-trash-incinerator-15605
Conclusion
https://www.wbrc.com/2019/05/08/growing-number-bhm-neighborhoods-oppose-concrete-plant-relocation/
https://www.sbsun.com/2019/10/10/rialto-company-that-recycles-hazardous-waste-gets-unacceptable-grade-from-state-regulators/
https://www.norwichhttps://
www.investigativepost.org/2https://www.politico.com/story/2018/05/14/emails-white-house-interfered-with-science-study-536950018/04/05/6836/bulletin.com/opinion/20190720/state-doesnt-hhttps:
/https://www.eenews.net/stories/1060119337
https://www.dispatch.com/news/20191023/environmentalists-question-use-of-radioactive-brine-waste-to-treat-roadshttp:
//www.investigativepost.orghttps:
/sentinelcolorado.com/0trending/171-acre-murphy-creek-development-back-for-another-aurora-city-council-discussion//2019/09/03/turning-a-blind-eye-to-environmental-risks/

https://www.lehighvalleylive.com/business/2020/02/developers-abandon-plans-for-controversial-sewage-sludge-plant-in-slate-belt.html

https://www.bostonglobe.com/metro/2019/10/22/new-epa-chief-new-england-barred-from-many-decisions-because-conflicts/5qR0Xop1bBPFqxWpy3wAxO/story.html

https://www.nhregister.com/business/article/Wheelabrator-confirms-end-of-Greene-County-Ash-13880102.php

https://www.salemreporter.com/posts/767/covanta-officials-say-trash-incinerator-could-close-if-energy-certificate-bill-fails

https://www.wbrc.com/2019/05/22/mayor-sherman-industries-not-coming-five-points-west-community/

https://triblive.com/local/south-hills/baldwin-zoning-board-denies-plans-for-hazardous-waste-disposal-company/

https://www.phoenixnewtimes.com/news/arizona-utility-aps-srp-customers-pay-lobbying-against-epa-regulations-11300926

https://helenair.com/opinion/columnists/if-you-build-it-radioactive-waste-will-come/article_48a70787-9971-5946-ae7e-3df9278cf97d.html

https://sentinelcolorado.com/0trending/171-acre-murphy-creek-development-back-for-another-aurora-city-council-discussion/

http://rochester.indymedia.org/node/108

https://www.eesi.org/papers/view/fact-sheet-fossil-fuel-subsidies-a-closer-look-at-tax-breaks-and-societal-costshttps://www.governing.com/topics/finance/gov-companies-receiving-largest-federal-subsidies.html

https://www.wastedive.com/news/stericycle-las-vegas-medical-waste-canceled-incinerator/569325/

https://www.cnn.com/2018/09/28/us/st-louis-landfill-epa/index.html

https://missoulacurrent.com/outdoors/2020/02/missoula-radioactive-waste/

https://www.salemreporter.com/posts/1278/marion-county-gives-trash-burner-another-year-to-win-state-tax-credits

https://www.fox8live.com/2019/08/08/central-city-residents-raise-concerns-high-lead-levels/

https://news.wbfo.org/post/165-million-clean-hickory-woods-neighborhood

http://www.investigativepost.org/2019/09/03/turning-a-blind-eye-to-environmental-risks/

About The Author

Fred Siegel spent three years living on top of a hazardous waste landfill in New Jersey. Unknowingly of course until many people began getting sick. That experience prompted him to begin thirty years doing environmental site assessments looking for Superfund sites, leaking coal ash dumps, old landfills and other environmental hazards. The book, Environmental Hazards - Are You Exposed?, brings his years of experience looking for and finding all types of dangerous environmental hazards to people concerned they too may be unknowingly exposed.

Fred's main area of interest was in hazardous waste management and disposal. He was offered a position by the state of New Jersey to sit on a citizen advisory board working on hazardous waste disposal issues but turned it down. Instead, Fred led a four year battle against the construction of a hazardous waste incinerator project culminating in the abandoning of the project and in finding safer solutions to the issue. Fred has been arrested twice over the years for standing up for his environmental beliefs and was once featured in a VH-1 Greenpeace video with Tony.

Each year, Fred gives many talks to various groups involved in their local environmental issues in addition to talking to civic and youth organizations. He reaches several thousand people through public speaking and many more by being actively involved with environmental issues across the country.